Handbook
of
Food, Drug, and Cosmetic Excipients

Susan C. Smolinske, B.S., R.Ph.

Manager
POISINDEX® AND IDENTIDEX® Information Systems
Micromedex, Inc.
Denver, Colorado

CRC Press
Taylor & Francis Group
Boca Raton London New York

CRC Press is an imprint of the
Taylor & Francis Group, an **informa** business

CRC Press
Taylor & Francis Group
6000 Broken Sound Parkway NW , Suite 300
Boca Raton, FL 33487-2742

First issued in paperback 2019

ISBN-13: 978-0-8493-3585-3 (hbk)
ISBN-13: 978-0-367-40281-5 (pbk)

Library of Congress Cataloging-in-Publication Data

Smolinske, Susan C., 1953–
Handbook of food, drug, and cosmetic excipients / author, Susan C. Smolinske.
    p. cm.
    Includes bibliographical references and index.
    ISBN 0-8493-3585-X
    1. Excipients—Handbooks, manuals, etc. I. Title.
    [DNLM: 1. Cosmetics—handbooks. 2. Excipients—handbooks.
    3. Food Additives—handbooks. QV 735 S666h]
RS201.E875S66   1992
615.9—dc20
DNLM/DLC
for Library of Congress

91-29427
CIP

Library of Congress Card Number 91-29427

Visit the Taylor & Francis Web site at
http://www.taylorandfrancis.com

and the CRC Press Web site at
http://www.crcpress.com

DEDICATION

To my husband,
Mark,
for adding your excitement, loving encouragement, and support to this project

To the Rocky Mountain Poison and Drug Center
for providing the clinical educational training background, and the opportunity to pursue
and develop my interest in excipient toxicology

ACKNOWLEDGMENTS

I wish to acknowledge Micromedex, Inc. for allowing the use of their equipment and resources, Priscilla Hall, L.P.N. for her expeditous library research assistance, and the Advisory Panel for providing many constructive comments and suggestions.

Portions of the text were adapted from the following articles published by Adis International Limited:

Golightly, L. K., Smolinske, S. C., Bennett, M. L., Sutherland, E. W., and Rumack, B. H., Pharmaceutical excipients: adverse effects associated with inactive ingredients in drug products (Part I), *Med. Toxicol.,* 3, 128, 1988.

Golightly, L. K., Smolinske, S. C., Bennett, M. L., Sutherland, E. W., and Rumack, B. H., Pharmaceutical excipients: adverse effects associated with inactive ingredients in drug products (Part II), *Med. Toxicol.,* 3, 209, 1988.

ADVISORY BOARD

TABLE OF CONTENTS

INTRODUCTION

I. REGULATORY ACTIVITY IMPACTING EXCIPIENTS

A. Summary of Regulations

The Pure Food and Drugs Act of 1906 was the first attempt to regulate the safety of additives in foods and pharmaceuticals. This act prohibited the use of any color additive in foods if the color would deceive the consumer, conceal inferiority or damage, or result in misbranding or adulteration. Under this statute, premarketing authority was not granted, and court proceedings had to be initiated to remove an adulterated or misbranded product. As a result of this act, 80 commonly used color additives were evaluated; only 7 of these were recommended as safe for use in foods. In 1907, these recommendations were published, along with a system for voluntary food color certification.

The voluntary color certification process was rendered mandatory by the Federal Food, Drug, and Cosmetic Act of 1938. This act was precipitated by a dramatic epidemic of excipient-related toxicity. Introduction of a new oral formulation of sulfanilamide in 1937, containing a vehicle composed of 72% diethylene glycol, resulted in 105 deaths from acidosis and renal failure by October 1938.[1-3]

The 1938 act required premarketing approval for the first time. Scientific evidence of safety of the submitted drug was necessary to allow approval and marketing. Drugs that were generally recognized as safe (GRAS) due to a long history of marketing were exempt from approval requirements.

The Food Additives Amendment was introduced in 1958, requiring demonstration of the safety of newly introduced food additives. Some 670 food additives with a long history of use were designated as GRAS. A review of scientific evidence on additives on the GRAS list began in the early 1970s, sparked by the discovery of suspected carcinogenicity of a widely used substance on the GRAS list, cyclamate.

The Color Additives Amendment to the Food, Drug, and Cosmetic Act in 1960 provided for color additives already in use to be "grandfathered" and allowed to be used on a provisional basis, while studies were completed to document safety and allow permanent listing. Color additives requiring certification included synthetic dyes made from coal tar and petroleum derivatives. Natural vegetable, animal, or mineral dyes were exempt from certification. New color additives were required to undergo an approval process with a demonstration of safety of the additive for its intended use. The permanent listing of the additive included a designation of the petitioned use listed in the approval application. Thus, dyes approved for drug and cosmetic use were designated either "D&C", allowing both internal and external use, or "Ext. D&C", allowing only external use. Dyes approved for unlimited use in foods, drugs, and cosmetics were designated "FD&C" dyes.

At the conclusion of the Food and Drug Administration (FDA) review of provisionally listed straight color additives in 1990, 90 of the 200 provisionally listed additives have been petitioned and permanently listed as safe. Many of these additives have insoluble derivatives, known as "lakes", which are still on the provisional list. A proposal to regulate these compounds is pending.

The dichotomy of color additives into "provisional" and "permanent" listings has created an inadvertent legal loophole, which limits the action that can be taken if a permanent-listed additive is later shown to be unsafe. This problem is illustrated by the recent banning of all provisional uses of erythrosine, which was found to be potentially carcinogenic in animals, while allowing continued use of the dye for permanent-listed uses. Thus, new oral animal studies indicating a potential ingestion hazard resulted in removing the color additive for use in cosmetics and externally applied drugs, while allowing its continued use in foods and oral pharmaceuticals.[4]

The 1958 Food Additive and 1960 Color Additive amendments included a provision known as the Delaney Clause, which assumes that cancer risks have no threshold dose; thus any amount of a carcinogenic agent is prohibited. This provision was upheld by a state Court of Appeals in the District of Columbia, which maintained that additives with documented "trivial" risks were not exempted from the Delaney Clause. Because this clause allows for evaluation of a "risk-benefit" ratio, it is unevenly applied to foods, drugs, and cosmetics. Naturally occurring food additives are exempt. Because cosmetics and foods have no proven "therapeutic" benefit to human health, any demonstrated risk results in invocation of the Delaney Clause. Drugs may continue to contain these risky excipients if the proven therapeutic benefits outweigh the risks.

In 1962, another amendment to the Federal Food, Drug, and Cosmetic Act established requirements for reporting to the FDA adverse effects, clinical experience, and data related to safety and efficacy of drugs with approved New Drug Applications (NDAs) or Abbreviated New Drug Applications (ANDAs). Pre-drug-law products without NDAs or ANDAs were exempt. Reporting requirements are confined to any serious and unexpected reaction or an increase in frequency of any serious expected adverse reaction.

Another component of the 1962 amendment required an extension of the premarketing approval process to include scientific evidence of efficacy for the intended use, as well as safety. Systematic review of drugs approved for safety considerations before 1962 was undertaken to confirm that these drugs were also efficacious. This evaluation of over 4000 products, known as Drug Efficacy Study Implementation (DESI), established panels of experts to review efficacy data and recommend acceptable marketing conditions for classes of drug products. Drugs falling outside of the DESI review included approximately 5000 products protected by the "grandfather" clause of the 1938 act.

Adverse drug reaction reporting requirements were extended to include all marketed prescription drug products in September 1986 as a result of over 38 deaths in premature infants attributed to a pre-drug-law vitamin product marketed without an ANDA.[5]

Over-the-counter drugs (OTCs) were excluded from this amendment, based on a presumption of an increased margin of safety in those OTC products that have attained NDA approval. Drugs switched from prescription to OTC status are subject to adverse drug reaction reporting.

An ongoing review of the safety and efficacy of pre-drug-law OTC drugs also requires adverse drug reaction data, establishing that the drug is generally recognized as safe, to allow continued marketing once the OTC final rule for that drug is promulgated and allow exemption from NDA submission.[5]

In 1984, in response to the E-Ferol®-related deaths in premature infants, the FDA also modified its regulations regarding permission of copies of old products to be marketed without specific new drug approval. Under this modification, new copies of old drugs can be marketed only if identical in directions, intended patient population, formulation, dosage form, route of administration, indications for use, and dosage or strength. Only drugs marketed after November 13, 1984 were affected.[6]

B. Excipient Labeling Requirements

1. Cosmetics

The Fair Packaging and Labeling Act (FPLA) of 1976 requires labeling of cosmetic ingredients present at a concentration of 1% or greater with ingredients listed in descending order of concentration. Individual components of flavor and fragrances are not required to be listed. A loophole in this law allows exemption of professional salon care products not intended for resale to the public.[7]

Salon care products exempt from the FPLA will be labeled on containers or package inserts under voluntary guidelines effective at the end of 1989. The voluntary program suggests listing ingredients in alphabetical order to prevent resale of the products to consumers.[7]

Cosmetic ingredients are registered on a voluntary basis with the FDA. It is estimated that there are about 8000 raw materials and fragrance ingredients available for use in cosmetics.[8] In response to a challenge by the FDA to regulate the safety of cosmetics, the Cosmetic, Toiletry and Fragrance Association established a voluntary safety review in 1976. The Cosmetic Ingredient Review Expert Panel was charged with systematically reviewing the published and unpublished scientific data and making a decision to place each ingredient in one of three categories:[9]

1. Safe for use, with or without limitations or restrictions
2. Unsafe for use
3. Insufficient data on which to base a conclusion

This committee has issued final reports on 310 ingredients of an estimated 2700 ingredients used extensively in cosmetics. Only two were found to be unsafe, *p*-hydroxyanisole and chloroacetamide.

2. Drugs

Labeling of pharmaceuticals generally falls under pharmacopoeial guidelines rather than regulatory statutes. The Drug Standards Division of the United States Pharmacopoeia (USP) requires labeling of inactive ingredients for topical, ophthalmic, and parenteral preparations and is seeking to expand this requirement to all drug dosage forms. These guidelines are enforceable under the Food, Drug, and Cosmetic Act.[10]

Voluntary guidelines have been published by the Pharmaceutical Manufacturers Association (PMA) and the Nonprescription Drug Manufacturers Association (NDMA). Unlike the proposed USP guidelines, these allow nondisclosure of ingredients if a trade secret would be violated by labeling the ingredient. Both of these proposed guidelines suggest labeling ingredients in alphabetical order, distinguished from active ingredients. Topical and parenteral products would continue to be labeled in descending order of ingredient concentration.

II. HISTORY OF EXCIPIENT TOXICITY

A. Historical Background

Significant Events in Food, Drug, and Cosmetic Excipient History

1906	Food and Drug Act
1937	Diethylene glycol/sulfanilamide toxicity epidemic
1938	Food, Drug and Cosmetic Act
1954	Miller Pesticides Amendment
1959	First report of adverse reaction to tartrazine
1960	Food Additive Amendment (Delaney Clause)
1966	Burn dressing excipient linked to coma in 51 patients
1969	Cyclamate banned
1980	Tartrazine labeling required for pharmaceuticals
1982	Outbreak of benzyl alcohol toxicity in neonates
1982	Polyethylene glycol linked to renal failure
1983	First report of propylene glycol hyperosmolality
1984	38 Deaths reported in infants receiving E-Ferol®
1984	Changes to FDA approval regulations
1986	Changes to FDA adverse reaction reporting regulations
1986	Sulfite GRAS status revoked
1989	Alupent reformulated to delete soya lecithin excipient

B. Food Additive Considerations

According to the United Nations Food and Agriculture Organization, the definition of a food additive is a "non-nutritive substance added intentionally to food, generally in small quantities to improve its appearance, texture or storage properties." Another definition, provided by the National Research Council, states "a substance added to foods either directly and intentionally for a functional purpose; or indirectly during some phase of production, processing, storage, or packaging without intending that it remain in the final product."

There are at least 2700 additives present in common foods.[11] Direct food additives include anticaking agents, emulsifying agents, preservatives, sequestrants, stabilizers, synthetic flavorings, and colorants. Indirect additives include pesticide residues, antibiotics, microorganisms, parasites, metals, radioactive compounds, and packaging residues.[12] Color additives are used to replace color lost in food processing, to inhibit natural color fading, and to ensure color uniformity in food products. The perception of goodness associated with coloring of food may be an innate response, which is illustrated by animal feeding chronic studies showing increased food consumption in animals fed highly colorized food.[13]

It has been estimated that the prevalence of hypersensitivity reactions to food additives in the general population is between 0.03 and 0.15%. Spices are common offenders, with 20% of atopic patients reacting to immediate skin test procedures.[14]

C. Cosmetic Excipient Considerations

It has been estimated that a dermatologist is consulted about an adverse reaction to a cosmetic by 210 individuals per million products used. In North American patients, reactions are most commonly caused by fragrance ingredients, followed by preservatives, *p*-phenylenediamine, lanolin, glyceryl monothioglycolate, and propylene glycol.[15]

The high degree of fragrance sensitivity presents a diagnostic and treatment problem. Fragrances are complex mixtures of natural products, which are difficult to isolate to a particular component. The voluntary disclosure of cosmetic excipients excludes disclosure of specific fragrance components, making it difficult for patients to avoid a particular allergen once it is identified. Only a few of the most clinically important fragrance components are represented here. A reasonable recommendation to these patients is avoidance of all perfumed products.

Atopic patients may be predisposed to develop contact dermatitis secondary to cosmetic use. Large case series of patients with cosmetic dermatitis have contained 16 to 31% atopics, compared to the incidence of atopy in the general population of about 10 to 20%.[15,16] The face is the most frequent site of dermatitis. The most commonly implicated products are skin care products (28%), hair preparations (24%), facial makeup (11%), nail preparations (8%), fragrance products (7%), and personal cleanliness products (6%).[15]

D. Drug Excipient Considerations

The term "excipient" has usually been defined in association with pharmaceutical products. The dictionary definition of an excipient is "any more or less inert substance added to an excipient in order to confer a suitable consistency or form to the drug." Excipients are a necessary component of pharmaceuticals, enabling delivery of medicinals in a variety of dosage forms. The necessity of colorants is less apparent. One possible significant benefit is the prevention of drug errors by the consumer, particularly the elderly or sight impaired.

Allergy to one or more components in topically applied medicaments accounted for one third of the cases of allergic contact dermatitis in a series of 4000 consecutive patients seen in one of five European clinics. Stasis dermatitis was more commonly associated with contact allergy to medicaments than dermatitis at other sites. Forty percent of women with lower leg dermatitis had documented allergy to applied medicaments, compared to 8% with hand dermatitis.[17]

As with most excipient-related problems, the relationship of a topically applied excipient to toxicity has been historically difficult to make. A mesh burn wound dressing impregnated with a supposedly inert and nonirritating component, hexylene glycol 80%, was associated with coma and renal failure in 51 cases. The excipient was almost the last ingredient evaluated.[18]

III. BOOK ORGANIZATION

Of the more than 8000 food, drug, and cosmetic excipients available, detailed monographs are presented on 77 of the most clinically important ingredients.

Attempts were made to include a table of representative pharmaceutical products in each of the excipients included. The dosage form was chosen in each case to reflect the route of administration associated with the adverse effect. The tables were compiled by reviewing at least three published databases,[19-21] relevant journal articles, and confirmation by the manufacturer in the event of a discrepancy between sources. To restrict the length of these tables to a manageable length, only brand name products were included. The intention is not to provide a complete listing of all products containing a particular excipient, but to provide an extensive listing of those products most likely to be encountered, as a starting place for investigation of the potential etiology of a suspected excipient-related problem in a given patient. The lists are published with the knowledge that inactive ingredients in pharmaceuticals may change without notification; therefore, the current ingredients should be verified by the manufacturer once the diagnosis is narrowed down to one or several choices.

REFERENCES

1. **Calvery, H. O. and Klumpp, T. G.,**The toxicity for human beings of diethylene glycol with sulfanilamide, *South. Med. J.,* 32, 1105, 1939.
2. **Leech, P. N.,** Elixir of sulfanilamide-Massengill: II, *JAMA,* 109, 1724, 1937.
3. **Geiling, E. M. K. and Cannon, P. R.,** Pathologic effects of elixir of sulfanilamide (diethylene glycol) poisoning, *JAMA,* 111, 1938.
4. **Blumenthal, D.,** Red No. 3 and other colorful controversies, *FDA Consumer,* 21, 18, 1990.
5. **Food and Drug Administration,** Adverse drug experience reporting requirements for marketed prescription drugs without approved new drug or abbreviated new drug applications, *Fed. Reg.,* 51, 24476, 1986.
6. **Food and Drug Administration,** Prescription drugs marketed without approved new drug applications; revised compliance policy, *Fed. Reg.,* 49, 38190, 1984.
7. **Rietschel, R. L. and Larsen, W. G.,** Salon care product labeling, *J. Am. Acad. Dermatol.,* 22, 309, 1990.
8. **Eiermann, H. J., Larsen, W. G., Maibach, H. I., and Taylor, J. S.,** Prospective study of cosmetic reactions: 1977–1980, *J. Am. Acad. Dermatol.,* 6, 909, 1982.
9. **Bergfeld, W. F., Elder, R. L., and Schroeter, A. L.,** The cosmetic ingredient review self-regulatory safety program, *Dermatologic Clin.,* 9, 105, 1991.
10. **United States Pharmacopeial Convention,** *USP XXII NF XVII,* Rockville, MD, 1990.
11. **Collins-Williams, C.,** Intolerance to additives, *Ann. Allergy,* 51, 315, 1983.
12. **Maher, T. J.,** Neurotoxicology of food additives, *Neurotoxicology,* 7, 183, 1986.
13. **Borzelleca, J. F. and Hallagan, J. B.,** Chronic toxicity/carcinogenicity studies of FD&C Yellow No. 5 (tartrazine) in rats, *Food Chem. Toxicol.,* 26, 179, 1988.
14. **Hannuksela, M. and Haahtela, T.,** Hypersensitivity reactions to food additives, *Allergy,* 42, 561, 1987.
15. **Adams, R. M. and Maibach, H. I.,** A five-year-study of cosmetic reactions, *J. Am. Acad. Dermatol.,* 13, 1062, 1985.
16. **De Groot, A. C., Liem, D. H., Nater, J. P., and van Ketel, W. G.,** Patch tests with fragrance materials and preservatives, *Contact Dermatitis,* 12, 87, 1985.
17. **Bandmann, H.-J., Calnan, C. D., Cronin, E., Fregert, S., Hjorth, N., Magnusson, B., Maibach, H., Malten, K. E., Meneghini, C. L., Pirila, V., and Wilkinson, D. S.,** Dermatitis from applied medicaments, *Arch. Dermatol.,* 106, 335, 1972.
18. **Procter, D. S. C.,** Coma in burns — the cause traced to dressings, *S. Afr. Med. J.,* 24, 1116, 1966.
19. **Barnhart, E. R.,** *Physicians' Desk Reference,* 44 ed., Medical Economics, Oradell, NJ, 1991.
20. **Rumack, B. H.,** *POISINDEX® Information System,* Micromedex, Denver, CO, edition expires 5/31/91.
21. **Olin, B.,** *Facts and Comparisons,* J.B. Lippincott, St. Louis, 1990.

ACACIA

I. REGULATORY CLASSIFICATION

Acacia is classified as an emulsifying and/or solubilizing agent, tablet binder, and a suspending and/or viscosity-increasing agent. Acacia syrup is a flavored and/or sweetened vehicle.

II. SYNONYMS

Gum arabic

III. AVAILABLE FORMULATIONS

A. Constituents

Acacia is a soluble hydrocolloid gum derived from the stems and branches of *Acacia senegal* or other related *Acacia* species. Constituents of acacia include arabin (a complex mixture of calcium, magnesium, and potassium salts of arabic acid), water 12 to 15%, and enzymes.[1]

B. Foods

Acacia has been on the FDA GRAS list since 1974 and has no assigned upper limit on the acceptable daily intake as a food additive.[2] It is found in many foods, including cake icing, frozen custard, diabetic foods, diet beverages, ice cream and pops, marshmallows, meringues, puddings, sherbets, and wheat cakes.[3]

C. Drugs

Acacia syrup NF contains acacia 10% w/w, sodium benzoate, vanilla tincture, sucrose, and water.

D. Cosmetics

Acacia is used as a viscosity increasing agent and hair fixative in aqueous formulations in cosmetic products, including hair products, bath soaps, cleansing products, and skin care products.[4]

E. Industrial Products

Workplace exposure may occur via contact with lithographic solutions, adhesive pastes, artificial flowers, cement, cigar manufacture, matches and fireworks, furniture polish, textile coating, metal polish, paints, drying and offset printing sprays, pottery manufacture, process engraving, shoe polish, textile sizing, varnish, and water colors.[5,6]

IV. TABLE OF COMMON PRODUCTS

A. Oral Drug Products

Trade name	Manufacturer
Aci-jel	Ortho
Advil	Whitehall
Afrinol repetab	Schering
Agoral	Parke-Davis
Amitriptyline tablet	Mylan
Aminophylline tablet	Searle
Apresoline tablet	Ciba
Apresoline-Esidrix	Ciba
Carters Little Pills	Carter-Wallace
Centrum Jr. plus extra C	Lederle
Cepastat lozenges	Lakeside
Chlortrimeton decongestant repetab	Schering
Chlortrimeton repetab	Schering
Choloxin tablet	Boots-Flint
Clusivol syrup	Whitehall
Compazine tablet	Smith Kline & French
Coricidin D	Schering
Coricidin tablet	Schering
Cytoxan tablet	Bristol-Myers
Darbid tablet	Smith Kline & French
Demazin tablet	Schering
Dexedrine capsule	Smith Kline & French
Disophrol tablet	Schering
Doxycycline tablet	Lederle
Dramamine tablet	Searle
Drixoral tablet	Schering
Dulcolax tablet	Boehringer Ingelheim
Enovid tablet	Searle
Erythromycin stearate	Mylan
Etrafon tablets	Schering
Evac-Q-Tabs	Adria
Ex-Lax extra gentle	Ex-Lax
Ex-Lax unflavored	Ex-Lax
Festalan tablet	Hoechst-Roussel
Ferro-Sequels	Lederle
Ibuprofen 400 mg tablet	Mylan
Ibuprofen 600 mg tablet	Mylan
Kaon CL tablet	Adria
Kaon CL-10 tablet	Adria

Trade name (cont'd)	Manufacturer (cont'd)
Kolantyl wafer	Lakeside
Leukeran tablet	Burroughs Wellcome
Levsin tablet	Kremers-Urban
Lomotil tablet	Searle
Mellaril tablet	Sandoz
Mephyton tablet	Merck Sharp & Dohme
Metandren linguet	Ciba
Methyldopa tablets	Mylan
Mintezol tablet	Merck Sharp & Dohme
Modane tablet	Adria
Modane mild tablet	Adria
Modane plus tablet	Adria
Mol-Iron	Schering
Motrin 300 mg tablet	Upjohn
Motrin 400 mg tablet	Upjohn
Naldecon tablet	Bristol
Neptazane tablet	Lederle
Ovcon 35-28 day	Mead Johnson
Ovcon 50-28 day	Mead Johnson
PBZ tablet	Geigy
Peganone tablet	Abbott
Persantine tablet	Boehringer Ingelheim
Polaramine tablet	Schering
Probenecid tablet	Mylan
Proventil repetab	Schering
Questran powder	Bristol
Rela tablet	Schering
Riopan	Whitehall
Riopan plus	Whitehall
Ser-as-es tablet	Ciba
Serentil tablet	Boehringer Ingelheim
Serutan granules	Beecham
Slow K tablet	Ciba
Sudafed tablet	Burroughs Wellcome
Synthroid tablet	Boots-Flint
Temaril tablet	Herbert
Theo-dur tablet	Key
Theragran hematinic	Squibb
Thermotabs	Beecham
Thioguanine tablet	Burroughs Wellcome
Throat Discs	Marion
Tindal tablet	Schering
Tine Test PPD	Lederle
Torecan tablets	Boehringer Ingelheim
Trendar tablet	Whitehall
Triaminic juvelets	Dorsey
Triaminic TR	Dorsey
Trilafon tablet	Schering
Trinalin repetab	Schering

V. ANIMAL TOXICITY DATA

Experiments in guinea pigs demonstrated anaphylaxis following intravenous injection in 63% of animals pretreated with intraperitoneal doses on four occasions.[7]

Subchronic feeding studies in rats showed a no-untoward effect level of 8.6% (5.2 g/kg/d) in the diet of male rats and 18.1% (13.8 g/kg/d) in female rats.[8]

VI. HUMAN TOXICITY DATA

A. Immediate Hypersensitivity

Acacia was formerly given intravenously for treatment of shock and was associated with many cases of severe anaphylactoid reactions after the first injection, including laryngeal stridor, cyanosis, dyspnea, pulmonary edema, and liver necrosis. [7,9-11] These were thought to be related to impurities in the acacia, but subsequent use of purified material produced similar adverse effects.[10] Anaphylaxis, occurring on the second dose, has also been described.[7]

Three kidney transplant patients receiving long-term therapy with prednisone tablets containing acacia and tragacanth were reported to develop hypersensitivity reactions consisting of rash, pruritus, fever, and arthralgia. One patient had a positive scratch test to acacia, one had a positive test to tragacanth, and the other patient was not tested.[12]

Occupational asthma and rhinitis related to acacia has been implicated by positive scratch and intradermal tests, demonstration of passive transfer in a plastic molder,[13] and in printers using an acacia-based offset or drying spray.[14-16] The onset of symptoms ranged from 2 weeks to 12 months following inhalation exposure.

An immediate systemic hypersensitivity reaction, consisting of local wheal, nausea, wheezing, and syncope, occurred within 5 min of receiving a Tine Test PPD in a 35-year-old woman. Although acacia could not be directly implicated, a RAST test for IgE-PPD was negative.[17]

B. Delayed Hypersensitivity

Allergic contact dermatitis has been reported in a litho-printer[18] and a flowermaker manipulating wet clay.[19]

VII. CLINICAL RELEVANCE

Although allergic reactions have been reported following ingestion of acacia, the incidence of reactions is believed to be comparable, but not greater than that elicited by hen ovalbumin. Inhalation exposure appears to carry a higher risk of sensitization. Contact dermatitis has been infrequently reported.

REFERENCES

1. **Tyler, V. E., Brady, L. R., and Robbers, J. E.,** *Pharmacognosy,* 9th ed., Lea & Febiger, Philadelphia, 1988, 47.
2. **Anderson, D. M. W.,** Evidence for the safety of gum arabic (*Acacia senegal* (L.) Willd.) as a food additive — a brief review, *Food Add. Contam.,* 3, 225, 1986.
3. **Nilsson, D. C.,** Sources of allergenic gums, *Ann. Allergy,* 18, 518, 1960.

4. **Nikitakis, J. M.,** *CTFA Cosmetic Ingredient Handbook,* 1st ed., The Cosmetic, Toiletry and Fragrance Association, Washington, D.C., 1988.

5. **Gelfand, H. H.,**The allergenic properties of the vegetable gums, *J. Allergy,* 14, 203, 1942.

6. *Hazardous Substances Data Bank,* National Library of Medicine, Bethesda, MD (CD-ROM Version), Micromedex Inc., Denver, CO, 1990.

7. **Maytum, C. K. and Magath, T. B.,** Sensitivity to acacia, *JAMA,* 99, 2251, 1932.

8. **Anderson, D. M. W., Ashby, P., Busuttil, A., Eastwood, M. A., Hobson, B. M., Ross, A. H. M., and Street, C. A.,** Sub-chronic effects of gum arabic in the rat, *Toxicol. Lett.,* 14, 221, 1982.

9. **De Kruif, P. H.,** Experimental research on the effect of intravenous injection of gum salt solutions, *Ann. Surg.,* 69, 297, 1919.

10. **Hanzlik, P. J. and Karsner, H. T.,** Anaphylactoid phenomena from the intravenous administration of various colloids, arsenicals and other agents, *J. Pharmacol. Exp. Ther.,* 14, 379, 1920.

11. **Studdeford, W. E.,** Severe and fatal reactions following the intravenous use of gum acacia glucose infusions, *Surg. Gynecol. Obstet.,* 64, 772, 1937.

12. **Rubinger, D., Friedlander, M., and Superstine, E.,** Hypersensitivity to tablet additives in transplant recipients on prednisone, *Lancet,* 2, 689, 1978.

13. **Spielman, A. D. and Baldwin, H. S.,** Atopy to acacia (gum arabic), *JAMA,* 101, 444, 1933.

14. **Bohner, C. B., Sheldon, J. M., and Trenis, J. W.,** Sensitivity to gum acacia, with a report of ten cases of asthma in printers, *J. Allergy,* 12, 290, 1940.

15. **Feinberg, S. M. and Schoenkerman, B. B.,** Karaya and related gums as causes of atopy, *Wis. Med. J.,* 39, 734, 1940.

16. **King, J. H.,** Asthma and allergic rhinitis due to gum arabic in non-offset spray, *J. Med.,* 22, 119, 1941.

17. **Wright, D. N., Ledford, D. K., and Lockey, R. F.,** Systemic and local allergic reactions to the Tine Test Purified Protein Derivative, *JAMA,* 262, 2999, 1989.

18. **van Ketel, W. G.,** Simultaneous sensitization to gum arabic and cobalt, *Contact Dermatitis,* 10, 180, 1984.

19. **Ilchyshyn, A. and Smith, A. G.,** Gum arabic sensitivity associated with epidemic hysteria dermatologica, *Contact Dermatitis,* 13, 282, 1985.

ACESULFAME

I. REGULATORY CLASSIFICATION

Acesulfame potassium is an artificial sweetener currently approved for use in certain foods.

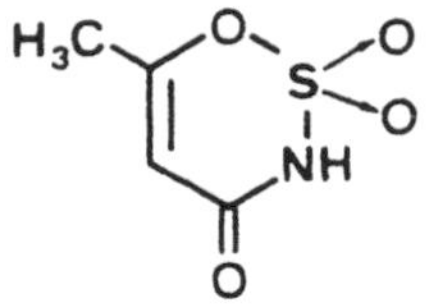

FIGURE 1. Acesulfame.

II. SYNONYMS

Acesulfame K
Acetosulfam
6-methyl-1,2,3-oxathiazine-4(3H)-one-2,2-dioxide potassium

III. AVAILABLE FORMULATIONS

Acesulfame is a non-nutritive sweetener, approximately 200 times sweeter than sucrose. It is approved for use as a table-top sweetener and as an ingredient in food products. Structurally, acesulfame has some resemblance to saccharin and shares the property of a bitter taste in high doses.

IV. TABLE OF COMMON PRODUCTS

Food products currently approved for use of acesulfame are[1]

Dry, free-flowing sugar substitutes in packets
Sugar substitute tablets

13

Chewing gum

Instant beverages, coffee, and tea

Dry bases for gelatins, puddings, and pudding deserts

Dry bases for dairy product analogs

A. Oral Drug Products

Rolaids Extra Strength (Warner-Lambert)

V. ANIMAL TOXICITY DATA

A. Carcinogenicity/Teratogenicity

A 2-year study in beagle dogs fed 90, 300, or 900 mg/kg/d in their diet did not show any evidence of toxic effects.[1]

Carcinogenicity was evaluated in 200 Swiss mice fed 0.3, 1, or 3% of acesulfame potassium in the diet. A consulting pathologist, the testing laboratory, and the FDA concluded that no association between neoplasms or any other adverse effects and acesulfame was documented. The high-dose group in female mice had an increased incidence of lymphocytic leukemia (4% compared to 1% in low-dose groups), but this was within the incidence of controls from this species at that laboratory (mean 5.2%).[1]

Carcinogenicity was also evaluated in 120 Wistar rats fed 0.3, 1, or 3% acesulfame potassium in the diet. A slightly higher incidence and early appearance of lymphoreticular pulmonary neoplasms (reticulum cell sarcomas) was reported. The lack of histopathologic examination in one third of the rats and the presence of chronic respiratory disease in the test animals (which is linked to sarcomas in this species) led to the conclusion that this study was inadequate and could not be used to assess carcinogencity of this compound.[1]

A second long-term rat study was performed to confirm the results in the previous study, using the same dosage levels and a different strain of Wistar rats. Evaluation of the results by a consultant pathologist did not show any evidence of a relationship to reticulum cell sarcomas or any adverse toxic, reproductive, or teratogenic effects. The human acceptable daily intake computed from this study was 15 mg/kg/d.[1]

B. Effect in Diabetes

An isolated pancreatic islet cell rodent model demonstrated a dose-dependent increase in insulin release in the presence of acesulfame, which was only evident with concomitant incubation with glucose. A direct potentiating action on the islet cells was demonstrated, which was independent of the neuroendocrine system.[2]

Injection of 150 mg/kg of acesulfame potassium to rats produced a transient approximately threefold elevation in plasma insulin concentrations with no effect on blood glucose levels; the insulin level returned to baseline by 15 min. No effect was seen with single doses of 50 mg/kg; while a doubling of plasma insulin occurred with 100 mg/kg. Continuous infusion of 20 mg/kg/min resulted in a sustained concomitant increase in plasma insulin and decrease in blood glucose (from a mean 103 to 72 mg/dl). Continuous infusion of 4 mg/kg/min had no effect on insulin secretion.[3]

VI. HUMAN TOXICITY DATA

Acesulfame potassium is reported to be excreted intact in the urine in humans with no detectable metabolism.[1]

VII. CLINICAL RELEVANCE

Based on an estimated 90th percentile estimated daily intake of 1.6 mg/kg, acesulfame is believed to be safe for human consumption.[1] The World Health Organization (WHO) has estimated the maximum acceptable daily intake to be 9 mg/kg.[4] Hypoglycemic effects demonstrated in animals occurred with a threshold of more than 30 times greater than the estimated daily amounts consumed by humans; therefore, acesulfame potassium can probably be safely used by diabetics.

REFERENCES

1. **Food and Drug Administration,** Food additives permitted for direct addition to food for human consumption; acesulfame potassium, *Fed. Reg.*, 53, 28379, 1988.
2. **Liang, Y., Maier, V., Steinbach, G., Lalic, L., and Pfeiffer, E. F.,** The effect of artificial sweetener on insulin secretion. II. Stimulation of insulin release from isolated rat islets by acesulfame K (in vitro experiments), *Horm. Metab. Res.*, 19, 285, 1987.
3. **Liang, Y., Steinbach, G., Maier, V., and Pfeiffer, E. F.,** The effect of artificial sweetener on insulin secretion. I. The effect of acesulfame K on insulin secretion in the rat (studies in vivo), *Horm. Metab. Res.*, 19, 233, 1987.
4. **World Health Organization,** Twenty-seventh report of the joint FAO/WHO expert committee on food additives, Tech. Rep. Ser. No. 696, WHO, Geneva, 1983.

ALUMINUM

I. REGULATORY CLASSIFICATION

Aluminum monostearate is classified as a suspending and/or viscosity-increasing agent.

Aluminum salts, such as alum solution, aluminum hydroxide, and aluminum phosphate are used in the manufacture of vaccines as an adjunct to enhance immunogenicity.

Parenteral products may contain aluminum as contaminants. The source of contamination is usually the drug product, but may also include leaching from glass containers and closures during autoclaving and storage. The FDA requires labeling of the aluminum content of parenterals intended for repeated use.[1] The FDA is considering regulations specifying an upper limit for aluminum content of large volume parenterals and labeling for small volume parenterals.[2]

II. AVAILABLE FORMULATIONS

A. Foods

Aluminum found in vegetation and vertebrate species contributes 2 to 3 mg/d from the normal diet.[3]

Tea (*Camellia sinensis*) accumulates aluminum in the leaves. Prepared teas may contain 2 to 6 mg/l.[4]

Aluminum contamination of foods is primarily a concern of products intended for use in infants, such as infant formulas. Higher amounts of aluminum are present in soy-based formulas, although the source has not been established. The method of preparation (i.e., use of tap water and aluminum cooking utensils) may contribute to contamination, although small amounts of aluminum salts are added to formula as contaminants of mineral supplements.[5]

In a study of British infant formulas, the aluminum content ranged from 0.03 to 0.2 mg/l for cow-based products and 0.64 to 1.34 mg/l for soy-based products.[5] In a similar Australian study, amounts of 0.09 to 10 mg/l were found in liquid formulas and 0.23 to 11.8 mg/l in powdered formula.[6]

Municipal drinking water deflocculated with aluminum sulfate may contain up to 1000 ng/ml of aluminum.[7] This is the major source of contamination of dialysis fluid.[8]

B. Drugs

Aluminum is found as a contaminant in drugs, such as plasma protein solutions, pediatric dialysis solutions, intravenous fluids, and human antihemophilic globulin. The source of contamination for intravenous and dialysis fluid preparations was the depth filter used to reduce the pyrogen content in one institution. Aluminum was present in the filter to bind endotoxin.[9] Factors contributing to filter contamination included high pH, high osmolarity, presence of divalent cations, and slow filtration rate.[9] Glass vials are a significant source of aluminum in human serum albumin injection. The amount of aluminum increased from 58.9 to 152.9 ppb after 28 d of storage in a glass container.[10]

C. Cosmetics

Aluminum salts are used in cosmetics as abrasives, anticaking agents, bulking agents, opacifying agents, emulsion stabilizers, viscosity-increasing agents, buffering agents, and absorbents. Aluminum powder is a color additive in some cosmetics. Preparations that may contain aluminum salts include eye shadows, eyebrow pencils, blushers, makeup products, mascara, dentifrices, lipsticks, rouges, moisturizing creams and lotions, permanent waves, face powders, mud packs, and various hair grooming aids. As an active ingredient, aluminum is used as an antiperspirant and astringent.[11]

III. TABLE OF COMMON PRODUCTS

A. Parenteral Vaccines

Adsorbed diphtheria toxoid USP
Adsorbed tetanus toxoid USP
Adsorbed diphtheria and tetanus toxoids USP (for pediatric use)
Adsorbed tetanus and diphtheria toxoids USP (for adult use)
Adsorbed diphtheria and tetanus toxoids and pertussis vaccine USP
Allpyral allergenic extracts
Recombivax HB

B. Parenteral Nutrition Solutions

Trade name	Aluminum content[12] (ng/ml)
Intralipid 10%	17–33
Intralipid 20%	10–27
Liposyn II 10%	19–50
Liposyn II 20%	6–33
Soyacal 10%	19–34
Soyacal 20%	3–21
Travamulsion 10%	45–60
Travamulsion 20%	31–59

C. Intravenous Solutions

Trade name	Aluminum content (ng/ml)
Calcium gluconate 10%	5,056[13]
Dextrose 5%	72[13]
Heparin 1,000 U/ml	684[13]
Heparin 5,000 U/ml	359[13]
Heparin 10,000 U/ml	468[13]
Normal serum albumin	163–1,108[14]
Normal serum albumin 25%	1,822[13]
Potassium chloride 3,000 mmol/l	6[13]
Potassium phosphate 3,000 mmol/l	16,598[13]
Sodium chloride 4,000 mmol/l	6[13]
Sodium phosphate 3,000 mmol/l	5,977[13]

IV. ANIMAL TOXICITY DATA

Studies in rats and piglets have shown accumulation of aluminum in the liver and resultant cholestasis after an intravenous load. The relevance of these studies to infants receiving aluminum via parenteral nutrition is unclear.[15,16]

V. HUMAN TOXICITY DATA

A. Summary
Manifestations of aluminum toxicity secondary to contamination of parenteral fluids, dialysis solutions, total parenteral nutrition, or infant formulas have included osteodystrophy (with associated bone pain, myopathy, and fractures), hypercalcemia, anemia, and progressive encephalopathy.[17-19]

B. Infant Formula
Freundlich et al.[17] reported a 3300 g neonate and a 1700 g premature neonate who developed fatal aluminum intoxication after receiving a ready-to-use infant formula for 3 and 1 months, respectively. Both infants were uremic, had no other identifiable source of aluminum, and were demonstrated to have elevated brain aluminum concentrations of 6.4 and 47.4 mcg/g, respectively. The features of aluminum toxicity were primarily neurologic, with sudden development of lethargy, hypotonia, and hypotension. A subsequent prospective study of 14 uremic infants receiving the same formula for 12 to 36 months failed to show an increase in baseline plasma aluminum levels, significant response to deferoxamine challenge, or histochemical aluminum bone deposition.[20]

C. Total Parenteral Nutrition
A 630 g premature neonate was reported to develop convulsions after receiving parenteral nutrition for 45 d. He died at age 93 d and was found to have a brain aluminum concentration of 40 mcg/g.[18]

A 620 g premature neonate who received total parenteral nutrition from age 7 weeks to 7 months was noted to have osteopenia. The serum aluminum level was 182 ng/ml.[19]

Bone pain, fractures, and elevated aluminum levels have been reported in adults with normal renal function receiving casein-containing total parenteral nutrition for 2 months to 3 years. The bone disease could not be totally attributed to aluminum and may have been partially a result of malnutrition prior to initiation of parenteral nutrition.[21]

A prospective study of eight children receiving long-term total parenteral nutrition demonstrated elevated plasma aluminum levels. Elevated bone levels were shown in one infant with osteopenic bone disease and a plasma aluminum level of 80 mcg/l.[22]

D. Plasma Exchange

Adult patients with normal renal function have been reported to have fivefold increases in serum aluminum levels following therapeutic plasma exchange. The increased levels (14 to 48 ng/ml) after one exchange were still below those considered to be toxic, and no manifestations of aluminum toxicity were observed.[23] The plasma aluminum level after two exchanges in one patient was 120 ng/ml, a potentially toxic level.[14]

Four patients with impaired renal function who received a single plasma exchange retained 60 to 74% of the aluminum given. Three of these patients had elevated bone aluminum levels and histological evidence of aluminum-related bone disease after a total of 29, 119, and 187 plasma exchanges over 6 to 53 months.[24]

E. Hypersensitivity Reactions

Two types of hypersensitivity dermatological reactions have been associated with aluminum salts. Granulomas or subcutaneous nodules may occur following vaccinations with aluminum-adsorbed toxoids or pollen extracts. These typically resolve within a few weeks,[25] but occasionally persist. Histology primarily supports a foreign-body reaction, but delayed-type hypersensitivity has been documented.[26] Persistent lesions contain a central eosinophilic necrosis surrounded by a mixed inflammatory cell infiltrate. Patch testing with aluminum hydroxide has been negative in three of five cases where testing was done and positive in one case.[25] Intradermal challenge may be necessary to prove causation.[26]

The other common dermatological problem is a typical delayed-type hypersensitivity eczematous lesion secondary to exposure to aluminum-containing antiperspirants or pollen extracts. Pruritic lesions following injections may persist for several years due to deposition of aluminum intracutaneously.[27-30]

VI. CLINICAL RELEVANCE

Three patient populations are at an increased risk of excessive exposure to aluminum excipients: patients with renal failure on chronic hemodialysis or CAPD, patients receiving long-term parenteral nutrition, especially those with compromised renal function, and premature neonates receiving parenteral nutrition.[2]

Contamination of infant formulas with aluminum has only been implicated in causing clinical problems in neonates with impaired renal function.[17] Infants with renal insufficiency or low birth weight should probably not receive soy-based formulas.[7]

Low-birth-weight premature infants are at high risk from aluminum toxicity because of prolonged parenteral nutrition therapy and decreased renal excretion capability.[18,19,31] Premature infants have been reported to excrete only 40 to 80% of a daily intravenous aluminum intake.[13,32]

Although aluminum-related osteodystrophy has mainly been reported in adult patients with renal insufficiency treated with dialysis, it has also been reported in patients with normal renal function who received total parenteral nutrition for 6 to 12 months.[33]

Patients with renal insufficiency are also at high risk of aluminum toxicity after repeated plasma exchange therapy. Although plasma aluminum levels may not be impressive, metabolic balance and bone biopsy studies have demonstrated accumulation of aluminum in these patients.[24]

Plasma aluminum levels of greater than 100 to 150 ng/ml are associated with the risk of aluminum toxicity.[7]

Dermatological reactions are more common in patients receiving aluminum-containing vaccines or hyposensitization dust and pollen extracts than in patients exposed to topical antiperspirants and other preparations.[30]

REFERENCES

1. **Food and Drug Administration,** Public Workshop: determination of aluminum in parenteral products, *Fed. Reg.,* 51, 36066, 1986.
2. **Food and Drug Administration,** Parenteral drug products containing aluminum as an ingredient or a contaminant: notice of intent and request for information, *Fed. Reg.,* 55, 20799, 1990.
3. **Koo, W. W. K. and Kaplan, L. A.,** Aluminum and bone disorders: with specific reference to aluminum contamination of infant nutrients, *J. Am. Coll. Nutr.,* 7, 199–214, 1988.
4. **Flaten, T. P. and Odegard, M.,** Tea, aluminum and Alzheimer's disease, *Food Chem. Toxicol.,* 26, 959–960, 1988.
5. **Fisher, C. E., Knowles, M. E., Massey, R. C., and McWeeny, D. J.,** Levels of aluminum in infant formulae, *Lancet,* 1, 1024, 1989.
6. **Weintraub, R., Hams, G., Meerkin, M., and Rosenberg, A. R.,** High aluminum content of infant milk formulas, *Arch. Dis. Child.,* 61, 914–916, 1986.
7. **Committee on Nutrition,** Aluminum toxicity in infants and children, *Pediatrics,* 78, 1150–1154, 1986.
8. **Parkinson, I. S., Ward, M. K., Feest, T. G., Fawcett, R. W. P., and Kerr, D. N. S.,** Fracturing dialysis osteodystrophy and dialysis encephalopathy:an epidemiological survey, *Lancet,* 1, 406–409, 1979.
9. **Vyth, A., Stolk, L. M. L., and Abbad, F. C. B.,** Aluminum contamination of dialysis and intravenous fluids after asbestos-free depth filtration, *Am. J. Hosp. Pharm.,* 43, 2390, 1986.
10. **Olson, W. P. and Kent, R. S.,** Aluminum from glass vials contaminates albumin, *Transfusion,* 29, 86–87, 1989.
11. **Nikitakis, J. M.,** *CTFA Cosmetic Ingredient Handbook,* 1st ed., The Cosmetic, Toiletry and Fragrance Association, Washington, D.C., 1988.
12. **Pesko, L. J. and Hudson, W. P.,** An evaluation of aluminum content in commonly prescribed fat emulsions, *Pharmacy Practice News,* September 1986.
13. **Sedman, A. B., Klein, G. L., Merritt, R. J., Miller, N. L., Weber, K. O., Gill, W. L., Anand, H., and Alfrey, A. C.,** Evidence of aluminum loading in infants receiving intravenous therapy, *N. Engl. J. Med.,* 312, 1337, 1985.
14. **Milliner, D. S., Shinaberger, J. H., Shuman, P., and Coburn, J. W.,** Inadvertent aluminum administration during plasma exchange due to aluminum contamination of albumin-replacement solutions, *N. Engl. J. Med.,* 312, 165, 1985.
15. **Klein, G. L., Sedman, A. B., Heyman, M. B., et al.,** Hepatic abnormalities associated with aluminum loading in piglets, *JPEN,* 11, 293, 1987.
16. **Klein, G. L., Heyman, M. B., Lee, T. C., et al.,** Aluminum-associated hepatobiliary dysfunction in rats: relationships to dosage and duration of exposure, *Pediatr. Res.,* 23, 275, 1988.
17. **Freundlich, M., Zillereulo, G., Abitol, C., and Strauss, J.,** Infant formula as a cause of aluminum toxicity in neonatal uraemia, *Lancet,* 2, 527, 1985.

18. **Bishop, N. J., Robinson, M. J., Lendon, M., Hewitt, C. D., Day, J. P., and O'Hara, M.,** Increased concentration of aluminum in the brain of a parenterally fed premature infant, *Arch. Dis. Child.,* 64, 1316, 1989.

19. **Klein, G. L., Snodgrass, W. R., Griffin, P., Miller, N. L., and Alfrey, A. C.,** Hypocalcemia complicating deferoxamine therapy in an infant with parenteral nutrition-associated aluminum overload: evidence for a role in the bone disease of infants, *J. Pediatr. Gastroenterol. Nutr.,* 9, 400, 1989.

20. **Salusky, I. B., Coburn, J. W., Nelson, P., and Goodman, W. G.,** Prospective evaluation of aluminum loading from formula in infants with uremia, *J. Pediatr.,* 116, 726, 1990.

21. **Ott, S. M., Maloney, N. A., Klein, G. L., Alfrey, A. C., Ament, M. E., Coburn, J. W., and Sherrard, D. J.,** Aluminum is associated with low bone formation in patients receiving chronic parenteral nutrition, *Ann. Intern. Med.,* 98, 910, 1983.

22. **Larchet, M., Chaumont, P., Galliot, M., Bourdon, R., Goulet, O., and Ricour, C.,** Aluminum loading in children receiving long-term parenteral nutrition, *Clin. Nutr.,* 9, 79, 1990.

23. **Monteagudo, F., Wood, L., Jacobs, P., Folb, F., and Cassidy, M.,** Aluminum loading during therapeutic plasma exchange, *J. Clin. Apheresis,* 3, 161, 1987.

24. **Maharaj, D., Fell, G. S., Boyce, B. F., Ng, J. P., Smith, G. D., Boulton-Jones, J. M., Cumming, R. L. C., and Davidson, J. F.,** Aluminum bone disease in patients receiving plasma exchange with contaminated albumin, *Br. Med. J.,* 295, 693, 1987.

25. **Bohler-Sommeregger, K. and Lindemayr, H.,** Contact sensitivity to aluminum, *Contact Dermatitis,* 15, 278, 1986.

26. **McFadden, N., Lyberg, T., and Hensten-Pettersen, A.,** Aluminum-induced granulomas in a tattoo, *J. Am. Acad. Dermatol.,* 20, 903, 1989.

27. **Clemmensen, O. and Kuudsen, H. E.,** Contact sensitivity to aluminum in patient hyposensitized with aluminum-precipitated grass pollen, *Contact Dermatitis,* 6, 305, 1980.

28. **Fisher, T. and Rystedt, I.,** A case of contact sensitivity to aluminum, *Contact Dermatitis,* 8, 343, 1982.

29. **Veien, N. K., Hattel, T., Justesen, O., and Norholm, A.,** Aluminum allergy, *Contact Dermatitis,* 15, 295, 1986.

30. **Castelain, P. Y., Castelain, M., Vervloet, D., Garbe, L., and Mallet, B.,** Sensitization to aluminum by aluminum-precipitated dust and pollen extracts, *Contact Dermatitis,* 19, 58, 1988.

31. **Broadbent, R. and Pybus, J.,** Aluminum contamination of intravenous fluids in neonates, *N.Z. Med. J.,* 99, 166, 1986.

32. **Koo, W., Kaplan, L., Bendon, R., Succop, P., Horn, J., Tsang, R., and Steichen, J.,** Response to aluminum in parenteral nutrition during infancy, *Pediatr. Res.,* 20, 352A, 1986.

33. **Klein, G. L., Alfrey, A. C., Miller, N. L., et al.,** Aluminum loading during total parenteral nutrition, *Am. J. Clin. Nutr.,* 35, 1425, 1982.

ANNATTO

I. REGULATORY CLASSIFICATION

Annatto is classified as a color additive.

II. SYNONYMS

Annotta
Arnatta
Arnatto
Arnotta
E160b

III. AVAILABLE FORMULATIONS

A. Foods
Annatto is a natural orange yellow pigment obtained from the seeds of *Bixa orellana*. The primary pigment is a carotenoid, beta-norbixin. Annatto extract was approved for use in foods in 1963 and is exempt from batch certification. In a local supermarket survey, 8 of 100 cereals and 9 of 20 cheese products listed annatto as an ingredient.[1]

B. Drugs
Annato extract was approved for use in ingested drugs in 1963 and for external drugs, including eye area use, in 1977.

C. Cosmetics
Annatto is a noncertified color additive approved for cosmetic use in 1977. It is used as a colorant, flavoring agent, or fragrance in cosmetic creams, lotions, suntan gels, and eye makeup products.[2]

IV. TABLE OF COMMON PRODUCTS

A. Foods That May Contain Annatto Include[1]

Cheeses
Snack foods
Beverages
Cereals
Ice creams
Margarines
Oils

V. HUMAN TOXICITY DATA

A. Urticaria/Angioedema

Among 56 patients with chronic urticaria and/or angioedema, 26% had adverse hypersensitivity reactions within 4 h following ingestion of annatto after single-blind oral challenge.[3] Oral single-blind provocation with 5 to 10 mg of annatto was performed in 112 patients with recurrent urticaria. Positive reactions were found in 10% of the patients tested.[4]

B. Anaphylaxis

Anaphylaxis was described in a 62-year-old man following ingestion of a new high fiber cereal containing annatto extract as a color additive. Generalized urticaria, lip and eye angioedema, and loss of consciousness occurred within 20 min after consuming the cereal. After being found with an unobtainable blood pressure by paramedics, he was transported to a local hospital and resuscitated with epinephrine and intravenous fluids. Five weeks later, he was skin prick tested to components of the cereal and had a strong response to the full strength annatto extract and a weaker reaction to a 1:1000 dilution. An IgE-binding band specific to an annatto protein was demonstrated on immunoblot assay. The protein responsible for the reaction was probably a contaminant in the extraction process during commercial production of the dye.[1]

VI. CLINICAL RELEVANCE

Annatto was not tolerated by 10 to 26% of patients with chronic urticaria using a single-blind challenge procedure. These results should be considered tentative until confirmed by double-blind, placebo-controlled studies in this patient population. Anaphylaxis has been reported in one case and may be related to a protein contaminant in the commercial extract.

REFERENCES

1. **Nish, W. A., Whisman, B. A., Goetz, D. W., and Ramirez, D. A.,** Anaphylaxis to annatto dye: a case report, *Ann. Allergy.* 66. 129, 1991.
2. **Nikitakis, J. M.,** *CTFA Cosmetic Ingredient Handbook.* 1st ed., The Cosmetic. Toiletry and Fragrance Association, Washington, D.C., 1988.
3. **Mikkelsen, H., Larsen, J. C., and Tarding, F.,** *Arch. Toxicol..* 1 (Suppl.), 141, 1978.
4. **Juhlin, L.,** Recurrent urticaria:clinical investigation of 330 patients, *Br. J. Dermatol..* 104, 369, 1981.

Aspartame

I. REGULATORY CLASSIFICATION

Aspartame is classified as a sweetening agent.

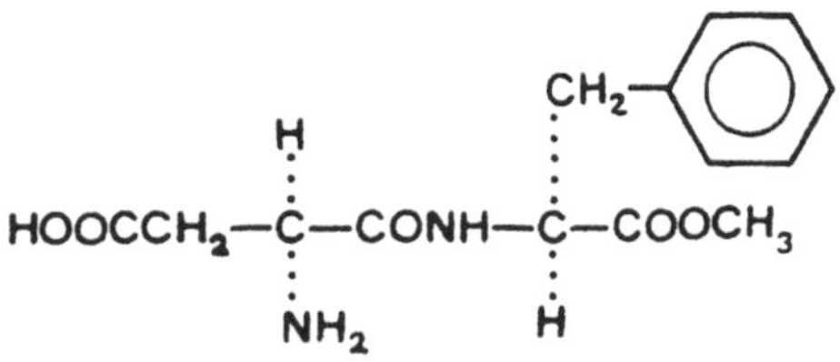

$$HOOCCH_2-\overset{\overset{H}{|}}{\underset{\underset{NH_2}{|}}{C}}-CONH-\overset{\overset{CH_2-\bigcirc}{|}}{\underset{\underset{H}{|}}{C}}-COOCH_3$$

FIGURE 2. Aspartame.

II. SYNONYMS

APM
Equal®
NutraSweet®

III. AVAILABLE FORMULATIONS

A. Foods

Aspartame is widely used as an artificial sweetener in foods and beverages. Some of the types of products containing aspartame include soft drinks, fruit juices, instant tea and coffee, powdered beverages, breakfast cereals, gelatin, puddings, pie fillings, whipped toppings, chewing gum, frozen stick-type confections, breath mints, and milk shakes. Since foods containing aspartame must be clearly labeled, it can easily be avoided by individuals who should not ingest these products.[1]

B. Drugs

Aspartame is used as an artificial sweetener in over-the-counter pharmaceuticals and in some prescription drugs, particularly in chewable tablets and sugar-free products. Both prescription and nonprescription products must be labeled with the phenylalanine content.[2]

IV. TABLE OF COMMON PRODUCTS

A. Foods/Beverages

Caffeine-free Diet Coke
Caffeine-free Tab
Diet Cherry Coca-Cola
Diet Coke
Diet Minute Maid Lemon-lime
Diet Minute Maid Orange
Diet Sprite
Diet Squirt Plus
Fresca
Tab

B. Drugs

Trade name	Manufacturer
Alka-Seltzer Advanced tablet	Miles
Anacin-3 Children's Chewable tablet	Whitehall
Children's CoTylenol Chewable tablet	McNeil
Dramamine Chewable tablet	Richardson-Vicks
Flintstones Children's Chewable Multivitamins	Miles
Metamucil (sugar-free)	Procter & Gamble
Pediacare Cough-Cold Formula Chewable tablet	McNeil
Questran Light	Bristol
Sunkist Children's Chewable Multivitamins	Ciba
Tempra Chewable tablet	Mead Johnson
Tempra Chewable double-strength tablet	Mead Johnson
Tylenol Chewable tablet	McNeil
Tylenol Chewable Jr. strength tablet	McNeil
Tylenol Cold and Flu hot medication	McNeil

V. ANIMAL TOXICITY DATA

Animal models have been used to study the mechanism for the alleged neurological adverse effects of aspartame. Rats given an oral dose of 200 mg/kg had a doubling of brain phenylalanine levels. Coingestion with glucose resulted in a further doubling. Brain tyrosine levels were also increased. It was speculated that these effects would produce an increase in adrenergic neurotransmitters.[3] These results, along with evidence of increased brain norepinephrine and dopamine, were confirmed in mice given 130 or 650 mg/kg.[4] Another study using larger doses in mice (up to 2500 mg/kg) refuted these conclusions and showed no alteration in seizure threshold nor in norepinephrine synthesis despite similar changes in levels of phenylalanine, tyrosine, and serotonin.[5] Aspartame, in amounts of 1 g/kg orally, did not increase the seizure potential of theophylline in rats.[6]

Prior to approval for marketing, several safety concerns related to aspartame and its major decomposition products were examined by the FDA. The first concern, the issue of potential brain damage from accumulation of phenylalanine and aspartate, has been adequately studied in humans, thus no animal data are presented. No impairment in the ability of individuals to metabolize aspartame and its metabolites has been demonstrated in groups of normal adults, children, infants, lactating women, diabetics, obese subjects, and glutamate-sensitive subjects.[7]

Another concern addressed the potential for causing brain tumors in animal experiments. In one rat study involving a major decomposition product, diketopiperazine, which is found in amounts of 3 to 4% of carbonated beverages stored at room temperature for 8 weeks, no relationship to brain tumors was found.[7]

In a 104-week rat study with aspartame, brain tumors were found in 3.75% of the treated group and in 0.8% of controls, but the FDA concluded that this was within the expected spontaneously occurring rate for these animals. Another 104-week study showed no increased incidence of tumors.[7]

A concern over possible methanol toxicity has also been addressed, since 10% of an aspartame dose is converted to methanol. Small amounts of methanol were found in infants given 34 mg/kg of aspartame.[8] The amounts of methanol formed are no greater than those found after consumption of other dietary sources of methanol, such as fruits and vegetables.

VI. HUMAN TOXICITY DATA

A. Headache

The most consistently reported adverse effect from aspartame has been headache. Following a case report of migraine shown to be related to aspartame by multiple placebo and aspartame challenges,[9] two double-blind trials were conducted. In a 4-week trial of 11 migraine patients, the incidence of headache increased from 1.55 to 3.55 per month during aspartame intake of 1200 mg/d.[10] In another trial involving 40 patients with self-reported aspartame-exacerbated headaches challenged with aspartame 3 doses of 10 mg/kg 2 h apart, no increased incidence of headaches was confirmed when compared to placebo.[11]

In an uncontrolled survey of 171 patients seen at a headache clinic, 8.2% of all patients and 11% of migraine patients reported that aspartame was a trigger for their headaches.[12] Headache was also reported in 45% of spontaneously submitted adverse drug reaction reports.[13] In a 6-month study in normal subjects, aspartame doses of 75 mg/kg/d were not associated with an increased incidence of headaches. Headaches were the most commonly reported complaint in both aspartame and placebo groups, underscoring the difficulty in attributing causation in individual case reports without placebo challenges.[14] Studies using capsule dosage forms may not mimic real-life usage of aspartame, and it is possible that breakdown products from storage of soft drinks are responsible for headaches.

B. Neuropsychiatric Symptoms

Several neuropsychiatric symptoms have been attributed to excessive aspartame intake, but few were confirmed with a controlled rechallenge and confounding or predisposing conditions were usually present. Drake et al.[15] reported a 33-year-old woman with asymptomatic mitral valve prolapse who developed panic attacks after ingestion of 20 cans of a diet cola containing aspartame and caffeine. Switching to noncaffeinated diet beverages did not result in decreased symptoms, while reducing the amount of aspartame did provide relief.

Wurtman[16] reported a 42-year-old woman with mood changes, headache, nausea, visual hallucinations, and a grand mal seizure after consumption of over 6 l/d of aspartame-containing beverages. Symptoms abated upon discontinuation and returned after resumption, but no double-blind challenge was done. The same author reported two other cases of seizures related temporally to aspartame intake.

A grand mal seizure, followed by an acute manic episode, was reported in a 54-year-old woman with a history of major depression following ingestion of up to 1 gal/d of aspartame-sweetened tea. The symptoms disappeared after switching to sugar-sweetened tea, but no rechallenge was done.[17]

There was no relationship between aggressive behavior and aspartame consumption in preschool children.[18]

C. Hypersensitivity

There are several case reports of hypersensitivity reactions to aspartame, confirmed by double-blind placebo-controlled challenge. In one case, erythema, pruritus, and urticaria developed. In another case, angioedema and urticaria were reported.[19] An unconfirmed case of a facial and chest pruritic dermatitis occurring 7 to 8 h after consumption of aspartame has been reported.[13]

One case of granulomatous panniculitis, resembling erythema nodosum, was reported in a 22-year-old woman who consumed 1080 to 1320 ml/d of aspartame-sweetened soft drink.[20] Lobular panniculitis was described in an insulin-dependent man who developed tender subcutaneous nodules on the back and arms after ingestion of aspartame 210 to 245 mg/d as a sugar substitute, along with aspartame-containing foods and beverages. Biopsy of one of the nodules revealed panniculitis. The nodules disappeared within 12 d after discontinuing use of aspartame. Double-blinded challenge with 300 mg twice daily resulted in recurrence of the nodules within 5 d.[21]

VI. CLINICAL RELEVANCE

A. Phenylketonuria

The major concern with aspartame ingestion is in patients with the autosomal recessive trait of phenylketonuria (PKU). This occurs in about one in 10,000 to 15,000 live births.[22,23] Heterozygous PKU carriers occur in about 1 in 60 individuals.[23] Ingestion of aspartame 34 mg/kg resulted in elevated phenylalanine levels in three 9-year-old children with classical PKU.[24] Two adolescents with PKU given the same aspartame load did not have markedly elevated levels.[25]

Patients, particularly children, with PKU and who have strict dietary restrictions should meticulously avoid consumption of aspartame. PKU patients who do not have dietary restrictions could safely consume a dose of 10 mg/kg (equivalent to 34.5 ounces of Kool-Aid® or three cans of soft drink).[26]

It is estimated that 4 million Americans are heterozygous for the PKU trait.[3] These patients do not appear to be at increased risk. Single ingestion of 100 mg/kg of aspartame, an amount equivalent to ingestion of 38 12-ounce cans of a diet carbonated beverage, did not increase plasma phenylalanine levels to toxic concentrations in such patients.[27] Repeated ingestion of smaller amounts, 10 mg/kg every 2 h for 3 doses, had a smaller effect,[28] and repeated ingestion of 30 mg/kg/h for 8 h, the equivalent of 24 12-ounce cans of aspartame-sweetened beverage, also produced modest but clinically insignificant increases in phenylalanine levels.[29]

B. Liver Disease

Patients with severe liver dysfunction (i.e., alcoholic cirrhosis) may have an impaired ability to metabolize phenylalanine. If the disease is severe enough to warrant dietary monitoring and modification by the physician, then aspartame should be considered as a source of aromatic amino acids.[2]

C. Migraine

Patients with chronic migraine headache may experience an increased frequency of headache after consumption of aspartame.[10,12] In some cases headaches occurred within 90 min of a 500 mg challenge,[9] but in others the association was only apparent after time. Because headache is a subjective symptom that may occur with equal frequency after placebo challenge, patients suspected of having aspartame-induced headache should undergo a placebo-controlled double-blind challenge.[11,14]

Capsules are the preferred challenge vehicle, since solutions are easily distinguished by taste. However, capsules may not mimic the typical use of aspartame beverages in terms of degradation products present. If a single-dose challenge is negative, a longer trial (i.e., 1 month) may be indicated. Aspartame and placebo capsules are available from the NutraSweet® Company.

D. Mitral Valve Prolapse

On the basis of a single case report of panic attack, it has been suggested that patients with mitral valve prolapse may be more sensitive to the effects of excessive aspartame intake. No double-blind rechallenge was done in this case, and it should be considered preliminary until confirmed under more controlled conditions.[15]

E. Glutamate Sensitivity

Because of the structural similarity of aspartame to glutamate, it has been suggested that patients with the glutamate-related "Chinese Restaurant Syndrome" might be similarly sensitive to aspartame. In a controlled study of six individuals shown to be sensitive to monosodium glutamate, no adverse effects were reported after challenge with aspartame 34 mg/kg.[30] Further study is needed before concluding that aspartame is safe in such individuals.

F. Diabetes

Aspartame has been convincingly not shown to present an increased risk to patients with either insulin or noninsulin-dependent diabetes or during pregnancy.[31,32]

G. Epilepsy

Although there are several case reports of seizures associated with aspartame consumption, there is no convincing evidence that epileptic patients are at increased risk. The Epilepsy Institute considers aspartame to be safe for people with epilepsy.[23]

REFERENCES

1. **Food and Drug Administration,** Food additives permitted for direct addition to food for human consumption: aspartame, *Fed. Reg.*, 48, 31378, 1983.
2. **Food and Drug Administration,** Aspartame as an inactive ingredient in human drug products: labelling requirements, *Fed. Reg.*, 48, 54993, 1983.
3. **Wurtman, R. J.,** Neurochemical changes following high-dose aspartame with dietary carbohydrates, *N. Engl. J. Med.*, 309, 429, 1983.

4. **Coulombe, R. A., Jr. and Sharma, R. P.,** Neurobiochemical alterations induced by the artificial sweetener aspartame (NutraSweet), *Toxicol. Appl. Pharmacol.,* 83, 79, 1986.
5. **Dailey, J. W., Lasley, S. M., Mishra, P. K., Bettendorf, A. F., Burger, R. L., and Jobe, P. C.,** Aspartame fails to facilitate pentylenetetrazol-induced convulsions in CD-1 mice, *Toxicol. Appl. Pharmacol.,* 98, 475, 1989.
6. **Zhi, J. and Levy, G.,** Aspartame and phenylalanine do not enhance theophylline-induced seizures in rats, *Res. Comm. Chem. Pathol. Pharmacol.,* 66, 171, 1989.
7. **Council on Scientific Affairs,** Aspartame review of safety issues, *JAMA,* 254, 400, 1985.
8. **Stegink, L. D., Brummel, M. C., Filer, L. J., Jr., and Baker, G. L.,** Blood methanol concentrations in one-year-old infants administered graded doses of aspartame, *J. Nutr.,* 113, 1600, 1983.
9. **Johns, D. R.,** Migraine provoked by aspartame, *N. Engl. J. Med.,* 315, 456, 1986.
10. **Koehler, S. M. and Glaros, A.,** The effect of aspartame on migraine headache, *Headache,* 28, 10, 1988.
11. **Schiffman, S. S., Buckley, C. E., III, Sampson, H. A., Massey, E. W., Baraniuk, J. N., Follett, J. V., and Warwick, Z. S.,** Aspartame and susceptibility to headache, *N. Engl. J. Med.,* 317, 1181, 1987.
12. **Lipton, R. B., Newman, L. C., Cohen, J. S., and Solomon, S.,** Aspartame as a dietary trigger of headache, *Headache,* 29, 90, 1989.
13. **Bradstock, M. K., Serdula, M. K., Marks, J. S., Barnard, R. J., Crane, N. T., Remington, P. L, and Trowbridge, F. L.,** Evaluation of reactions to food additives:the aspartame experience, *Am. J. Clin. Nutr.,* 43, 464, 1986.
14. **Leon, A. S., Hunninghake, D. B., Bell, C., Rassin, D. K., and Tephyl, T. R.,** Safety of long-term large doses of aspartame, *Arch. Intern. Med.,* 149, 2318, 1989.
15. **Drake, M. E.,** Panic attacks and excessive aspartame ingestion, *Lancet,* 2, 631, 1986.
16. **Wurtman, R. J.,** Aspartame: possible effect on seizure susceptibility, *Lancet,* 2, 1060, 1985.
17. **Walton, R. G.,** Seizure and mania after high intake of aspartame, *Psychosomatics,* 27, 218, 1986.
18. **Kruesi, M. J. P., Rapoport, J. L., Cummings, E. M., Berg, C. J., Ismond, D. R., Flament, M., Yarrow, M., and Zahn-Waxler, C.,** Effects of sugar and aspartame on aggression and activity in children, *Am. J. Psychiatry,* 144, 1487, 1987.
19. **Kulczycki, A., Jr.,** Aspartame-induced urticaria, *Ann. Intern. Med.,* 104, 207, 1986.
20. **Novick, N. L.,** Aspartame-induced granulomatous panniculitis, *Ann. Intern. Med.,* 102, 206, 1985.
21. **McCauliffe, D. P. and Poitras, K.,** Aspartame-induced lobular panniculitis, *J. Am. Acad. Dermatol.,* 24, 298, 1991.
22. **Kempe, C. H., Silver, H. K., O'Brien, D., and Fulginiti, V. A.,** *Current Pediatric Diagnosis & Treatment,* 9th ed., Appleton & Lange, East Norwalk, CT, 1987.
23. **Garriga, M. M. and Metcalfe, D. D.,** Aspartame intolerance, *Ann. Allergy,* 61, 63, 1988.
24. **Guttler, F. and Lou, H.,** Aspartame may imperil dietary control of phenylketonuria, *Lancet,* 1, 525, 1985.
25. **Koch, R., Schaeffler, G., and Shaw, K. N. F.,** Results of loading doses of aspartame by two phenylketonuric (PKU) children compared with two normal children, *J. Toxicol. Environ. Health,* 2, 459, 1976.
26. **Caballero, B., Mahon, B. E., Rohr, F. J., Levy, H. L., and Wurtman, R. J.,** Plasma amino acid levels after single-dose aspartame consumption in phenylketonuria, mild hyperphenylalaninemia, and heterozygous state for phenylketonuria, *J. Pediatr.,* 109, 668, 1986.
27. **Stegink, L. D., Filer, L. J., Jr., Baker, G. L., and McDonnell, J. E.,** Effect of an abuse dose of aspartame upon plasma and erythrocyte levels of amino acids in phenylketonuric heterozygous and normal adults, *J. Nutr.,* 110, 2216, 1980.
28. **Stegink, L. D., Filer, L. J., Baker, G. L., Bell, E. F., Ziegler, E. E., Brummel, M. C., and Krause, W. L.,** Repeated ingestion of aspartame-sweetened beverage: effect on plasma amino acid concentrations in individuals heterozygous for phenylketonuria, *Metabolism,* 38, 78, 1989.
29. **Stegink, L. D., Filer, L. J., Bell, E. F., Ziegler, E. E., Tephly, T. R., and Krause, W. L.,** Repeated ingestion of aspartame-sweetened beverages: further observations in individuals heterozygous for phenylketonuria, *Metabolism,* 39, 1076, 1990.
30. **Stegink, L. D., Filer, L. J., Jr., and Baker, G. L.,** Effect of aspartame and sucrose loading in glutamate-susceptible subjects, *Am. J. Clin. Nutr.,* 34, 1899–1905, 1981.
31. **Nehrling, J. K., Kobe, P., McLane, M. P., Olson, R. E., Kamath, S., and Horwitz, D. L.,** Aspartame use by persons with diabetes, *Diabetes Care,* 8, 415, 1985.
32. **Gupta, V., Cochran, C., Parker, T. F., Long, D. L., Ashby, J., Gorman, M. A., and Liepa, G. U.,** Effect of aspartame on plasma amino acid profiles of diabetic patients with chronic renal failure, *Am. J. Clin. Nutr.,* 49, 1302, 1989.

BENZALKONIUM CHLORIDE

I. REGULATORY CLASSIFICATION

Benzalkonium chloride is classified as an antimicrobial preservative. It is an effective bacteriostatic agent against most Gram-positive and some Gram-negative organisms.

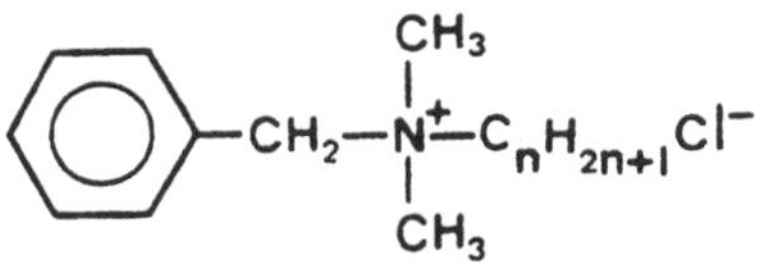

$$\langle\bigcirc\rangle - CH_2 - \overset{\overset{\displaystyle CH_3}{|}}{\underset{\underset{\displaystyle CH_3}{|}}{N^+}} - C_nH_{2n+1} \quad Cl^-$$

FIGURE 3. Benzalkonium chloride.

II. SYNONYMS

Alkylbenzyldimethylammonium chlorides
Cloreto de Benzalconio

III. AVAILABLE FORMULATIONS

A. Drugs

Manufacturers have been replacing thimerosal-preserved ophthalmic products with other antimicrobials due to increasing reports of adverse reactions to that excipient.[1] In many cases, the preservative has been replaced with benzalkonium chloride, which has become the most widely used ocular excipient. In January 1988, the FDA listed 92 approved ophthalmic products containing benzalkonium chloride and 12 products containing thimerosal.[2]

A side benefit of the toxic properties of benzalkonium to cell membranes is increased epithelial permeability, resulting in enhanced ocular drug penetration.[3]

B. Cosmetics

Benzalkonium chloride is used as an antimicrobial in cosmetics, such as baby products, hair conditioners and rinses, shampoos, personal cleanliness products, face creams and lotions, hand creams and lotions, deodorants, mouthwashes, dentifrices, and skin fresheners.[4,5] Benzalkonium may bleach the color out of dyed hair when used in shampoo formulations.[6]

31

IV. TABLE OF COMMON PRODUCTS

A. Ophthalmic Drug Products

Trade name	Manufacturer	Therapeutic class
Akarpine	Akorn	Miotic
Al-Cide susp	Akorn	Steroid/antibiotic
Ak-Con	Akorn	Decongestant
Ak-Con A	Akorn	Decongestant/antihistamine
Ak-Dilate	Akorn	Mydriatic
Ak-Nefrin	Akorn	Decongestant
Ak-Pentolate	Akorn	Cycloplegic
Ak-Pred	Akorn	Steroid
Ak-Rinse	Akorn	Irrigant
Ak-Spore H.C. susp	Akorn	Steroid/antibiotic
Ak-Taine	Akorn	Anesthetic
Ak-Trol susp	Akorn	Steroid/antibiotic
Akwa Tears	Akorn	Artificial tear
Albalon	Allergan	Decongestant
Albalon A	Allergan	Decongestant/antihistamine
Alcaine	Alcon	Anesthetic
Baldex soln	Bausch & Lomb	Steroid
Betagan	Allergan	Beta-blocker
Betoptic	Alcon	Beta-blocker
Blephamide	Allergan	Steroid/antibiotic
Clear Eyes	Ross	Decongestant
Collyrium Fresh	Wyeth-Ayerst	Decongestant
Comfort Drops	Barnes-Hind	Artificial tear
Cyclogyl	Alcon	Cycloplegic
Cyclomydril	Alcon	Cycloplegic
Decadron soln	Merck Sharp & Dohme	Steroid
Dacriose	IOLAB	Irrigant
Degest 2	Barnes-Hind	Decongestant
Dexacidin susp	IOLAB	Steroid/antibiotic
Dual Wet	Alcon	Artificial tear
Efricel	Professional	Decongestant
Econopred	Alcon	Steroid
Enuclene	Alcon	Cleaner/lubricant
Epifrin	Allergan	Antiglaucoma
Eye Stream	Alcon	Irrigant
Eyewash	Allergan	Irrigant
Fluor-Op	IOLAB	Steroid
FML susp	Allergan	Steroid
FML Forte	Allergan	Steroid
FML-S	Allergan	Steroid/antibiotic
Garamycin soln	Schering	Antibiotic

Trade name (cont'd)	Manufacturer (cont'd)	Therapeutic class (cont'd)
Gentak soln	Akorn	Antibiotic
Genoptic soln	Allergan	Antibiotic
Gentacidin soln	IOLAB	Antibiotic
Gentrasul soln	Bausch & Lomb	Antibiotic
Gonak	Akorn	Demulcent
Herplex	Allergan	Antiviral
HMS susp	Allergan	Steroid
Humorsol	Merck Sharp & Dohme	Antiglaucoma
Hypotears	IOLAB	Artificial tear
Infectrol susp	Bausch & Lomb	Steroid/antibiotic
Inflamase Mild	IOLAB	Steroid
Inflamase Forte	IOLAB	Steroid
Iopidine	Alcon	Antiglaucoma
Isopto Alkaline	Alcon	Artificial tear
Isopto Atropine	Alcon	Cycloplegic
Isopto Carbachol	Alcon	Miotic
Isopto Carpine	Alcon	Miotic
Isopto Cetapred	Alcon	Steroid
Isopto Frin	Alcon	Decongestant
Isopto Homatropine	Alcon	Cycloplegic
Isopto Hyoscine	Alcon	Cycloplegic
Isopto Plain	Alcon	Artificial tear
Isopto Tears	Alcon	Artificial tear
Lavoptik	Lavoptik	Irrigant
Lyteers	Barnes-Hind	Artificial tear
Maxidex susp	Alcon	Steroid
Maxitrol susp	Alcon	Steroid/antibiotic
Metimyd susp	Schering	Steroid/antibiotic
Murine	Ross	Decongestant
Murine Plus	Ross	Decongestant
Murocoll 2	Bausch & Lomb	Cycloplegic
Mydfrin	Alcon	Decongestant
Mydriacyl	Alcon	Cycloplegic
Naphcon	Alcon	Decongestant
Naphcon A	Alcon	Decongestant/antihistamine
Naphcon Forte	Alcon	Decongestant
Natacyn	Alcon	Antifungal
Neodecadron soln	Merck Sharp & Dohme	Steroid/antibiotic
Ocuclear	Schering	Decongestant
Opcon	Bausch & Lomb	Decongestant
Opcon A	Bausch & Lomb	Decongestant/antihistamine
Ophthetic	Allergan	Anesthetic
Opticrom	Fisons	Antiallergic
Pilocar	IOLAB	Antiglaucoma

Trade name (cont'd)	Manufacturer (cont'd)	Therapeutic class (cont'd)
Pilopine HS gel	Alcon	Miotic
Pilostat	Bausch & Lomb	Miotic
Pred Forte	Allergan	Steroid
Pred G susp	Allergan	Steroid/antibiotic
Pred Mild	Allergan	Steroid
Prefrin	Allergan	Decongestant
Prefrin A	Allergan	Decongestant/antihistamine
Propine	Allergan	Antiglaucoma
Soothe	Alcon	Artificial tear
Tear-Efrin	Cooper	Decongestant
Tearisol	IOLAB	Artificial tear
Tears Naturale	Alcon	Artificial tear
Tears Renewed	Akorn	Artificial tear
Timoptic Ocumeter	Merck Sharp & Dohme	Beta-blocker
Tobradex susp	Alcon	Steroid/antibiotic
Tobrex soln	Alcon	Antibiotic
Tropicacyl	Akorn	Cycloplegic
Vasocon A	IOLAB	Decongestant/antihistamine
Visine	Leeming	Decongestant
Visine A.C.	Leeming	Decongestant
Visine Extra	Leeming	Decongestant
Zincfrin	Alcon	Decongestant

B. Ophthalmic Contact Lens Products

Trade name	Manufacturer
Liquifilm Wetting solution	Allergan
Soakare soaking solution	Allergan
Total solution	Allergan
Wet-N-Soak Wetting/Soaking	Allergan
Wet-N-Soak Plus	Allergan

C. Inhalation Drug Products

Trade name	Manufacturer
Alupent solution for inhalation 5%	Boehringer Ingelheim
Proventil solution for inhalation	Schering
Ventolin solution for inhalation 5%	Allen & Hanburys

V. ANIMAL TOXICITY DATA

A. Ototoxicity

Instillation of 0.1% benzalkonium chloride into the inner ear of guinea pigs for 10 to 60 min produced vestibular and cochlear damage.[7]

B. Ocular Toxicity

Application of benzalkonium chloride 0.02% completely inhibited corneal epithelial healing in rabbits.[8]

VI. HUMAN TOXICITY DATA

A. Bronchoconstriction

In 1982 and 1983, cases of ipratropium bromide-induced paradoxical bronchoconstriction were first reported,[9-11] and the search for the causative ingredient began. At first, the hypotonicity of the product (Atrovent) was demonstrated to be a factor, and it was reformulated as an isotonic solution.[12] Reports of bronchoconstriction continued despite the reformulation.[13]

Beasley[14] suggested that benzalkonium chloride was responsible for these paradoxical reactions, based on *in vitro* work demonstrating histamine release.[15,16] He subsequently demonstrated dose-related bronchoconstriction in 6 of 20 stable asthmatic patients exposed to nebulized benzalkonium chloride alone in the concentration found in the ipratropium product, the ipratropium bromide solution itself, containing both EDTA and benzalkonium, and following EDTA alone, only in concentrations more than twice that found in the product.[17] A case report in a 3-year-old boy who demonstrated paradoxical bronchoconstriction secondary to a benzalkonium chloride-preserved beclomethasone nebulizer solution strengthened the association with this preservative, but was not conclusive.[18]

Paradoxical deterioration was also reported in a 16-month-old child receiving nebulized terbutaline preserved with benzalkonium chloride. Inhalation challenge with nonpreserved terbutaline produced a favorable response.[19] Because the paradoxical reaction is dose related, patients may experience an additive effect when receiving combined therapy with preserved inhalation products.

A study in 30 asthmatic patients showed significant bronchoconstriction following Atrovent nebulization in 5 patients when compared to a nonpreserved isotonic ipratropium preparation. Atopic patients had more severe and prolonged responses than nonatopic patients, suggesting that sensitization plays a role in the pathogenesis of this reaction. The author speculated that the mast cell secretagogue activity of benzalkonium chloride was responsible, perhaps compounded by enhanced airway smooth muscle contractility by EDTA.[20]

O'Callaghan et al. proposed two other possible etiologies for paradoxical bronchoconstriction observed in infants given nebulized preservative-free ipratropium treatments; first that the acidic pH (3.6) may be responsible, and second that the nonselective muscarinic antagonist properties of ipratropium could overwhelm the bronchodilator effect by blocking prejunctional receptors.[21]

A reproducible dose-related bronchoconstrictive response was demonstrated in a study of 28 stable asthmatics challenged with benzalkonium inhalation. The peak response occurred 1 min after inhalation with a return to baseline within 60 min. Patients with a documented increased sensitivity to histamine were also more sensitive to benzalkonium. Inhalation of benzalkonium also enhanced the response to histamine challenges given 1 h later. The reactions were inhibited by prior treatment with cromolyn sodium.[22]

B. Nasal Toxicity

Nasal congestion and ocular pruritus were reported by a 33-year-old woman following use of a benzalkonium chloride-preserved nasal decongestant drop. The reaction was confirmed by rechallenge with an aqueous benzalkonium chloride solution. The patient had preexisting asthma and had previously demonstrated the aspirin/tartrazine triad hypersensitivity reaction.[23]

A mild Loffler's syndrome, consisting of nonproductive cough, mild dyspnea, and pleuritic pain, was temporally associated with the use of a benzalkonium chloride-preserved nasal decongestant in one patient; however, the excipient was not confirmed to be causative.[24]

C. Ocular toxicity

Benzalkonium chloride is a commonly used preservative in ophthalmic solutions. The eye appears to be more sensitive to irritant effects than the skin. Concentrations as low as 0.015% will produce a 60% incidence of corneal cell lysis within 1 h.[25]

Similar to other ophthalmic preservatives, benzalkonium chloride may bind to contact lens materials, providing continuous exposure to the cornea. In a 36-year-old woman with an aphakic contact lens, a corneal ulcer developed 3 d after switching from a thimerosal-preserved to a benzalkonium-preserved saline solution.[26]

Patients with disrupted corneal epithelium are more sensitive to the toxic effects of benzalkonium chloride. A 56-year-old man with extensive ocular surface disease developed corneal epithelial damage requiring corneal transplantation after using benzalkonium-preserved ophthalmic products over an 8-year period.[27]

Contact hypersensitivity reactions may also occur following ocular exposure. Application of two different prednisolone ophthalmic products produced eyelid edema, conjunctivitis, tearing, and burning in one patient. Cutaneous patch testing demonstrated strong positive reactions to benzalkonium chloride, the preservative in common with both ophthalmic products.[28]

Severe allergic conjunctivitis was reported in five patients sensitized via occupational exposure to medical products.[5] Eyelid contact dermatitis has also been reported following use of a benzalkonium-preserved hard contact lens wetting solution.[29] Contact lens exposure generally results in a pruritic conjunctivitis with no involvement of the eyelids.[30]

The chronic use of benzalkonium-preserved artificial tears may create an unstable tear film and aggravate the dry eye condition.[31,32]

D. Direct Irritant Reactions

Benzalkonium aqueous solutions of 0.5% and greater are directly irritating. Moderate to severe reactions occurred in 66% of the subjects.[33] A concentration of 0.1% is recommended for patch testing.

E. Delayed Hypersensitivity Reactions

Allergic contact dermatitis has been reported infrequently to cosmetics containing benzalkonium chloride. In one reported case, daily application of a preserved roll-on deodorant for several years resulted in axillary dermatitis.[34]

Occupational contact dermatitis has been described in surgical personnel following application of skin disinfectants or use of instruments soaked in benzalkonium chloride.[35]

Topical exposure to benzalkonium as a preservative in plaster of Paris casts and adhesive strip bandages has resulted in erythema, pruritus, and eczematous eruptions.[5,36,37]

A benzalkonium chloride solution used to disinfect a tracheostomy catheter was implicated in causing severe hemorrhagic contact dermatitis of the tracheal mucosa at the tracheostomy site, resulting in repeated episodes of breathing difficulty and spasmodic coughing in a 3-month-old infant with croup. Patch testing confirmed hypersensitivity to the disinfectant solution.[38]

VI. CLINICAL RELEVANCE

A. Delayed Hypersensitivity

The incidence of positive patch test reactions occurred in two patients out of several thousand tested in North America.[39] In a smaller series of 157 eczema patients, 5% had positive reactions.[40] Similarly, in a series of 142 otitis externa patients chronically exposed to benzalkonium-preserved otic solutions, 6.3% tested positive.[41] The higher incidence in this series may reflect a heavily exposed population with underlying otic pathology.

In a population of 100 patients with chronic conjunctivitis, the incidence of positive patch tests to 0.07% benzalkonium chloride was 6%. Positive reactions were confirmed by serial dilutions of the test material, with two patients reacting to concentrations as low as 0.005%. Again, this population had been continuously exposed to various ophthalmic preparations for at least 3 months.[42]

Patients with occupational exposure to benzalkonium chloride, such as physicians, nurses, dentists, and veterinarians, may be more likely to develop contact dermatitis from subsequent exposure to preserved ophthalmic or cosmetic products.[5]

B. Bronchoconstriction

Asthmatic patients with moderate or severe hyper-responsive airways as documented by histamine challenge may develop a paradoxical bronchoconstriction after inhalation of benzalkonium chloride-preserved antiasthmatic medications. The reactions are rapid in onset and may be prolonged for 30 to 60 min in most sensitive individuals. The response is dose related, and combination therapy with two or more preserved preparations may increase the likelihood of an adverse response. Administration of bronchodilator medications via a spacer device may minimize contact with the excipients and prevent this reaction.[21] Although the majority of recent evidence implicates benzalkonium chloride as the causative agent, an additive reaction to other inhaled excipients, such as EDTA and the osmolarity and pH of the solution, should be considered.

REFERENCES

1. **Olson, R. J. and White, G. L.,** Preservatives in ophthalmic topical medications: a significant cause of disease, *Cornea,* 9, 363, 1990.
2. **Weiner, M. and Bernstein, I. L.,** *Adverse Reactions to Drug Formulation Agents,* Marcel Dekker, New York, 1989.
3. **Mondino, B. J. and Groden, C. R.,** Conjunctival hyperemia and corneal infiltrates with chemically disinfected soft contact lenses, *Arch. Ophthalmol.,* 98, 1767, 1980.
4. **Nikitakis, J. M.,** *CTFA Cosmetic Ingredient Handbook,* 1st ed., The Cosmetic, Toiletry and Fragrance Association, Washington, D.C., 1988.
5. **Fisher, A. A.,** Allergic contact dermatitis and conjunctivitis from benzalkonium chloride, *Cutis,* 39, 381, 1987.
6. **Spoor, H. J.,** Shampoos and hair dyes, *Cutis,* 20, 189, 1977.
7. **Aursnes, J.,** Ototoxic effect of quaternary ammonium compounds, *Acta Otolaryngol.,* 93, 421, 1982.
8. **Collin, H. B. and Grabsch, B. E.,** The effect of ophthalmic preservatives on the healing rate of the rabbit corneal epithelium after keratectomy, *Am. J. Optom. Physiol. Opt.,* 59, 215, 1982.
9. **Connolly, C. K.,** Adverse reaction to ipratropium bromide, *Br. Med. J.,* 285, 934, 1982.
10. **Jolobe, O. M. P.,** Adverse reaction to ipratropium bromide, *Br. Med. J.,* 285, 1425, 1982.
11. **Howarth, P. H.,** Bronchoconstriction in response to ipratropium bromide, *Br. Med. J.,* 286, 1825, 1983.

12. **Mann, J. S., Howarth, P. H., and Holgate, S. T.,** Bronchoconstriction induced by ipratropium bromide in asthma: relation to hypertonicity, *Br. Med. J.,* 289, 469, 1984.

13. **Patel, K. R. and Tullet, W. M.,** Bronchoconstriction in response to ipratropium bromide, *Br. Med. J.,* 286, 1318, 1983.

14. **Beasley, C. R. W., Rafferty, P., and Holgate, S.,** Benzalkonium chloride and bronchoconstriction, *Lancet,* 2, 1227, 1986.

15. **Coleman, J. W., Holgate, S. T., Church, M. K., and Godfrey, R. C.,** Immunoglobulin E decapeptide-induced 5-hydroxytryptamine release from rat peritoneal mast cells: comparison with corticotripin-(1-24)peptide, polyarginine, polylysine and antigen, *Biochem. J.,* 198, 615, 1981.

16. **Read, G. W. and Kiefer, E. F.,** Benzalkonium chloride: selective inhibitor of histamine release induced by compound 48/80 and other polyamines, *J. Pharmacol. Exp. Ther.,* 211, 711, 1979.

17. **Beasley, C. R. W., Rafferty, P., and Holgate, S. T.,** Bronchoconstrictor properties of preservatives in ipratropium bromide (Atrovent) nebuliser solution, *Br. Med. J.,* 294, 1197, 1987.

18. **Clark, R. J.,** Exacerbation of asthma after nebulised beclomethasone diproprionate, *Lancet,* 2, 574, 1986.

19. **Menendez, R., Lowe, R. S., and Kersey, J.,** Benzalkonium chloride and bronchoconstriction, *J. Allergy Clin. Immunol.,* 84, 272, 1989.

20. **Rafferty, P., Beasley, R., and Holgate, S. T.,** Comparison of the efficacy of preservative free ipratropium bromide and Atrovent nebuliser solution, *Thorax,* 43, 446, 1988.

21. **O'Callaghan, C., Milner, A. D., and Swarbrick, A.,** Paradoxical bronchoconstriction in wheezing infants after nebulised preservative free iso-osmolar ipratropium bromide, *Br. Med. J.,* 299, 1433, 1989.

22. **Zhang, Y. G., Wright, W. J., Tam, W. K., Nguyen-Dang, T. H., Salome, C. M., and Woolcock, A. J.,** Effect of inhaled preservatives on asthmatic subjects. II. Benzalkonium chloride, *Am. Rev. Respir. Dis.,* 141, 1405, 1990.

23. **Hillerdal, G.,** Adverse reaction to locally applied preservatives in nose drops, *Oto-Rhino-Laryngology,* 47, 278, 1985.

24. **Cohen, H. P. and Israel, R. H.,** Loffler's syndrome secondary to "NTZ" nose drops: a self-limited illness, *Respiration,* 38, 168, 1979.

25. **Neville, R., Dennis, P., Sens, D., and Crouch, R.,** Preservative cytotoxicity to cultured corneal epithelial cells, *Curr. Eye Res.,* 5, 367, 1986.

26. **Gasset, A. R.,** Benzalkonium chloride toxicity to the human cornea, *Am. J. Ophthalmol.,* 84, 169, 1977.

27. **Lemp, M. A. and Zimmerman, L. E.,** Toxic endothelial degeneration in ocular surface disease treated with topical medications containing benzalkonium chloride, *Am. J. Ophthalmol.,* 105, 670, 1988.

28. **Fisher, A. A. and Stillman, M. A.,** Allergic contact sensitivity to benzalkonium chloride, *Arch. Dermatol.,* 106, 169, 1972.

29. **Yorav, S., Ronnen, M., and Suster, S.,** Eyelid contact dermatitis due to Liquifilm wetting solution of hard contact lenses, *Contact Dermatitis,* 17, 314, 1987.

30. **Fisher, A. A.,** Allergic reactions to contact lens solutions, *Cutis,* 36, 209, 1985.

31. **Gobbels, M. and Spitznas, M.,** Influence of artificial tears on corneal epithelium in dry eye syndrome, *Arch. Clin. Exp. Ophthalmol.,* 227, 139, 1989.

32. **Gilbard, J. P., Rossi, S. R., and Heyda, K. G.,** Ophthalmic solutions, the ocular surface, and a unique therapeutic artificial tear formation, *Am. J. Ophthalmol.,* 107, 348, 1989.

33. **Holst, R. and Moller, H.,** One hundred twin pairs patch tested with primary irritants, *Br. J. Dermatol.,* 93, 145, 1975.

34. **Shmunes, E. and Levy, E. J.,** Quaternary ammonium compound contact dermatitis from a deodorant, *Arch. Dermatol.,* 105, 91, 1972.

35. **Wahlberg, J. E.,** Two cases of hypersensitivity to quaternary ammonium compounds, *Acta Derm. Venereol.,* 42, 230, 1962.

36. **Lovell, C. R. and Staniforth, P.,** Contact allergy to benzalkonium chloride in plaster of Paris, *Contact Dermatitis,* 7, 343, 1981.

37. **Staniforth, P.,** Allergy to benzalkonium chloride in plaster of Paris after sensitisation to cetrimide: a case report, *J. Bone Joint Surg.,* 62B, 500, 1980.

38. **Padnos, E., Horwitz, I. D., and Wunder, G.,** Contact dermatitis complicating tracheostomy: causative role of aqueous solution of benzalkonium chloride, *Am. J. Dis. Child.,* 109, 90, 1965.

39. **Adams, R. M., Maibach, H. I., Clendenning, W. E., et al.,** North American Contact Dermatitis Group: a five year study of cosmetic reactions, *J. Am. Acad. Dermatol.,* 13, 1062, 1985.

40. **Garcia-Perez, A. and Moran, M.,** Dermatitis from quaternary ammonium compounds, *Contact Dermatitis,* 1, 316, 1975.

41. **Fraki, J. E., Kalimo, K., Tuohimaa, P., and Aantaa, E.,** Contact allergy to various components of topical preparations for treatment of external otitis, *Acta Otolaryngol.,* 100, 414, 1985.

42. **Afzelius, H. and Thulin, H.,** Allergic reactions to benzalkonium chloride, *Contact Dermatitis,* 5, 60, 1979.

BENZOIC ACID

I. REGULATORY CLASSIFICATION

Benzoic acid is classified as an antimicrobial preservative agent.

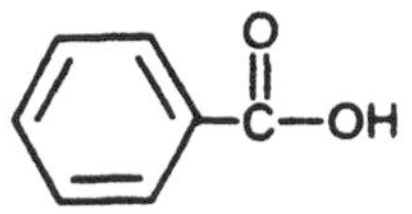

FIGURE 4. Benzoic acid.

II. SYNONYMS

Acidum benzoicum
Benzenecarboxylic acid
Benzoesaure
Carboxybenzene
Dracylic acid
Flowers of benzoin
Phenylcarboxylic acid
Phenylformic acid

III. AVAILABLE FORMULATIONS

A. Foods
The GRAS status of benzoic acid has been confirmed by the FDA. It occurs naturally in some foods and may be added as a preservative in concentrations not exceeding 0.1% for combined benzoic acid and sodium benzoate.[1]

B. Drugs
Benzoic acid is a moderately effective bacteriostatic and fungistatic agent at a pH of 5 or less. A concentration of 0.1% is used to preserve oral liquid and parenteral products. It is present as an active ingredient in the topical antifungal compound, Compound Benzoic Acid Ointment, in a concentration of 5%.

Benzoic acid is listed as an inactive ingredient in 16 injectable products in concentrations of 0.17 to 5%; 32 oral concentrates, solutions, elixirs, suspensions, and syrups in concentrations of 0.01 to 0.5%; 11 topical creams, lotions, and solutions in concentrations of 0.1 to 0.2%; and 6 vaginal foams, creams, or sponges.[2]

C. Cosmetics

Benzoic acid was among the top 20 preservatives used in cosmetics in 1982 and was present in 116 of 20,183 (0.6%) products surveyed by the FDA.[3] It is also used in cosmetics to adjust the pH. Types of products include bath products, sachets, shampoos, hair conditioners and rinses, cleansing products, skin fresheners, creams, and lotions.[4]

IV. TABLE OF COMMON PRODUCTS

A. Foods[1]
1. Natural Sources

Anise
Cinnamon
Cloves
Cranberries
Prunes
Raspberries
Tea

2. Food Additive Sources

Alcoholic beverages
Baked goods
Cheese
Chewing gum
Condiments
Fats and oils
Frozen dairy products
Gelatins
Imitation dairy products
Nonalcoholic beverages
Puddings
Relishes
Soft candy
Sugar substitutes

B. Parenteral Drug Products

Trade name	Manufacturer
Valium injectable	Roche

C. Topical Drug Products

Trade name	Manufacturer
Balmex ointment	Macsil
Delfen foam	Ortho
Gynol II jelly	Ortho
Micatin cream	Ortho
Monistat-Derm	Ortho
Ortho-Dienestrol cream	Ortho
Oxistat cream	Glaxo
Spectrazole cream	Ortho
Today sponge	Whitehall

D. Oral Liquid Drug Products

Trade name	Manufacturer
Agoral	Parke-Davis
Aldomet oral suspension	Merck Sharp & Dohme
Aludrox suspension	Wyeth-Ayerst
Cheracol D	Upjohn
Children's Tylenol liquid	McNeil
Decadron elixir	Merck Sharp & Dohme
Demerol syrup	Winthrop
Diuril oral suspension	Merck Sharp & Dohme
Gantrisin syrup	Roche
Geritol liquid	Beecham
Neo-Calglucon syrup	Sandoz
Pediacare cough-cold formula liquid	McNeil
Pediacare Nightrest cough-cold formula	McNeil
Pediacare infants oral decongestant drop	McNeil
Pepto-Bismol maximum strength liquid	Procter & Gamble
Rynatan suspension	Wallace
Rynatuss suspension	Wallace
Tacaryl syrup	Westwood-Squibb
Taractan syrup	Roche
Triaminic expectorant with codeine	Sandoz
Triaminic expectorant DH	Sandoz
Triaminic oral infant drops	Sandoz

V. ANIMAL TOXICITY DATA

Based on animal toxicology data, the WHO has recommended a maximum acceptable daily intake of up to 5 mg/kg as the sum of benzoic acid, potassium benzoate, and sodium benzoate.[5]

VI. HUMAN TOXICITY DATA

A. Nonimmunologic Contact Urticaria

Topical exposure to benzoic acid 5% in petrolatum induced nonimmunologic contact urticaria in 52% of 105 subjects tested.[6] The typical response to open patch testing in both atopic and nonatopic individuals is tingling, burning, and itching, lasting about 30 min. A limited study of six subjects reported positive open patch tests in all six tested with 0.25% benzoic acid in aqueous solution and in 50% tested with 5% in petrolatum.[7]

Contact urticaria secondary to perioral contact from foods containing benzoic acid has also been described. This was observed in 18 of 20 children, aged 1 to 4 years, who smeared a mayonnaise and fruit salad around their mouths. Subsequent testing of ten healthy adults with the salad dressing and a 0.1% aqueous solution of benzoic acid applied in a closed 20-min patch test to the perioral region, disclosed erythema and stinging after 10 to 30 min in 1 subject. A dose-response curve was shown in 41 subjects, which plateaued at 1%, with 34% reacting to 0.1% and 63.4% reacting to 1, 5, or 10% concentrations.[8]

There are differences in the sensitivity of various application sites to the urticariogenic activity of benzoic acid. The most sensitive area is the face, followed by the antecubital space, inner forearm, lower back, and leg.[9]

Pretreatment with indomethacin 50 mg three times a day completely inhibits the urticarial reaction to benzoic acid. Although this was suggestive of a prostaglandin effect, the specific mechanism has not been proven.[10]

B. Chronic Urticaria

Sodium benzoate has been shown to exacerbate symptoms in patients with recurrent urticaria and angioedema in open challenge procedures.[11] The incidence of positive challenges was 5.8% in 308 patients tested by one investigator.[12] A placebo effect was suggested by Lahti and Hannuksela, who administered benzoic acid 500 mg in a double-blind placebo-controlled fashion (using lactose as the placebo) to 150 dermatological inpatients. Objective symptoms (rash, lip or throat edema, rhinitis, and urticaria) were seen in 3 to 7% of the benzoic acid group and in 6 to 14% in the placebo group.[13]

C. Purpura

Allergic vascular purpura was diagnosed in a 60-year-old woman 1 month after taking a cough medicine containing benzoic acid. Oral provocation with sodium benzoate resulted in pronounced worsening of the purpura. Aspirin, tartrazine, and 4-hydroxybenzoic acid were also implicated. The purpura resolved after a diet eliminating salicylates, benzoates, and dyes.[14]

D. Asthma

Oral provocation with benzoic acid resulted in positive reactions (asthma, rhinitis, or urticaria) in 47% of 100 patients with chronic, steroid-dependent asthma. The incidence of positive challenges was similar for tartrazine (54%).[15] A German study found intolerance to benzoic acid in 11.5% of 96 asthma patients and intolerance to tartrazine in 7.3%.[12] Most subsequent studies of this phenomenon involved sodium benzoate, which has better gastrointestinal tolerance by the oral route.

The mechanism of action of sodium benzoate was studied in an *in vitro* model using blood from healthy volunteers. The presence of 5 to 15 mM of sodium benzoate inhibited formation of thromboxane B2 by noradrenaline-activated platelets to a degree about one half that seen with 0.1 mM of aspirin.[16] These concentrations are comparable to those achieved in neonates receiving 500 mg/kg/d intravenously.[17] The inhibition of the cyclooxygenase pathway of arachidonic acid metabolism is predicted to cause preferential metabolism via the lipoxygenase pathway, resulting in formation of bronchospasm-inducing leukotrienes. This activity remains to be proven in serum levels achievable by excipient doses.

VII. CLINICAL RELEVANCE

A. Urticaria

Nonimmunologic contact urticaria occurs in about 34% of normal subjects exposed to preservative concentrations. Symptoms of tingling, accompanied by a pronounced flare, appear within 30 min and generally last less than 2 h.

Exacerbation of chronic urticaria has also been described in patients who ingest foods containing this additive. A large placebo effect has been observed in adults, making the open challenge studies difficult to interpret.

B. Asthma

Unselected groups of asthmatic patients have demonstrated exacerbation of symptoms in about 11.5 to 47% in unblinded open challenge studies. The incidence observed with sodium benzoate of 2.2 to 3.6% following double-blind placebo-controlled challenges may be a more reasonable estimate. The mechanism is unclear, but may be related to inhibition of cyclooxygenase.

REFERENCES

1. **Lecos, C.,** Food preservatives: a fresh report, *FDA Consumer,* U.S. Department of Health and Human Services, Washington, D.C., April 1984.
2. **Food and Drug Administration,** *Inactive Ingredients Guide,* FDA, Washington, D. C., March 1990.
3. **Decker, R. L. and Wenninger, J. A.,** Frequency of preservative use in cosmetic formulas as disclosed to FDA-1982 update, *Cosmet. Toilet.,* 97, 57, 1982.
4. **Nikitakis, J. M.,** *CTFA Cosmetic Ingredient Handbook,* 1st ed., The Cosmetic, Toiletry and Fragrance Association, Washington, D.C., 1988.
5. **World Health Organization,** Seventeenth report of the joint FAO/WHO expert committee on food additives, Tech. Rep. Ser. No. 539, WHO, Geneva, 1974.
6. **Lahti, A.,** Non-immunologic contact urticaria, *Acta Derm. Venereol.,* 50 (Suppl. 91), 1, 1980.
7. **Nethercott, J. R., Lawrence, M. J., Roy, A., and Gibson, B. L.,** Airborne contact urticaria due to sodium benzoate in a pharmaceutical manufacturing plant, *J. Occup. Med.,* 26, 734, 1984.
8. **Clemmensen, O. and Hjorth, N.,** Perioral contact urticaria from sorbic acid and benzoic acid in a salad dressing, *Contact Dermatitis,* 8, 1, 1982.
9. **Gollhausen, R. and Kligman, A. M.,** Human assay for identifying substances which induce non-allergic contact urticaria: the NICU-test, *Contact Dermatitis,* 13, 98, 1985.
10. **Lahti, A., Oikarinen, A., Viinikka, L., Ylikorkala, O., and Hannuksela, M.,** Prostaglandins in contact urticaria induced by benzoic acid, *Acta Derm. Venereol.,* 63, 425, 1983.
11. **Ros, A., Juhlin, L., and Michaelsson, G.,** A follow-up study of patients with recurrent urticaria and hypersensitivity to aspirin, benzoates and azo dyes, *Br. J. Dermatol.,* 95, 19, 1976.
12. **Wuthrich, B. and Fabro, L.,** Acetylsalicylsaure-und Lebensmitteladditiva — Intoleranz bei urtikaria, asthma bronchiale und chronischer rhinopathie, *Schweiz. Med. Wochenschr.,* 111, 1445, 1981.
13. **Lahti, A. and Hannuksela, M.,** Is benzoic acid really harmful in cases of atopy and urticaria?, *Lancet,* 2, 1055, 1981.

14. **Michaelsson, G., Pettersson, L., and Juhlin, L.,** Purpura caused by food and drug additives, *Arch. Dermatol.,* 109, 49, 1974.

15. **Rosenhall, L. and Zetterstrom, O.,** Asthmatic patients with hypersensitivity to aspirin, benzoic acid and tartrazine, *Tubercle,* 56, 168, 1975.

16. **Williams, W. R., Pawlowicz, A., and Davies, B. H.,** Aspirin-like effects of selected food additives and industrial sensitizing agents, *Clin. Exp. Allergy,* 19, 533.

17. **Green, T. P., Marchessault, R. P., and Freese, D. K.,** Disposition of sodium benzoate in newborn infants with hyperammonemia, *J. Pediatr.,* 102, 785, 1983.

Benzyl Alcohol

I. REGULATORY CLASSIFICATION

Benzyl alcohol is classified as an antimicrobial preservative. It is bacteriostatic and moderately active against Gram-positive bacteria, molds, fungi, and yeasts in concentrations of 0.9 to 2. It is used as a solubilizing agent in aqueous or oily preparations in concentrations of 5 or more. In concentrations of about 10, it is used as a local anesthetic and disinfectant, however concentrations of 0.9 also produce local anesthesia and reduce the pain produced by intradermal injections.[1]

Following reports of toxicity in neonates due to benzyl alcohol, the FDA published an intent to require labeling of multiple-dose parenterals containing any antimicrobial preservative to include a caution against use in neonates, but ruled in 1989 not to require labeling, since no new cases of toxicity had been reported.[2] The USP has required labeling of bacteriostatic water or sodium chloride for injection with the phrase "Not for use in newborns" since 1983.

II. SYNONYMS

Benzenemethanol
Phenylcarbinol
Phenylmethanol

III. AVAILABLE FORMULATIONS

A. Drugs
Benzyl alcohol is most commonly used in large volume or multiple-use parenterals.

B. Cosmetics
Various cosmetic products may contain benzyl alcohol as a preservative, fragrance component, or solvent. Types of products include hair conditioners, shampoos, sprays, dyes, rinses, and makeup such as foundations, facial creams and lotions, moisturizing creams and lotions, perfumes, and aftershave lotions.[3] The commonly used hydrophilic cosmetic and pharmaceutical cream base Aquatain® contains benzyl alcohol.

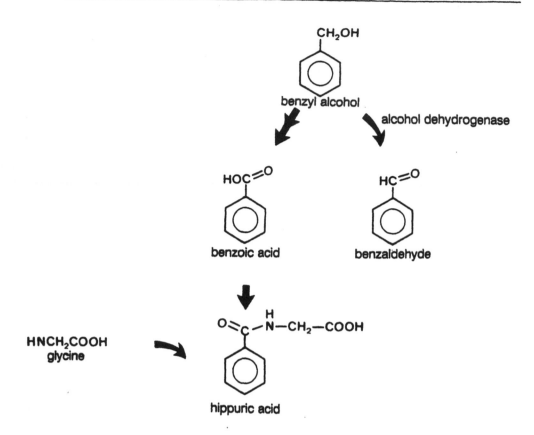

FIGURE 5. Benzyl alcohol metabolic pathway.

IV. TABLE OF COMMON PRODUCTS AND ALTERNATIVES

A. Parenteral Drug Products

Trade name/(manufacturer)	Concentration (%)	Alternative manufacturer
A-Hydrocort (Abbott)	0.9	None
A-Methapred (Abbott)	0.9	None
Amicar (Lederle)	0.9	None
Aminophylline	2	None
AquaMephyton neonatal (Merck Sharp & Dohme)	0.9	None
Ativan (Wyeth-Ayerst)	2	None
Atropine sulfate (various)	1.5	Abbott, LyphoMed
Bacteriostatic saline (various)	0.9	Abbott, Baxter, Kendall, Elkins-Sinn
Bacteriostatic water for injection (various)	0.9	Elkins-Sinn 10 ml ampule
Bactrim IV (Roche)	1	None
Berocca Parenteral Nutrition (Roche)	1	MVI (Rorer)
Bumex injection (Roche)	1	None
Compazine multidose vial/disp syringe (Smith Kline & French)	0.75	2 ml ampules

Trade name/(manufacturer) (cont'd)	Concentration (%) (cont'd)	Alternative manufacturer (cont'd)
Decadron LA (Merck Sharp & Dohme)	0.9	None
Dexamethasone (Elkins-Sinn)	1	Merck Sharp & Dohme
Diazepam	1.5	None
Dopram (Robins)	0.9	None
Dramamine injection (Searle)	5	None
Factrel injection (Wyeth-Ayerst)	2	None
Folate sodium	1.5	None
Glycopyrrolate (various)	0.9	None
Haldol Decanoate (McNeil)	1.2	None
Heparin 1000 U/m (various)	1	LyphoMed, Squibb, Abbott, Winthrop
Hydeltra TBA (Merck Sharp & Dohme)	0.9	None
Hydrocortone acetate (Merck Sharp & Dohme)	0.9	None
Kenalog-10 (Squibb)	0.9	None
Kenalog-40 (Squibb)	0.75	None
Levsin injection (Schwarz)	1.5	None
Librium injectable (Roche)	1.5	None
Lupron (TAP)	0.9	Lupron depot
Myochrysine (Merck Sharp & Dohme)	0.5	None
Navane IM (Roerig)	NS	None
Netromycin (Schering)	1	None
Norcuron (Organon)	0.9	Reconstitute with sterile water for injection
Pavulon (Organon)	1	None
Pepcid IV multidose vial (Merck Sharp & Dohme)	0.9	2ml vial (Merck Sharp & Dohme)
Phenobarbital (Elkins-Sinn)	1.5	Lilly
Physostigmine (various)	2	None
Premarin IV (Wyeth-Ayerst)	2	None
Prolixin Decanoate (Princeton)	1.2	None
Prolixin Enanthate (Princeton)	1.5	None
Pyridostigmine (Organon)	1	ICN
Robinul (Robins)	0.9	None
Septra IV (Burroughs Wellcome)	1	None
Solu-Cortef (Upjohn)	0.9	None
Solu-Medrol (Upjohn)	0.9	None
Stelazine (Smith Kline & French)	0.75	None
Succinylcholine (Organon)	1	Abbott, Burroughs Wellcome, Squibb
Thorazine multidose vial (Smith Kline & French)	2	1, 2 ml ampules
Tracrium multidose vial (Burroughs Wellcome)	0.9	5 mL ampules
Tubocurarine (Abbott, Squibb)	0.9	Lilly

Trade name/(manufacturer) (cont'd)	Concentration (%) (cont'd)	Alternative manufacturer (cont'd)
Valium injectable (Roche)	1.5	None
Vasotec IV (Merck Sharp & Dohme)	0.9	None
Vepesid (Bristol-Myers)	0.6	None
Versed (Roche)	1	None
Vitamin E	2	None

B. Topical Drug Products

Trade name	Manufacturer
Aristocort A Cream	Lederle
Capitrol Shampoo	Westwood
Cordran Lotion	Dista
Cyclocort Cream, Lotion, Ointment	Lederle
Estar Gel	Westwood
Fluroplex Topical Cream	Herbert
Loprox Cream, Lotion	Hoechst
Lotrimin Cream, Lotion	Schering
Lotrisone Cream	Schering
Moisturel Lotion	Westwood
Mycelex Cream	Miles
Mytrex Cream	Savage
Naftin Cream	Herbert
Penecort Cream, Topical Solution	Herbert
Pramegel	GenDerm
Zostrix Cream	GenDerm

C. Nebulizers/Aerosols

Nebulizer solutions administered with bacteriostatic saline instead of nonpreserved saline have caused toxicity believed due to benzyl alcohol.[4]

V. ANIMAL TOXICITY DATA

A. Neonatal Toxicity

A comparison of the intraperitoneal toxicity of benzyl alcohol in adult and term neonatal mice did not demonstrate any difference in lethal dose or metabolite profile. Premature mice were not studied. Benzyl alcohol, and not its metabolites, was shown to be responsible for the primary toxic effects, including sedation, dyspnea, and loss of motor function, with an LD_{50} of about 1000 mg/kg for both groups.[5]

B. Intrathecal Toxicity

During investigation of reports of paraparesis in humans following intrathecal administration of benzyl alcohol, studies in rats showed demyelination and active nerve fiber degeneration in animals exposed to 1.5% benzyl alcohol by this route. Prolonged exposure to 0.9% solutions showed less dramatic results with only scattered demyelination.[6]

VI. HUMAN TOXICITY DATA

A. Neonatal "Gasping Syndrome"

The most widely publicized benzyl alcohol-related excipient reaction began with simultaneous reports from two institutions of a fatal syndrome in 16 very low birth weight premature neonates, primarily weighing less than 1250 g. A constellation of symptoms beginning 2 to 14 d postnatally was described, including severe metabolic acidosis unresponsive to bicarbonate therapy, seizures, encephalopathy, thrombocytopenia, increased serum bilirubin, renal dysfunction, progressive bradycardia, and cardiovascular collapse.[7,8]

The hallmark of this syndrome was reported as a peculiar "gasping" respiration in one of these reports.[8] The common denominator in all of these cases was the administration of large volumes of intravascular flush solutions, such as bacteriostatic saline, preserved with benzyl alcohol. Endotracheal tube lavage solutions were later implicated as a source of exposure. Gasping respirations were later found not to be present universally, but indicative of benzyl alcohol toxicity if present.[9]

Neonatal intensive care units (ICU) discontinued the use of these preserved flush solutions, which were subsequently labeled with a caution against use in neonates. Subsequent studies comparing the incidence of mortality in the neonatal ICU before and after discontinuation of benzyl alcohol exposure showed a marked decrease in mortality in the latter groups.[9,10]

Other clinical symptoms, such as intraventricular hemorrhage and kernicterus, were also markedly decreased after discontinuation of benzyl alcohol solutions.[9-12] There is no difference in the incidence of these symptoms in neonates receiving total benzyl alcohol doses of less than or greater than 270 mg/kg/d.[12]

Morbidity, including cerebral palsy and developmental delay, has been observed in surviving infants with similar high-dose benzyl alcohol exposure.[11]

Removing the large volume flush solutions did not remove all sources of exposure in this population. Many commonly used parenteral medications in the neonatal ICU contain benzyl alcohol, such as diazepam and phenobarbital. Low-dose exposure from parenteral medications has not been demonstrated to produce toxicity, although a minimum toxic dose has not been established.

B. Neurologic Toxicity

Administration of pharmaceuticals preserved with benzyl alcohol by the intrathecal or epidural route has been associated with development of paraparesis.[6,13-17] Onset of paraparesis was immediate in some cases and delayed for up to 36 h in others. Recovery was similarly rapid in some cases and gradual over 1 to 6 weeks in others. Permanent impairment was reported in two cases.[16,17]

C. Inhalation Toxicity

Dilution of an albuterol nebulizer solution with bacteriostatic saline was associated with severe bronchitis and hemoptysis in a 64-year-old man. This reaction appeared to be dose related, clinically evident only after increasing the saline dose to 2 ml four times daily for about 10 d and resolving after substitution with nonpreserved saline.[4]

D. Allergic Contact Dermatitis

Only rarely has benzyl alcohol been attributed as a cause of delayed contact dermatitis.[18-22] Usually, scratch and intradermal tests are negative.[22]

E. Contact Urticaria

Immediate contact urticaria and delayed contact dermatitis were demonstrated simultaneously by patch testing to 1% benzyl alcohol in a 44-year-old woman.[19]

F. Systemic Hypersensitivity

Three case reports describe systemic reactions. In one case, fever and maculopapular rash occurred repeatedly after subcutaneous and intravenous administration of several preparations containing benzyl alcohol. Intradermal testing was positive.[23] In another case, arthralgia and urticaria were described after administration of intramuscular benzyl alcohol-preserved cyanocobalamin.[24] A third case described angioedema, fatigue, and nausea following intramuscular injection of a similar product. Subcutaneous injection of benzyl alcohol was positive; skin prick tests were negative.[25] The positive immediate skin reactions are suggestive of an IgE-mediated hypersensitivity, but the delayed onset in one case (4 to 5 h after subcutaneous injection and minutes after intravenous injection) is not totally consistent with this.

VII. CLINICAL RELEVANCE

A. Neonates

While the complete syndrome with a fatal outcome was seen primarily in premature infants weighing less than 1500 g, exclusively in those infants weighing 2500 g or less, and was associated with total daily doses of 99 to 234 mg/kg of benzyl alcohol; the effects of exposure to smaller amounts and in larger premature neonates or term neonates is not completely understood. A critically ill infant weighing 3350 g developed high levels of benzyl alcohol metabolites and signs of toxicity following doses of 32 to 105 mg/kg/d.[26] Infants with renal insufficiency and other critical organ dysfunction are probably at higher risk.

Low dose medication-related exposure was studied in 14 term and 9 premature neonates receiving benzyl alcohol-preserved phenobarbital intravenously or intramuscularly. Although peak benzoic acid levels were ten times higher in the premature infants, despite comparable doses when adjusted for weight, no signs of toxicity were observed.[27]

It is recommended that intravascular and intratracheal flush solutions containing benzyl alcohol should be avoided in any neonate, regardless of birth weight.[28-30] Exposure to other preserved pharmaceuticals generally results in total daily doses of 1 to 20 mg/kg benzyl alcohol. The use of any of these agents cannot be totally avoided in neonates, since there is no source of nonpreserved preparations in many cases. In many instances, nonpreserved diluents can be substituted for the supplied benzyl alcohol preparation. Benzyl alcohol should only be used in neonates if the drug is clearly indicated and there is no alternative nonpreserved product available.

B. Hypersensitivity

Benzyl alcohol is a rare sensitizer and is replacing compounds associated with a higher incidence of sensitivity in preparations intended for atopic populations, such as allergenic extracts. The local anesthetic effect may also be beneficial in patients prone to irritant dermatitis.[31] Patients displaying evidence of benzyl alcohol-induced contact dermatitis may tolerate parenteral preparations, but dual immediate and delayed hypersensitivity has rarely occurred. Cross-reactions may occur in patients sensitized to balsam of Peru.[32]

REFERENCES

1. **Thomas, D. V.,** Pain of skin infiltration with local anaesthetics, *Anaesth. Intensive Care,* 13, 101, 1984.
2. **Food and Drug Administration,** Parenteral drug products containing benzyl alcohol or other antimicrobial preservatives: withdrawal of notice of intent, *Fed. Reg.,* 54, 49772, 1989.
3. **Nikitakis, J. M.,** *CTFA Cosmetic Ingredient Handbook,* 1st ed., The Cosmetic, Toiletry and Fragrance Association, Washington, D.C., 1988.
4. **Reynolds, R. D.,** Nebulizer bronchitis induced by bacteriostatic saline, *JAMA,* 264, 35, 1990.
5. **McCloskey, S. E., Gershanik, J. J., Lertora, J. J. L., White, L., and Geroge, W. J.,** Toxicity of benzyl alcohol in adult and neonatal mice, *J. Pharm. Sci.,* 75, 702, 1986.
6. **Hahn, A. F., Feasby, T. E., and Gilbert, J. J.,** Paraparesis following intrathecal chemotherapy, *Neurology,* 33, 1032, 1983.
7. **Brown, W. J., Buist, N. R. M., Gipson, H. T. C., Huston, R. K., and Kennaway, N. G.,** Fatal benzyl alcohol poisoning in a neonatal intensive care unit, *Lancet,* 1, 1250, 1982.
8. **Gershanik, J., Boecler, B., Ensley, H., McCloskey, S., and George, W.,** The gasping syndrome and benzyl alcohol poisoning, *N. Engl. J. Med.,* 307, 1384, 1982.
9. **Menon, P. A., Thach, B. T., Smith, C. H., Landt, M., Roberts, J. L., Hillman, R. E., and Hillman, L. S.,** Benzyl alcohol toxicity in a neonatal intensive care unit, *Am. J. Perinatol.,* 1, 288, 1984.
10. **Hiller, J. L., Benda, G. I., Rahatzad, M., Allen, J. R., Culver, D. H., Carlso, C. V., and Reynolds, J. W.,** Benzyl alcohol toxicity: impact on mortality and intraventricular hemorrhage among very low birth weight infants, *Pediatrics,* 77, 500, 1986.
11. **Benda, G. I., Hiller, J. L., and Reynolds, J. W.,** Benzyl alcohol toxicity: impact on neurologic handicaps among surviving very low birth weight infants, *Pediatrics,* 77, 507, 1986.
12. **Jardine, D. S. and Rogers, K.,** Relationship of benzyl alcohol to kernicterus, intraventricular hemorrhage, and mortality in preterm infants, *Pediatrics,* 83, 153, 1989.
13. **Bagshawe, K. D., Magrath, I. T., and Golding, P. R.,** Intrathecal methotrexate, *Lancet,* 2, 1258, 1969.
14. **Saiki, J. H., Thompson, S., Smith, F., and Atkinson, R.,** Paraplegic following intrathecal chemotherapy, *Cancer,* 29, 370, 1972.
15. **Gagliano, R. G. and Costanzi, J. J.,** Paraplegia following intrathecal methotrexate: report of a case and review of the literature, *Cancer,* 37, 1663, 1976.
16. **Luddy, R. E. and Gilman, P. A.,** Paraplegia following intrathecal methotrexate, *J. Pediatr.,* 83, 988, 1973.
17. **Craig, D. B. and Habib, G. G.,** Flaccid paraparesis following obstetrical epidural anesthesia: possible role of benzyl alcohol, *Anesth. Analg.,* 56, 219, 1977.
18. **Lazzarini, S.,** Contact allergy to benzyl alcohol and isopropyl palmitate, ingredients of topical corticosteroid, *Contact Dermatitis,* 8, 349, 1980.
19. **Edwards, E. K.,** Allergic reactions to benzyl alcohol in a sunscreen, *Cutis,* 28, 332, 1981.
20. **Shoji, A.,** Allergic reaction to benzyl alcohol in an antimycotic preparation, *Contact Dermatitis,* 9, 510, 1983.
21. **Shmunes, E.,** Allergic dermatitis to benzyl alcohol in an injectable solution, *Arch. Dermatol.,* 120, 1200, 1984.
22. **Fisher, A. A.,** Allergic reactions to the parabens and benzyl. The relationship of the "delayed" reaction to the immediate variety, *Contact Dermatitis,* 1, 281, 1975.
23. **Wilson, J. P., Solimando, D. A., and Edwards, M. S.,** Parenteral benzyl alcohol-induced hypersensitivity reaction, *Drug Intell. Clin. Pharm.,* 20, 689, 1986.
24. **Lagerholm, B., Lodin, A., and Gentele, H.,** Hypersensitivity to phenylcarbinol preservative in Vitamin B12 for injection, *Acta Allergol.,* 12, 295, 1958.
25. **Grant, J. A., Bilodeau, P. A., Guernsey, B. G., and Gardner, F. H.,** Unsuspected benzyl alcohol hypersensitivity, *N. Engl. J. Med.,* 306, 108, 1982.
26. **Anderson, C. W., Ng, K. J., Andresen, B., and Cordero, L.,** Benzyl alcohol poisoning in a premature newborn infant, *Am. J. Obstet. Gynecol.,* 148, 344, 1984.
27. **LeBel, M., Ferron, L., Masson, M., Pichette, J., and Carrier, C.,** Benzyl alcohol metabolism and elimination in neonates, *Dev. Pharmacol. Ther.,* 11, 347, 1988.
28. **American Academy of Pediatrics,** Benzyl alcohol: toxic agent in neonatal units, *Pediatrics,* 72, 356, 1983.
29. **Centers for Disease Control,** Neonatal deaths associated with use of benzyl alcohol-United States, *MMWR,* 31, 290, 1982.

30. **Food and Drug Administration,** *FDA Bull.,* May 28, 1982.
31. **Fisher, A. A.,** *Contact Dermatitis,* 3rd ed., Lea & Febiger, Philadelphia, 1986.
32. **Schultheiss, E.,** Benzyl alcohol sensitivity, *Dermatol. Wochenschrift.,* 135, 629, 1957.

BRONOPOL

I. REGULATORY CLASSIFICATION

Bronopol is a water-soluble antimicrobial preservative that is especially effective against *Pseudomonas* species. Activity can be demonstrated against Gram-negative bacteria, Gram-positive bacteria, yeasts, and fungi.[1]

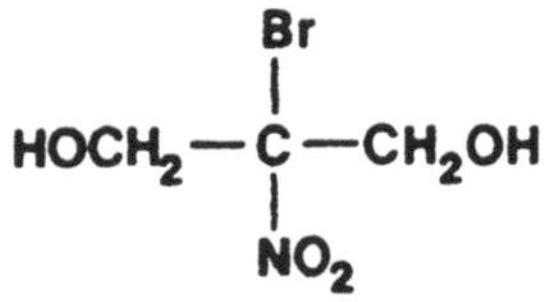

$$HOCH_2 - \underset{\underset{NO_2}{|}}{\overset{\overset{Br}{|}}{C}} - CH_2OH$$

FIGURE 6. Bronopol.

II. SYNONYMS

BNPD
2-Bromo-2-nitro-1,3-propanediol
2-Bromo-2-nitropropane-1,3-diol
Myacide BT®
Onyxide 500®

III. AVAILABLE FORMULATIONS

A. Cosmetics
Bronopol is considered a formaldehyde-releasing antimicrobial agent. In neutral and alkaline pHs, it decomposes to formaldehyde and bromo compounds. A yellow or brown color develops on exposure to light. Bronopol was present in about 2.7% of 20,183 cosmetics surveyed in 1982.[2]

Common types of cosmetic products containing bronopol include shampoos, makeup bases, hair conditioners, blushers, cleansing products, oil-in-water emulsion creams and lotions, rinses, and eye makeup. Concentrations in cosmetics usually range from 0.01 to 0.1%.[3]

IV. TABLE OF COMMON PRODUCTS

Trade name	Manufacturer
Balmex lotion	Macsil

V. HUMAN TOXICITY DATA

A. Irritant Reactions

Primary irritation is often observed with 1% concentrations.[4] The irritant threshold was determined in normal subjects to be between 0.5 and 1% in petrolatum.[5] Concentrations of 0.25 to 0.5% concentration in petrolatum have been recommended for patch testing. Repeated patch tests and usage tests should be performed to confirm suspected reactions.[6]

B. Allergic Contact Dermatitis

Contact dermatitis to bronopol in cosmetics was first reported in patients who used a recently reformulated popular emollient lubricant, Eucerin® ointment. Bronopol was added to the formulation in 1978 to control growth of *Pseudomonas aeruginosa* in the product. In a series of seven cases, all patients had used the product on abnormal skin for 5 weeks to 2 years.[3]

Occupational dermatitis has been reported secondary to exposure to an undiluted 8% bronopol solution used in milk testing[7,8] and in a veterinarian exposed via contact with a bronopol-preserved lubricant jelly to abraded skin.[9]

Patients with hypersensitivity to formaldehyde often have positive patch tests to bronopol, and patients with allergic cosmetic dermatitis related to bronopol invariably react to formaldehyde.[10] A bronopol concentration of 0.02% releases between 10 and 15 ppm of formaldehyde. Cosmetics with higher concentrations may be more likely to approach the threshold level of 30 ppm in formaldehyde-sensitive individuals. Cross-reactions with a related biocide used in cutting oils, Tris Nitro®, have been reported.[6]

VI. CLINICAL RELEVANCE

Clinically relevant bronopol-related contact dermatitis was reported in 3.5% of 228 patients tested in the U.S. in 1979 and 1980,[3] and in 2.2% of 713 patients with cosmetic dermatitis tested over a 64-month period.[11] Positive patch tests to 0.5 or 1% in petrolatum, which included some irritant responses, were found in 0.21 to 0.8% of British patients in two series involving over 8800 patients.[12,13]

In a population of patients referred for severe, extensive dermatitis, 9 of 72 (12.5%) had clinically relevant bronopol-related dermatitis. Discontinuation of the use of Eucerin® in this population reduced the incidence of bronopol reactivity to 2.3%. Two patient series demonstrated that one third of the bronopol reactors had positive patch tests to formaldehyde.[13,14]

The use of bronopol-preserved leave-on cosmetic products has not been associated with sensitization in patients with normal skin. The prolonged use of these cosmetics on damaged or dermatitic skin carries a significant risk of sensitization to both bronopol and formaldehyde. Individuals with occupational exposures to higher concentrations should wear protective gloves and clothing.

REFERENCES

1. **Stretton, R. J. and Manson, T. W.,** Some aspects of the mode of action of the antibacterial compound Bronopol (2-bromo-2-nitropropane-1,3-diol), *J. Appl. Bacteriol.,* 36, 61, 1973.
2. **Decker, R. L. and Wenninger, J. A.,** Frequency of preservative use in cosmetic formulas as disclosed to FDA-1982 update, *Cosmet. Toilet.,* 97, 57, 1982.
3. **Storrs, F. J. and Bell, D. E.,** Allergic contact dermatitis to 2-bromo-2-nitropropane-1,3-diol in a hydrophilic ointment, *J. Am. Acad. Dermatol.,* 8, 157, 1983.
4. Final report of the safety assessment for 2-bromo-2-nitropropane-1,3-diol, *J. Environ. Pathol. Toxicol.,* 4, 47, 1980.
5. **Maibach, H. I.,** Dermal sensitization potential of 2-bromo-2-nitropropane-1,3-diol (Bronopol), *Contact Dermatitis,* 3, 99, 1977.
6. **Robertson, M. H. and Storrs, F. J.,** Allergic contact dermatitis in two machinists, *Arch. Dermatol.,* 118, 997, 1982.
7. **Grattan, C. E. H., Harman, R. R. M., and Tan, R. S. H.,** Milk recorder dermatitis, *Contact Dermatitis,* 14, 217, 1986.
8. **Herzog, J., Dunne, J., Aber, R., Claver, M., and Marks, J. G.,** Milk tester's dermatitis, *J. Am. Acad. Dermatol.,* 19, 503, 1988.
9. **Wilson, C. L. and Powell, S. M.,** An unusual case of allergic contact dermatitis in a veterinary surgeon, *Contact Dermatitis,* 23, 42, 1990.
10. **Fisher, A. A.,** Cosmetic dermatitis Part II. Reactions to some commonly used preservatives, *Cutis,* 26, 136, 1980.
11. **Adams, R. M. and Maibach, H. I.,** A five-year-study of cosmetic reactions, *J. Am. Acad. Dermatol.,* 13, 1062, 1985.
12. **Ford, G. P. and Beck, M. H.,** Reactions to quaternium 15, bronopol and Germall 115 in a standard series, *Contact Dermatitis,* 14, 271, 1986.
13. **Frosch, P. J., White, I. R., Rycroft, R. J. G., Lahti, A., Burrows, D., Camarasa, J. G., Ducombs, G., and Wilkinson, J. D.,** Contact allergy to bronopol, *Contact Dermatitis,* 22, 24, 1990.
14. **Peters, M. S., Connolly, S. M., and Schroeter, A. L.,** Bronopol allergic contact dermatitis, *Contact Dermatitis,* 9, 397, 1983.

Butylated Hydroxyanisole/ Butylated Hydroxytoluene

I. REGULATORY CLASSIFICATION

Butylated hydroxytoluene (BHT) and butylated hydroxyanisole (BHA) are classified as antioxidants. They are on the GRAS list, but have not yet been reaffirmed as safe.

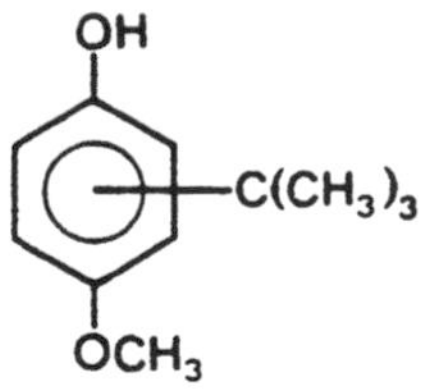

FIGURE 7. BHA.

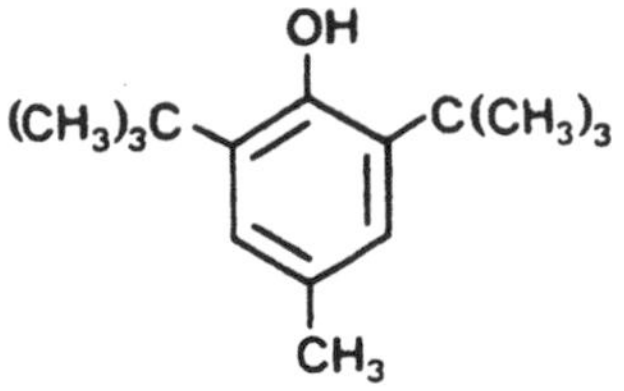

FIGURE 8. BHT.

II. SYNONYMS

    BHT
    2-*tert*-Butyl-4-methoxyphenol (BHA)
    2,6-Di-*tert*-butyl-*p*-cresol (BHT)
    Tenox® (BHA)
    Tenox BHT®

BHA is a mixture of approximately 90% 3-*tert*-butyl-4-hydroxyanisole and 8% 2-*tert*-butyl-4-hydroxyanisole.

III. AVAILABLE FORMULATIONS

A. Foods

BHA and BHT are widely used antioxidants in foods, particularly in products containing fats and oils. Antimicrobial activity is best against molds and Gram-positive bacteria. BHT has some activity against lipid-containing virus *in vitro*, which has been exploited in health food supplement products promoted for the treatment of *Herpes* infections.

59

These antioxidants are used both as direct additives and indirect additives in defoaming agents, food packaging materials, adhesives, and lubricants. The FDA limits the total combined amounts of these compounds to not more than 0.02% of the total fat and oil content. Flavorings and essential oils may contain up to 0.5%. BHA is more stable at high temperatures and is more frequently used than BHT.

B. Drugs
BHA and BHT are used in topical pharmaceutical products in concentrations of up to 0.2%.

C. Cosmetics
BHA and BHT are among the top ten most commonly used preservatives in cosmetics. In 1982, BHA was listed in 618 and BHT in 475 formulations of a total of 20,183 products filed with the FDA.[1] Concentrations of 0.01 to 0.1% may be used.

D. Other
Other consumer and industrial products that may contain BHA or BHT as antioxidants include petroleum products, such as cutting oils, jet fuels, rubber latex gloves, paints, lacquers, adhesive hardeners, cleaning agents, printing products, thinners, and plastics.[2]

IV. TABLE OF COMMON PRODUCTS

A. Foods

Active dry yeast
Breakfast cereals
Beverage dry mixes
Cake mixes
Candies
Chewing gum
Enriched rice
Margarine
Mixed, diced, or glazed fruits
Potato chips
Potato flakes
Salted peanuts
Shortening
Vegetable oils

B. Topical Drug Products

Trade name	Manufacturer
Bactine HC cream	Miles
Eucerin lotion	Beiersdorf
Micatin cream	Ortho
Vagistat	Fujisawa
Vanoxide HC	Dermik

V. ANIMAL TOXICITY DATA

A. Carcinogenicity

When combined with some mutagens, BHT accentuates the tumor-promoting effects. Inhibition of the carcinogenic or mutagenic effects of many other chemical carcinogens has been demonstrated in the presence of BHT. The possible mechanisms for the anticarcinogenic effects of BHT include inhibition of carcinogen metabolism by impairing hepatic hydroxylation reactions and augmenting the activity of enzymes involved in detoxifying carcinogens.[3]

The WHO has revised its recommended acceptable dietary intake from 0.3 to 0.5 mg/d, based on a no-observed-effect level of 50 mg/kg/d of dietary intake in rats. Continuous exposure to higher levels produced squamous cell carcinoma in the rat forestomach after 6 to 12 months.[4]

Forestomach tumors were inducible in rats with dietary levels of 2% BHA. Although there is no precise human equivalent of this structure, similar tissue is found in human esophagus, pharynx, larynx, oral cavity, and anorectal junction.[5]

B. Pulmonary Toxicity

BHT induces dose-related, species-specific toxicity to pulmonary pneumocytes in certain strains of mice. The mechanism has been determined to involve metabolism to BHT-quinone methide, an electrophilic reactive compound that binds covalently to pulmonary tissue. BHA enhances the formation of the BHT-quinone methide, enabling production of BHT-induced lung damage with lower doses. Doses greater than 250 mg/kg are required.[6,7]

C. Liver Toxicity

Centrilobular necrosis has been produced following administration of 500 mg/kg of BHT to mice depleted of hepatic glutathione by concomitant administration of buthionine sulfoximine, an inhibitor of glutathione synthesis. The species-specific BHT-quinone methide has been implicated in causing liver damage.[8]

D. Hematologic Toxicity

Vitamin K-dependent hypoprothrombinemia has been produced in animals receiving dietary levels of 0.017% BHT; this is equivalent to 14.7 mg/kg/d in humans.[9] This is dose related, with a mean prothrombin index of 28% following doses of 250 mg/kg/d and 10% following doses of 640 mg/kg/d.[10] The primary mechanism may be inhibition of vitamin K epoxide reductase in the liver by an active metabolite of BHT.

VI. HUMAN TOXICITY DATA

A. Contact Dermatitis

1. Cosmetics

Between 1977 and 1983, there were three cases of cosmetic-related contact dermatitis attributed to BHA among 713 patients (0.4%) evaluated.[11] During a 2-year period, 7 of 1096 (0.6%) British patients with facial dermatitis were diagnosed with BHA-related allergic contact dermatitis. The patch test concentration used was 1% in petrolatum. In both of these series, relevance of BHA to clinical dermatitis was established.[12]

No cases of hypersensitivity to BHT were found in a series of 1336 consecutive Danish patients tested with 2% in petrolatum.[2]

2. Drugs

Contact dermatitis to BHA in a topical pharmaceutical was first reported in 1975 in a 20-year-old patient treated with a miconazole cream containing BHA.[13]

Two cases were described by Tosti et al. in 1987.[14] The first case was a 67-year-old woman with chronic psoriasis. Topical application of a coal tar ointment containing 0.1% BHA to the legs was followed by itching and vesicular eczema. Patch tests were positive for BHA 2% in petrolatum and negative for BHT. The lesions cleared within 3 weeks after dietary antioxidant restriction and topical steroid therapy. The second case was a 69-year-old woman with chronic hand eczema exacerbated by use of a topical antifungal cream containing 0.2% BHA. Patch test results were similar to the first case.

Topical application of an antifungal cream in a 31-year-old man with no history of skin or atopic diseases resulted in red scaly lesions of the right palm and soles and sensitization to BHT present in the cream in a concentration of 0.02%. Patch tests were positive to BHT, as well as to several other components of the international standard series.[15]

Two elderly patients with venous stasis were reported to develop dermatitis following the use of support bandages containing BHT. Both had positive patch tests to BHT.[16]

Contact dermatitis of the cheek and lip occurring 2 d after a dental appointment was attributed to latex examining gloves containing the antioxidant 2-*tert*-butyl 5 methyl phenol (Lowinox® 44S36 monomer). Patch tests were positive to this antioxidant, BHA, and Puritee and Travenol Ultraderm gloves, which contain BHA. The monomer of Lowinox® is closely structurally related to BHA; both have a hydroxyl group in the *ortho* position to the *tert* butyl group. This patient was a baker with occupational exposure to these antioxidants.[17]

Oral challenge with BHA and BHT, in doses of 10 to 40 mg, produced exacerbations of contact dermatitis, manifested by vesicular eczema localized to the fingers and lower lip border, in two patients with positive patch tests. Immediate urticaria was produced in one patient after application of 5% BHA or BHT in ethanol. The eczematous dermatitis cleared when the antioxidants were eliminated from the diet.[18]

B. Urticaria

1. Foods

Intolerance to BHA and BHT in foods was suspected in seven patients with asthma or rhinitis. Symptoms reported after challenges with 125 or 250 mg of BHA and BHT included rhinitis, headaches, asthma, flushing, conjunctivitis, diaphoresis, and somnolence. This intolerance was presumably verified by a doubling of earlobe bleeding time, a questionable procedure that has since been refuted.[19,20]

An investigation of 330 patients with chronic urticaria, 70% with concomitant angioedema, given single-blind oral challenges with 1 to 50 mg of a combined capsule of BHA and BHT, found positive reactions in 15% of 156 patients tested. Repeat placebo-controlled challenges in eight of the positive reactors within 1 year of the initial challenge were positive in three patients.[21]

Chewing gum is a common source of BHT. An urticarial disseminated eruption involving the legs, forearms, abdomen, and back was attributed to daily consumption of a BHT-preserved chewing gum for 3 months in a 30-year-old woman. Biopsy of the skin lesion revealed signs of vasculitis with endothelial deposits of IgM and complement factors C1, C3, and C9. Single-blind oral challenges with BHT resulted in appearance of the lesions within a few hours. An immune complex mechanism was suggested.[22]

Double-blind, placebo-controlled prospective oral challenges with a capsule containing 25 mg each of BHA and BHT produced equivocal results in 2 of 44 chronic urticaria patients and 2 of 91 atopic dermatitis patients. The reactions could not be confirmed when retested 4 d later.[23]

Two patients with chronic urticaria had exacerbations 1 to 6 h after double-blind, placebo-controlled oral challenges with BHA and BHT (administered simultaneously) in doses of 125 and 250 mg each. Reactions were confirmed by repeated double-blind challenges. The previously described serial earlobe bleeding time procedure was done on both patients with negative results. Attempts to identify an immunologic etiology in an extensive evaluation of one patient failed. These patients were part of a group of 271 patients referred with chronic urticaria. Of 11 who responded favorably to elimination diets, these 2 were the only ones with positive challenges to BHA and BHT.[24]

2. Industrial Products

A 19-year-old woman with a history of skin eruptions provoked by plastic was shown to react to both BHA and BHT in 20-min patch tests using 1% in ethanol; she did not react to lower concentrations. An immediate contact urticarial reaction to oleylamide, a structurally unrelated plastic additive, was also documented.[25]

C. Overdose

A 22-year-old woman ingested 4 g of BHT, purchased as 250 mg capsules in a health food store for the treatment of genital herpes simplex infection. Adverse reactions included severe epigastric cramping, weakness, nausea, vomiting, dizziness, confusion, and a brief loss of consciousness. Symptoms resolved within a few days.[26]

Another report of overdose of 80 g of BHT in safflower oil produced dizziness, ataxia, and slurred speech within 1 h.[27]

VII. CLINICAL RELEVANCE

BHA and BHT are associated with a low incidence of Type IV allergic delayed contact dermatitis, accounting for about 0.5% of cosmetic dermatitis reactions. Isolated positive patch tests to either of these compounds has occurred without positive reactions to the other. Cross-reactions between them may also occur; reactions to other structurally related compounds are possible.

One case of an apparent Type III immune complex urticarial reaction has been described. Immediate contact urticarial reactions and exacerbation of chronic urticaria are presumed to be nonimmunologic reactions. These reactions appear to be infrequent when evaluated by careful double-blind, placebo-controlled challenges, and often cannot be reproduced on subsequent testing.

REFERENCES

1. **Decker, R. L. and Wenninger, J. A.,** Frequency of preservative use in cosmetic formulas as disclosed to FDA-1982 update, *Cosmet. Toilet.,* 97, 57, 1982.
2. **Flyvholm, M. and Menne, T.,** Sensitizing risk of butylated hydroxytoluene based on exposure and effect data, *Contact Dermatitis,* 23, 341, 1990.

3. **Busch, J.,** Final report on the safety assessment of butylated hydroxyanisole, *J. Am. Coll. Toxicol.*, 3, 83, 1984.

4. **World Health Organization,** Evaluation of certain food additives and contaminants: thirty-third report of the joint FAO/WHO expert committee on food additives, Tech. Rep. Ser. No. 776, WHO, Geneva, 1989.

5. **Grice, H. C.,** Safety evaluation of butylated hydroxyanisole from the perspective of effects on forestomach and oesophageal squamous epithelium, *Food Chem. Toxicol.*, 26, 717, 1988.

6. **Thompson, D. C. and Trush, M. A.,** Enhancement of butylated hydroxytoluene-induced mouse lung damage by butylated hydroxyanisole, *Toxicol. Appl. Pharmacol.*, 96, 115, 1988.

7. **Yamamoto, K., Tajima, K., Okino, N., and Mizutani, T.,** Enhanced lung toxicity of butylated hydroxytoluene in mice by coadministration of butylated hydroxyanisole, *Res. Commun. Chem. Pathol. Pharmacol.*, 59, 219, 1988.

8. **Mizutani, T., Nomura, H., Nakanishi, K., and Fujita, S.,** Hepatotoxicity of butylated hydroxytoluene and its analogs in mice depleted of hepatic glutathione, *Toxicol. Appl. Pharmacol.*, 87, 166, 1987.

9. **Takahashi, O. and Hiraga, K.,** Dose-response study of hemorrhagic death by dietary butylated hydroxytoluene (BHT) in male rats, *Toxicol. Appl. Pharmacol.*, 43, 399, 1978.

10. **Takahashi, O., Ichikawa, H., and Sasaki, M.,** Hemorrhagic toxicity of d-a-tocopherol in the rat, *Toxicology*, 63, 157, 1990.

11. **Adams, R. M. and Maibach, H. I.,** A five-year-study of cosmetic reactions, *J. Am. Acad. Dermatol.*, 13, 1062, 1985.

12. **White, I. R., Lovell, C. R., and Cronin, E.,** Antioxidants in cosmetics, *Contact Dermatitis*, 11, 265, 1984.

13. **Degreef, H. and Verhoeve, L.,** Contact dermatitis to miconazole nitrate, *Contact Dermatitis*, 1, 269, 1975.

14. **Tosti, A., Bardazzi, F., Valeri, F., and Russo, R.,** Contact dermatitis from butylated hydroxyanisole, *Contact Dermatitis*, 17, 257, 1987.

15. **Bardazzi, F., Misciali, C., Borrello, P., and Capobianco, C.,** Contact dermatitis due to antioxidants, *Contact Dermatitis*, 19, 385, 1988.

16. **Dissanayake, M. and Powell, S. M.,** Allergic contact dermatitis from BHT in leg ulcer patients, *Contact Dermatitis*, 21, 195, 1989.

17. **Rich, P., Belozer, M. L., Norris, P., and Storrs, F. J.,** Allergic contact dermatitis to two antioxidants in latex gloves: 4,4-thio*bis*(6-*tert*-butyl-*meta*-cresol)(Lowinox 44S36) and butylhydroxyanisole, *J. Am. Acad. Dermatol.*, 24, 37, 1991.

18. **Roed-Petersen, J. and Hjorth, N.,** Contact dermatitis from antioxidants: hidden sensitizers in topical medications and foods, *Br. J. Dermatol.*, 94, 233, 1986.

19. **Fisherman, E. W. and Cohen, G. N.,** Chemical intolerance to butylated hydroxyanisol (BHA) and butylated hydroxytoluene (BHT) and vascular response as an indicator and monitor of drug tolerance, *Ann. Allergy*, 31, 126, 1973.

20. **Cloninger, P. and Novey, H. S.,** The acute effects of butylated hydroxyanisole ingestion in asthma and rhinitis of unknown etiology, *Ann. Allergy*, 32, 131, 1974.

21. **Juhlin, L.,** Recurrent urticaria: clinical investigation of 330 patients, *Br. J. Dermatol.*, 104, 369, 1981.

22. **Moneret-Vautrin, D. A., Faure, G., and Bene, M. C.,** Chewing-gum preservative induced toxidermic vasculitis, *Allergy*, 41, 546, 1986.

23. **Hannuksela, M. and Lahti, A.,** Peroral challenge tests with food additives in urticaria and atopic dermatitis, *Int. J. Dermatol.*, 25, 178, 1986.

24. **Goodman, D. L., McDonnell, J. T., Nelson, H. S., Vaughan, T. R., and Weber, R. W.,** Chronic urticaria exacerbated by the antioxidant food preservatives, butylated hydroxyanisole (BHA) and butylated hydroxytoluene (BHT), *J. Allergy Clin. Immunol.*, 86, 570, 1990.

25. **Osmundsen, P. E.,** Contact urticaria from nickel and plastic additives (butylhydroxytoluene, oleylamide), *Contact Dermatitis*, 6, 452, 1980.

26. **Shlian, D. M. and Goldstone, J.,** Toxicity of butylated hydroxytoluene, *N. Engl. J. Med.*, 314, 648, 1986.

27. **Grogan, W. A.,** Toxicity from BHT ingestion, *West. J. Med.*, 145, 245, 1986.

CANTHAXANTHIN

I. REGULATORY CLASSIFICATION

Canthaxanthin is a color additive approved for use in foods and drugs since 1969 in amounts not exceeding 30 mg per pound or pint. It is a natural color, exempt from batch certification.

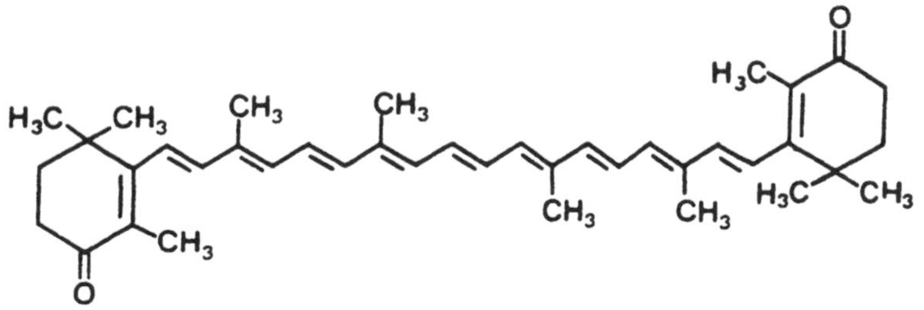

FIGURE 9. Canthaxanthine.

II. SYNONYMS

    4,4-diketo-beta-carotene
    C.I. Food Orange 8
    C.I. 40850

III. AVAILABLE FORMULATIONS

A. Foods
Canthaxanthin is a naturally occurring orange-red carotenoid that lacks vitamin A activity. It is used both as a direct human food additive and as an animal feed additive to enhance the color of chicken skin and egg yolks.

B. Drugs
Despite discouragement from the FDA, canthaxanthin continues to be available as an oral tanning preparation in tanning salons and by mail order. It is used experimentally, in combination with beta-carotene, for the treatment of vitiligo and other photodermatologic conditions. As

an excipient, canthaxanthin may be used as a color additive in sugar-coated tablets to impart a peach to red color.

IV. TABLE OF COMMON PRODUCTS

A. Oral Suntan Products

Trade name	Manufacturer
Bronz Glo tablet 30mg	Bronz Glo
Easytan tablet 30 mg	David Carter Products
Tanamin tablet	Standard Pharmacal

B. Foods
1. Natural Sources

Mushrooms (Chanterelle)
Salmon
Seaweed
Sunflowers
Trout

2. Foods that May Contain Canthaxanthin as an Additive

Baked goods
Barbecue sauces
Butter
Catsup
Cheese
Fruit drinks
Pizza
Pudding
Salad dressings
Sauces
Soups
Spaghetti sauces
Tomato products
Vegetable oils

V. ANIMAL TOXICITY DATA

The temporary acceptable daily intake of canthaxanthin in foods is up to 0.05 mg/kg (decreased from the previous ADI of 25 mg/kg), pending further information on the threshold dose, dose-duration relationship, reversibility, and animal models for development of eye pigmentation. This amount is based on a tenfold safety margin for the minimal eye pigmentation effect in humans.[1]

Several unsuccessful attempts have been made to reproduce the retinal pigmentation in animals, including cats fed up to 16 mg/kg/d, rats fed 28 mg/kg of feed, rabbits, and rhesus monkeys. Crystal formation was documented in one monkey in a glaucomatous eye.[2]

VI. HUMAN TOXICITY DATA

A. Retinopathy

Canthaxanthin ingestion results in a dose-related decrease of the scotopic electroretinogram rod b- and a-wave amplitudes; no effect was seen after a dose of 15 mg/d for 1 month, while 60 mg/d for an additional month had a demonstrated effect. Retinal changes are associated with the development of glistening, golden-yellow paramacular pigmentation, believed to occur from active transport of canthaxanthin into retinal Muller (glial) cells, forming crystals which are then exocytosed into the extracellular space.[3]

In a review of 259 published cases of ophthalmic investigations in patients ingesting canthaxanthin, 92 (35%) were found to have retinal crystals.[3] The incidence was reported to be 100% after a cumulative dose of 60 g and 50% after a total dose of 37 g.[4] Risk factors for accelerated development of crystals include increasing age, high intraocular pressure, and retinitis pigmentosa, indicating underlying pigment epitheliopathy.[5]

Once crystal formation develops, gradual reversal occurs following discontinuation of canthaxanthin ingestion, which is usually only partial. In one long-term, follow-up study, a significant decrease in retinal deposits was only seen at 26 months after discontinuation in 74% of the patients; some patients still had significant deposits 7 years after discontinuation.[4]

The pigmentary changes are invariably not associated with a loss of visual sensitivity, visual acuity, or color vision, even after high-dose usage for up to 13 years. Occasional patients with mild dysfunction in dark adaptation, static perimetry, and ERG have been reported, but are still within clinically normal ranges. Glare sensations are reported rarely.[2]

B. Aplastic Anemia

One case of fatal aplastic anemia has been described in a 20-year-old woman who ingested a tanning salon product containing canthaxanthin for several weeks at an unknown dosage. She presented to the hospital 4 months after the onset of this therapy and was noted to have yellow or reddish-brown skin. She refused supportive care based on religious reasons and died of severe anemia.[6]

C. Amenorrhea

There are several anecdotal, poorly documented cases of amenorrhea in women who have taken canthaxanthin as a tanning agent.[6,7]

D. Discoloration

Canthaxanthin ingestion in sufficient amounts imparts a bright orange to red color to body tissues and fluids, including skin, blood, feces, and sweat.[8]

E. Other

Adverse reactions spontaneously reported to the FDA include nausea, diarrhea, stomach cramps, hepatitis, pruritus, and generalized urticaria.[9]

VII. CLINICAL RELEVANCE

Canthaxanthin has not been associated with adverse effects when used as a food or drug excipient. It produces a dose-related characteristic orange color to the skin and body fluids and golden deposits in the retina, which are usually not associated with any functional disturbance. The total reversibility and clinical significance of the retinal changes are debatable, and the pigmentary changes appear to very slowly, but not completely, resolve in most patients.

Patients who appear to be at increased risk for retinopathy (glaucoma, elderly) should avoid excessive use. The promotion of this natural dye for cosmetic "tanning" purposes should be discouraged.

REFERENCES

1. **World Health Organization,** Evaluation of certain food additives and contaminants: sixteenth report of the joint FAO/WHO expert committee on food additives, Tech. Rep. Ser. No. 759, WHO, Geneva, 1987.
2. **Barker, F. M.,** Canthaxanthin retinopathy, *J. Toxicol. Cut. Ocular Toxicol.,* 7, 223, 1988.
3. **Arden, G. B., Oluwole, J. O. A., Polkinghorne, P., Bird, A. C., Barker, F. M., Norris, P. G., and Hawk, J. L. M.,** Monitoring of patients taking canthaxanthin and carotene: an electroretinographic and ophthalmological survey, *Hum. Toxicol.,* 8, 439, 1989.
4. **Harnois, C., Samson, J., Malenfant, M., and Rousseau, A.,** Canthaxanthin retinopathy, *Arch. Ophthalmol.,* 107, 538, 1989.
5. **Cortin, P., Boudreault, G., Rousseau, A. P., Tardif, Y., and Malenfant, M.,** La retinopathie a la canthaxanthine. 2. Facteurs predisposants, *Can. J. Ophthalmol.,* 19, 215, 1984.
6. **Bluhm, R., Branch, R., Johnston, P., and Stein, R.,** Aplastic anemia associated with canthaxanthin ingested for "tanning" purposes, *JAMA,* 264, 1141, 1990.
7. **Mathews-Roth, M.,** Amenorrhea associated with carotenemia, *JAMA,* 250, 731, 1983.
8. **Rock, G. A., Decary, F., and Cole, R. S.,** Orange plasma from tanning capsules, *Lancet,* 1, 1419, 1981.
9. **Jones, B.,** Indications and safety of canthaxanthine, *DICP,* 21, 173, 1987.

CASTOR OIL

I. REGULATORY CLASSIFICATION

Castor oil is classified as an oleaginous vehicle.

II. SYNONYMS

Aceite de ricino
Huile de ricini
Oleum ricini
Ricini oleum

$$CH_3(CH_2)_5CH(OH)CH_2-CH=CH(CH_2)_7COOH$$

FIGURE 10. Ricinoleic acid.

III. AVAILABLE FORMULATIONS

A. Constituents
Castor oil is a fixed oil obtained from the seeds of *Ricinus communis* and contains 80% ricinoleic acid triglyceride.

B. Drugs
Castor oil is present in 6 FDA-approved intramuscular injectables, 26 solid oral dosage forms in amounts of 0.08 to 23 mg, and in 4 topical pharmaceuticals in concentrations of 5 to 12.5%. Hydrogenated castor oil is also included in the U.S.N.F. and is present in 27 solid oral dosage forms and 1 topical cream.[1]

C. Cosmetics
Castor oil is used as an occlusive skin conditioner and solvent in products such as lipsticks, eye and face makeup, and nail polish and enamels.[2]

IV. HUMAN TOXICITY DATA

A. Allergic Contact Dermatitis

Contact sensitization to the sulfonated castor oil component of a commerical hair conditioner spray was demonstrated in 4 of 53 normal volunteers undergoing a repeated insult occlusive patch test procedure. Rechallenge with the individual components of the hair product 4 weeks later showed allergic responses only to sulfonated castor oil.[3]

Castor oil is commonly found in lipsticks in concentrations of 10 to 67%. Two cases of lipstick-induced cheilitis were shown to be related to allergic hypersensitivity to castor oil.[3,4] The reaction spread to the entire face and neck in one case. This patient had a severe spreading reaction to pure ricinoleic acid, an impurity in poor quality castor oil.[5] Cosmetic facial dermatitis was attributed to castor oil in a makeup remover in the case of a 23-year-old French woman.[6]

V. CLINICAL RELEVANCE

Castor oil is a rare cause of delayed-Type IV allergic contact dermatitis. Patch testing with this excipient should be included in the evaluation of patients with allergic cheilitis related to lipstick products.

REFERENCES

1. **Food and Drug Administration,** *Inactive Ingredients Guide.* FDA, Washington, D.C., March 1990.
2. **Nikitakis, J. M.,** *CTFA Cosmetic Ingredient Handbook,* 1st ed., The Cosmetic, Toiletry and Fragrance Association, Washington, D.C., 1988.
3. **Fisher, L. B. and Berman, B.,** Contact allergy to sulfonated castor oil, *Contact Dermatitis,* 7, 339, 1981.
4. **Sai, S.,** Lipstick dermatitis caused by castor oil, *Contact Dermatitis,* 9, 75, 1983.
5. **Andersen, K. E. and Nielsen, R.,** Lipstick dermatitis related to castor oil, *Contact Dermatitis,* 11, 253, 1984.
6. **Brandle, I., Boujnah-Khouadja, A., and Foussereau, J.,** *Contact Dermatitis,* 9, 424, 1983.

CELLULOSE

I. REGULATORY CLASSIFICATION

Microcrystalline cellulose is classified as a tablet and capsule diluent, tablet disintegrant, and suspending and/or viscosity-increasing agent. Carboxymethylcellulose is used as a suspending agent and emulsifying agent in the preparation of gels and is widely used in foods.

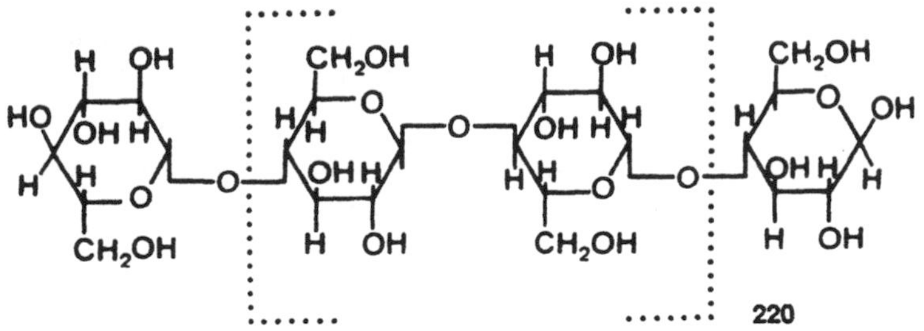

FIGURE 11. Microcrystalline cellulose.

II. SYNONYMS

Alpha-cellulose
Cellulose gel (microcrystalline cellulose)
Cellulose powder (alpha-cellulose)
Crystalline cellulose
Avicel®
E460

III. AVAILABLE FORMULATIONS

A. Constituents

Cellulose is a natural polysaccharide obtained from plant fibers. Microcrystalline cellulose is the colloidal crystalline portion obtained by hydrolysis of cellulose with dilute mineral acid solutions. Carboxymethylcellulose sodium (carmellose sodium) is the sodium salt of a polycarboxymethyl ether of cellulose.

B. Drugs

Microcrystalline cellulose is available in two pharmaceutical grades: a colloidal, used in conjunction with other cellulose derivatives as a suspending agent, and a non-water-dispersible form, used in tablet manufacturing. Concentrations of 5 to 20% are used as tablet binders, disintegrants, glidants, and antiadherents. Up to 30% may be present as a capsule diluent.

Cellulose was listed as an inactive ingredient in 96 oral pharmaceutical products, including drops, suspensions, oral tablets, and sublingual tablets; amounts ranged from 6 to 100 mg. Well over 2000 products were listed containing microcrystalline cellulose, including nasal inhalation metered sprays, intravenous injections, capsules, suspensions, syrups, tablets, sublingual tablets, topical powders, and vaginal tablets.[1]

C. Cosmetics

Cellulose and microcrystalline cellulose are used in cosmetics as an absorbent, anticaking agent, emulsion stabilizer, bulking agent, opacifying agent, and viscosity-increasing agent. Types of products include makeup, douches, and cleansing products.[2]

D. Other

Cellulose is an ingredient used in the preparation of hemodialysis membranes. The cuprammonium cellulose type is made by combining raw cellulose with copper, ammonia, and other additives, which allow the cellulose to be spun into a porous membrane. Another process involves melting a partially acetylated cellulose to form hollow fibers. Regenerated cellulose membranes include saponified cellulose ester membranes and cellulose acetate membranes.[3]

IV. TABLE OF COMMON PRODUCTS

A. Oral Drug Products with Abuse Potential

Trade name	Manufacturer
Darvon-N tablet	Lilly
Dolophine hydrochloride tablet	Lilly
Empirin with codeine tablet	Burroughs Wellcome
Fiorinal with codeine capsule	Sandoz
Methadone tablet	Roxane
Percocet tablet	DuPont
Percodan tablet	DuPont
Phenaphen with codeine capsule	A. H. Robins
Roxicodone tablet	Roxane
Soma compound with codeine tablet	Wallace
Talacen tablet	Winthrop
Talwin Nx tablet	Winthrop
Vicodin tablet	Knoll

V. HUMAN TOXICITY DATA

A. Hemodialysis Membranes

Cellulose-based membranes are widely used in hemodialysis treatments. The cuprammonium hollow-fiber cellulose type is well known to cause transient activation of the alternate pathway of the complement system, resulting in symptoms due to C3a and C5a anaphylatoxins. A fourfold increase in C3a levels occurred with a peak at 15 min in asymptomatic patients undergoing dialysis with new membranes. In patients with mild first-use syndrome, the peak occurred at 10 min and was 13 times the predialysis level.[4] Profound leukopenia accompanies complement activation and can be used as a marker for bioincompatibility with the membrane. Leukopenia is caused by sequestration of leukocytes in the lung and may lead to pulmonary dysfunction. Reuse of the same membrane in the same patient is associated with a diminished response, presumably due to saturation of complement binding sites on the membrane surface.[5]

Clinical findings in patients with dialysis membrane-related complement activation include dyspnea, angioedema, severe bronchospasm, diaphoresis, facial flushing, a burning sensation at the vascular access site, hypotension, cardiopulmonary collapse, and rarely urticaria. The incidence of severe reactions was 21 in 260,000 dialysis treatments in a retrospective study.[3] Eosinophilia may accompany these reactions, which may improve after switching to a polyacrylonitrile membrane.[6] Less severe symptoms may occur in 3 to 5% of dialysis patients exposed to new cellulose membranes, including shortness of breath, chest tightness, back pain, nausea, vomiting, hypotension, and a feeling of malaise.

Although most of these symptoms are confined to the first use of the cuprammonium dialyzer, recurrent episodes have been described, necessitating a change to a polyacrilonitrile membrane.[3,7,8] Reprocessing the dialyzer with sodium hypochlorite may restore the reactivity of the membrane.[9]

B. Intravenous Drug Abuse

Deliberate intravenous injection of aqueous suspensions made from crushed tablets containing microcrystalline cellulose excipients induces foreign body granulomas when the suspension extravasates. Pulmonary intravascular and perivascular granulomas have been described in eight subjects who died after intravenous injection of pentazocine tablets or other narcotics. Crystalline deposits were also found in the hepatic portal triads, Kupffer cells, splenic macrophages, renal glomeruli, and cardiac interstitium. Large cellulose emboli in the heart, kidneys, brain, liver, pancreas, and spleen were described in one case.[10,11]

Sequelae of cellulose embolism have included thrombosis and dry gangrene. A 37-year-old man inadvertently injected 240 mg of codeine as tablets dissolved in water intra-arterially. Three days later gangrene of the fingers and thumb of the right hand developed. Amputation of the hand was required. A study of dogs injected with the individual components of the tablets, which included lactose, gelatin, carboxymethyl cellulose, calcium stearate, talc, and microcrystalline cellulose, demonstrated gangrene only after the latter compound.[12]

Microcrystalline cellulose can be differentiated from talc during histologic study of the granulomas by pale to dark violet staining with PAS, deep black staining with methenamine silver, and brilliant orange stain with yellow-green birefringence with Congo Red. Talc appears as smaller, needle-shaped crystals that are achromatic with these stains.[10]

Pulmonary cellulose granulomas were reported in a 26-year-old university student with a history of cocaine sniffing who denied intravenous drug abuse.[13]

VI. CLINICAL RELEVANCE

The usual uses of cellulose in pharmaceutical and food products present no reported adverse effects. Deliberate abuse of drugs containing cellulose as a diluent or binder is known to cause granulomas, pulmonary emboli and thrombosis, and gangrene. Cellulose dialyzers are a cause of significant morbidity with the first use of the membrane, which usually diminishes with subsequent use. Persistent reactions can be eliminated by changing to a polyacrylonitrile membrane. Ethylene oxide sensitivity produces a similar clinical picture in dialysis patients and should be excluded before attributing the reaction to cellulose.

REFERENCES

1. **Food and Drug Administration,** *Inactive Ingredients Guide,* FDA, Washington, D.C., March 1990.
2. **Nikitakis, J. M.,** *CTFA Cosmetic Ingredient Handbook,* 1st ed., The Cosmetic, Toiletry and Fragrance Association, Washington, D.C., 1988.
3. **Daugirdas, J. T., Ing, T. S., Roxe, D. M., Ivanovich, P. T., Krumlovsky, F., Popli, S., and McLaughlin, M. M.,** Severe anaphylactoid reactions to cuprammonium cellulose hemodialyzers, *Arch. Intern. Med.,* 145, 489, 1985.
4. **Hakim, R. M., Breillatt, J., Lazarus, J. M., and Port, F. K.,** Complement activation and hypersensitivity reactions to dialysis membranes, *N. Engl. J. Med.,* 311, 878, 1984.
5. **Eknoyan, G.,** Side effects of hemodialysis, *N. Engl. J. Med.,* 311, 915, 1984.
6. **Vanherweghem, J.-L., Leon, M., Goldman, M., Lietar, N., Gamar, N., and Thayse, C.,** Membrane-related eosinophilia in hemodialysis, *Kidney Int.,* 33 (Suppl. 24), 73, 1988.
7. **Cruz, I. A., Dillard, M. G., Malveaux, F. J., Herry, V. E., and Hosten, O.,** Acute bronchospasm associated with a cuprophane capillary dialyzer, *Clin. Nephrol.,* 22, 53, 1984.
8. **Bhat, K., Lee, S. M. K., and Lozano, J.,** Anaphylactic reaction on subsequent exposure to cuprophan-hollow fiber dialyzer: a case report, *Ann. Allergy,* 52, 282, 1984.
9. **Rancourt, M., Senger, K., and DeOreo, P.,** Cellulosic membrane induced leukopenia after reprocessing with sodium hypochlorite, *Trans. Am. Soc. Artif. Intern. Organs,* 30, 49, 1984.
10. **Tomashefski, J. F., Hirsch, C. S., and Jolly, P. N.,** Microcrystalline cellulose pulmonary embolism and granulomatosis, *Arch. Pathol. Lab. Med.,* 105, 89, 1981.
11. **Zeltner, T. B., Nussbaumer, U., Rudin, O., and Zimmermann, A.,** Unusual pulmonary vascular lesions after intravenous injections of microcrystalline cellulose, *Virchows Arch.,* 395, 207, 1982.
12. **Goldberg, I., Bahar, A., and Yosipovitch, Z.,** Gangrene of the upper extremity following intra-arterial injection of drugs, *Clin. Orthop.,* September, 223, 1984.
13. **Cooper, C. B., Bai, T. R., Heyderman, E., and Corrin, B.,** Cellulose granulomas in the lungs of a cocaine sniffer, *Br. Med. J.,* 286, 2021, 1983.

CETYL ALCOHOL

I. USP CLASSIFICATION

II. AVAILABLE FORMULATIONS

A. Cosmetics/Drugs

Cetyl alcohol is used as an emulsifying agent, opacifying agent, emollient, viscosity-increasing agent, and foam booster in cosmetics and topical pharmaceuticals. Types of products include eye products, sachets, hair conditioners, lipsticks, makeup, creams and lotions, and cleansing products.[1] Cetostearyl alcohol is a mixture of equal parts of cetyl alcohol and stearyl alcohol. Stearyl alcohol is a component present in commercial cetyl alcohol in concentrations of up to 30%.[2]

$$CH_3(CH_2)_{14}CH_2OH$$

FIGURE 12. Cetyl alcohol.

III. SYNONYMS

Hexadecyl alcohol
Palmityl alcohol

IV. TABLE OF COMMON DRUG PRODUCTS

Trade name	Manufacturer
Anthranil lotion	Young
Bactine hydrocortisone cream	Miles
Chloromycetin cream	Parke-Davis
Decadron cream	Merck Sharp & Dohme
Exelderm cream	Westwood-Squibb
Femstat vaginal cream	Syntex
Florone E cream	Dermik
Fungizone lotion	Squibb

Trade name (cont'd)	Manufacturer (cont'd)
Halog cream	Westwood-Squibb
Kenalog lotion	Westwood-Squibb
Lac-Hydrin lotion	Westwood-Squibb
Lac-Hydrin five fragrance free lotion	Westwood-Squibb
Lacticare HC lotion	Stiefel
Lubriderm cream, lotion	Warner-Lambert
Masse breast cream	Ortho
Maxivate cream	Westwood-Squibb
Moisturel cream, lotion	Westwood-Squibb
NeoDecadron cream	Merck Sharp & Dohme
Nutracort lotion	Owen/Galderma
Nutraderm cream, lotion	Owen/Galderma
Oxistat cream	Glaxo
Shepard's cream lotion	Dermik
Shepard's skin cream	Dermik
SSD cream	Boots
Sulfamylon cream	Winthrop
Sween cream	Sween
Vioform Hydrocortisone cream, mild cream, lotion	Ciba

V. HUMAN TOXICITY DATA

A. Contact Dermatitis

Cetyl alcohol is a well-known cause of allergic delayed-type hypersensitivity reactions in patients with stasis dermatitis or leg ulcers documented in 5.4% of 116 cases.[3] Four patients with stasis dermatitis were reported to have positive patch tests to cetyl alcohol in a series of 23 patients with contact dermatitis from topical medications. Cross-reactions were documented to lanolin and stearyl alcohol.[2]

Cross-reactions to lanolin were also reported in three patients with reactions to lanolin, wool alcohol, and cetyl alcohol 10%.[3] Cross-reactions to cetostearyl alcohol and to cetomacrogolis cream have also been confirmed.[3] Contact urticaria has also been reported.[4]

Patch testing is usually done with a 30% concentration in petrolatum;[4] however, some investigators believe that this concentration may produce irritant reactions and recommend testing with 5%.[3]

It has been suggested that hypersensitivity reactions to cetyl alcohol are related to impurities in the product, since testing of patients with clearly positive patch tests to a commercial mixture containing 90 to 95% cetyl alcohol could not be reproduced by testing with a purified preparation (99.5%).[5]

VI. CLINICAL RELEVANCE

In the absence of stasis dermatitis, cetyl alcohol is an infrequent cause of cosmetic dermatitis, responsible for 0.2% of reactions in a prospective study of 487 patients in the U.S.,[6] in 0.7% of eczema patients in Finland,[5] and in 0.4% of 737 patients in Italy.[7]

REFERENCES

1. **Nikitakis, J. M.,** *CTFA Cosmetic Ingredient Handbook,* 1st ed., The Cosmetic, Toiletry and Fragrance Association, Washington, D.C., 1988.
2. **Degreef, H. and Dooms-Goossens, A.,** Patch testing with silver sulfadiazine cream, *Contact Dermatitis,* 12, 33, 1985.
3. **van Ketel, W. G. and Wemer, J.,** Allergy to lanolin and "lanolin-free" creams, *Contact Dermatitis,* 9, 420, 1983.
4. **Fisher, A. A.,** *Contact Dermatitis,* 3rd ed., Lea & Febiger, Philadelphia, 1986.
5. **Hannuksela, M. and Salo, H.,** The repeated open application test (ROAT), *Contact Dermatitis,* 14, 221, 1986.
6. **North American Contact Dermatitis Group,** Prospective study of cosmetic reactions 1977–1980, *J. Am. Acad. Dermatol.,* 6, 909, 1982.
7. **Tosti, A., Guerra, L., Morelli, R., and Bardazzi, F.,** Prevalence and sources of sensitization to emulsifiers: a clinical study, *Contact Dermatitis,* 23, 68, 1990.

CHLOROACETAMIDE

I. REGULATORY CLASSIFICATION

Chloroacetamide is an antimicrobial preservative used in cosmetics and pharmaceuticals. This preservative is not registered or approved for use in the U.S.

$$Cl-CH_2\overset{\overset{\displaystyle O}{\|}}{C}-NH_2$$

FIGURE 13. Chloroacetamide.

II. SYNONYMS

Chloracetamide
2-Chloroacetamide

III. AVAILABLE FORMULATIONS

A. Cosmetics
Chloroacetamide is present in face, hand, and body creams and lotions; wrinkle removers; cleansing products; and skin fresheners in cosmetics marketed outside of the U.S.[1] The usual concentration is 0.4%.[2]

B. Drugs
Chloroacetamide is used as an antimicrobial preservative in topical pharmaceuticals in Europe.

IV. HUMAN TOXICITY DATA

A. Allergic Contact Dermatitis
Sensitization to chloracetamide is easily achieved. In two studies attempting to sensitize human volunteers, application of 0.5% aqueous concentrations induced sensitization in 17 and 31% of the subjects.[3,4]

All of the cases reported have been to products used in European countries or exported to the U.S. Sensitization to a cosmetic cream containing plant extracts and 0.015% chloroacetamide was reported. Patch testing confirmed sensitivity to 0.1% chloroacetamide in petrolatum.[5] A pruritic facial dermatitis was attributed to the chloroacetamide preservative in an astringent spray imported from France and sold in the U.S.[6]

Occupational dermatitis from chloroacetamide in a cutting oil has been described.[7] Other industrial uses for this preservative include glues, adhesives, inks, paints and varnishes, textile finishes, leather and tanning agents, and packaging materials.[5]

Shoe dermatitis was attributed to leather tanned with chloroacetamide-preserved products. The reaction occurred while wearing new shoes. Testing of the shoes 6 months later did not reveal the presence of chloroacetamide. A patch test was positive only to 0.2% chloroacetamide. The authors concluded that trace amounts of this preservative remained after the tanning process and dissipated with time.[8]

V. CLINICAL RELEVANCE

Chloroacetamide is a potent sensitizer in concentrations as low as 0.015 to 0.07%.[2,5] The incidence of positive patch tests to chloroacetamide was 0.6% of 501 Dutch patients.[9] While not used in products manufactured in the U.S., imported European cosmetics may contain this preservative. Patch testing with an 0.2% aqueous solution is recommended.

REFERENCES

1. **Nikitakis, J. M.,** *CTFA Cosmetic Ingredient Handbook,* 1st ed., The Cosmetic, Toiletry and Fragrance Association, Washington, D.C., 1988.
2. **Detmar, U. and Agathos, M.,** Contact allergy to chloroacetamide, *Contact Dermatitis,* 19, 66, 1988.
3. **Marzulli, F. N. and Maibach, H. I.,** Antimicrobials: experimental contact sensitization in man, *J. Soc. Cosmet. Chem.,* 24, 399, 1971.
4. **Jordan, W. P. and King, S. E.,** Delayed hypersensitivity in females, *Contact Dermatitis,* 3, 19, 1977.
5. **Dooms-Goossens, A., Degreef, H., Vanhee, J., Kerkhofs, L., and Chrispeels, M. T.,** Chlorocresol and chloracetamide: allergens in medications, glues, and cosmetics, *Contact Dermatitis,* 7, 51, 1981.
6. **Koch, S. E., Mathias, T., and Maibach, H. I.,** Chloracetamide: an unusual cause of cosmetic dermatitis, *Arch. Dermatol.,* 121, 172, 1985.
7. **Lama, L., Vanni, D., Barone, M., Patrone, P., and Antonelli, C.,** Occupational dermatitis to chloroacetamide, *Contact Dermatitis,* 15, 243, 1986.
8. **Jelen, G., Cavelier, C., Protois, J. P., and Fousserequ, J.,** A new allergen responsible for shoe allergy: chloroacetamide, *Contact Dermatitis,* 21, 110, 1989.
9. **De Groot, A. C., Bos, J. D., Jagtman, B. A., Bruynzel, D. P., Van Joost, T., and Weyland, J. W.,** Contact allergy to preservatives-II, *Contact Dermatitis,* 15, 218, 1986.

CHLOROBUTANOL

I. REGULATORY CLASSIFICATION

Chlorobutanol is classified as an antimicrobial preservative.

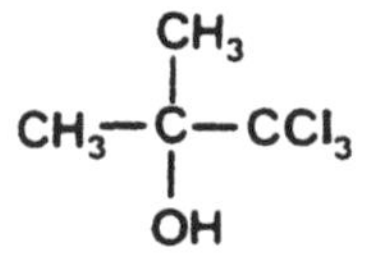

$$CH_3-\underset{\underset{OH}{|}}{\overset{\overset{CH_3}{|}}{C}}-CCl_3$$

FIGURE 14. Chlorobutanol.

II. SYNONYMS

Chlorbutol
Chloretone
2-Propanol,1,1,1-trichloro-2-methyl
1,1,1,-Trichloro-2-methyl-2-propanol

III. AVAILABLE FORMULATIONS

A. Drugs

Chlorobutanol has bacteriostatic properties against both Gram-positive and Gram-negative organisms, including *Pseudomonas aeruginosa* and *Staphylococcus albus*. It has some activity against yeasts and fungi, such as *Candida albicans*. Its primary use is in ophthalmic and parenteral drug products.

The FDA lists 13 parenteral products containing concentrations of 0.0001 to 0.5% and 15 ophthalmic products containing 0.2 to 0.65%.[1]

81

IV. TABLE OF COMMON PRODUCTS

A. Ophthalmic Products

Trade name	Manufacturer
AK-Chlor solution	Akorn
AK-Sulf solution	Akorn
Atropine sulfate solution, ointment	Allergan
Blink-N-Clean contact lens solution	Allergan
Chloroptic solution, ointment	Allergan
Epitrate solution	Wyeth-Ayerst
Fluor-I-Strip applicator	Wyeth-Ayerst
Fluress	Sola/Barnes-Hind
Homatropine solution	Allergan
Isopto Eserine solution	Alcon
Lacril	Allergan
Lacri-Lube S.O.P. ointment	Allergan
Liquifilm Tears	Allergan
Mycitracin ointment	Upjohn
Neo-Cortef ointment	Upjohn
Ophthalgan solution	Wyeth-Ayerst
PE solution	Alcon
Phospholine iodide solution	Wyeth-Ayerst
Pilagan solution	Allergan
Pilofrin solution	Allergan
Polyspectrin ointment	Allergan
Pred-G ointment	Allergan
PV Carpine solution	Allergan
Scopolamine ointment	Allergan
Soquette soaking and storing solution	Sola/Barnes-Hind
Sulten-10 solution	Bausch & Lomb
Tear Plus Lubricant	Allergan
Tobradex ointment	Alcon
Tobrex ointment	Alcon

B. Parenteral Drug Products

Trade name	Manufacturer
Adrenalin chloride injection steri-vial	Parke-Davis
Ana-Kit epinephrine syringe	Hollister-Stier
Bentyl injection	Marion Merrell Dow
Calciferol in oil	Schwarz Pharma
Codeine phosphate injection	Elkins-Sinn
DDAVP injection	Rhone-Poulenc Rorer
Dolophine hydrochloride	Lilly
Morphine sulfate injection	Lilly

Trade name (cont'd)	Manufacturer (cont'd)
Novocain injection 2%	Winthrop
Oxytocin	Lyphomed
Pitocin injection	Parke-Davis
Priscoline injection	Ciba
Pyridoxine hydrochloride	Lyphomed
Syntocinon	Sandoz
Thiamine hydrochloride	Lyphomed

V. ANIMAL TOXICITY DATA

An *in vitro* study using cultured human and rat corneal epithelial cells demonstrated 40% lysis after incubation with chlorobutanol 0.5% and less than 5% lysis at lower concentrations. This degree of cytotoxicity was less than that observed for other commonly used ophthalmic preservatives such as benzalkonium chloride, chlorohexidine, and thimerosal.[2]

Continuous bathing of rabbit eyes with a 0.4% chlorobutanol solution for 20 min produces keratitis epithelialis, manifested as a superficial corneal haze.[3]

VI. HUMAN TOXICITY DATA

A. Parenteral Administration

Both hypersensitivity and toxic reactions have been described following parenteral administration of chlorobutanol-preserved drugs. A delayed hypersensitivity reaction with red, indurated eruptions at the injection site and a raised erythematous rash over both thighs was reported in a 26-year-old postpartum woman receiving heparin preserved with chlorobutanol. Intradermal skin testing and macrophage migration inhibition tests confirmed a cell-mediated reaction to chlorobutanol.[4]

Severe pruritus and a maculopapular rash was attributed to the chlorobutanol preservative in a nasal DDAVP solution in a 54-year-old woman with partial diabetes insipidus. The reaction began 8 h after administration and lasted for several hours. Pruritus was reproduced after challenge with chlorobutanol-preserved saline and not by a preservative-free DDAVP solution. Intradermal challenge was negative.[5]

A systemic hypersensitivity reaction, described as severe anaphylactic shock and confirmed by positive scratch test, was reported in a 34-year-old woman following injection of a chlorobutanol-preserved oxytocin.[6]

Pharmacological effects attributed to chlorobutanol include myocardial contractility depression and decreased systemic vascular resistance.[7,8] A consistent decrease in systolic blood pressure (mean decrease 15.4 mmHg) was documented in 20 patients given a trial of chlorobutanol-preserved heparin (containing 2.5 mg/kg of chlorobutanol) prior to elective coronary bypass surgery in a randomized, double-blind, placebo-controlled study. Four patients required intervention (intravenous fluid, calcium gluconate, discontinuation of enflurane) to reverse systolic blood pressures that had fallen to 85 to 90 mmHg. Nonpreserved heparin did not produce a significant effect.[9]

Chlorobutanol is used as a nonprescription hypnotic in Australia with a usual dose of 150 mg. Unexplained somnolence was attributed to chlorobutanol, infused as a preservative during high-dose morphine therapy, in a 19-year-old woman with Ewing's sarcoma. At the highest dose, delivering 90 mg/h for 4 d of chlorobutanol (2.16 g/d), CNS depression and respiratory rates as low as 6 breaths per minute were observed.[10]

Lethargy, respiratory depression, and nystagmus may have been related to chlorobutanol given as a preservative during high-dose pyridoxine therapy for the treatment of *Gyromitra aesculenta* mushroom ingestion in a 27-year-old woman and a 33-year-old man. The total dose of chlorobutanol given over a 3-d period, assuming a 0.5% concentration, was 2.2 and 3.05 g/d, respectively. These doses are comparable to that associated with somnolence in the previously described case.[11]

B. Ophthalmic Administration

A transitory stinging sensation and conjunctival erythema was reported in 9 of 16 subjects tested with one drop of an ophthalmic solution containing polyvinyl alcohol 1.4% and chlorobutanol 0.5%. Testing with polyvinyl alcohol alone did not produce irritation. It was suggested that the irritation was due to crystallization of chlorobutanol, which is close to saturation at this concentration. Artificial tear preparations are unbuffered and may be more likely to produce crystals with fluctuations in pH.[12]

Gonioscopy solutions containing 0.4% chlorobutanol, applied under the contact lens, have been associated with visual fogging, haloes around lights, and a foreign body sensation, which intensifies over several hours and recovers spontaneously in 1 or 2 d.[3]

Generalized seizures and loss of consciousness were observed in a 28-year-old man less than 60 s after instillation of a chlorobutanol-preserved fluorescein solution. No effort was made to confirm a relationship to any of the active or inactive components of this product.[13]

VII. CLINICAL RELEVANCE

Isolated case reports have documented a variety of reactions to parenteral administration of chlorobutanol, including Type IV delayed hypersensitivity, Type I anaphylaxis, and nonimmunologic cardiovascular and CNS toxicity. Hypersensitivity reactions appear to be extremely rare. A survey of over 30,000 million units of chlorobutanol-preserved heparin administered over 3 years did not reveal any hypersensitivity reactions.[14]

CNS toxicity may occur in patients receiving high-dose therapy with continuous infusions of drugs preserved with chlorobutanol. Chlorobutanol has a long half-life, in the range of 13 d after chronic use, and may accumulate in patients receiving long-term parenteral therapy.[15]

REFERENCES

1. **Food and Drug Administration**, *Inactive Ingredients Guide*, FDA, Washington, D.C., March 1990.
2. **Neville, R., Dennis, P., Sens, D., and Crouch, R.**, Preservative cytotoxicity to cultured corneal epithelial cells, *Curr. Eye Res.*, 5, 367, 1986.
3. **Grant, W. M.**, *Toxicology of the Eye*, 3rd ed., Charles C Thomas, Springfield, IL, 1986.
4. **Dux, S., Pitlik, S., Perry, G., and Rosenfeld, J. B.**, Hypersensitivity reaction to chlorbutol-preserved heparin, *Lancet*, 1, 149, 1981.
5. **Itabashi, A., Katayama, S, and Yamaji, T.**, Hypersensitivity to chlorobutanol in DDAVP solution, *Lancet*, 1, 108, 1982.
6. **Hofmann, H., Goerz, G., and Plewig, G.**, Anaphylactic shock from chlorobutanol-preserved oxytocin, *Contact Dermatitis*, 15, 241, 1986.

7. **Hermsmeyer, K. and Aprigliano, O.,** Effects of chlorobutanol and bradykinin on myocardial excitation, *Am. J. Physiol.,* 230, 306, 1976.
8. **Barrigon, S., Tejerina, T., Delgrado, C., and Tamargo, J.,** Effects of chlorbutol on Ca movements and contractile responses of rat aorta and its relevance to the actions of Syntocinon, *J. Pharm. Pharmacol.,* 36, 521, 1984.
9. **Bowler, G. M. R., Galloway, D. W., Meiklejohn, B. H., and Macintyre, C. C. A.,** Sharp fall in blood pressure after injection of heparin containing chlorbutol, *Lancet,* 1, 848, 1986.
10. **DeChristoforo, R., Corden, B. J., Hood, J. C., Narang, P. K., and Magrath, I. T.,** High-dose morphine infusion complicated by chlorobutanol-induced somnolence, *Ann. Intern. Med.,* 98, 335, 1983.
11. **Albin, R. L., Albers, J. W., Greenberg, H. S., Townsend, J. B., Lynn, R. B., Burke, J. M., and Alessi, A. G.,** Acute sensory neuropathy-neuronopathy from pyridoxine overdose, *Neurology,* 37, 1729, 1987.
12. **Fassihi, A. R. and Naidoo, N. T.,** Irritation associated with tear-replacement ophthalmic drops, *S. Afr. Med. J.,* 75, 233, 1989.
13. **Cohn, H. C. and Jocson, V. L.,** A unique case of grand mal seizures after Fluress, *Ann. Ophthalmol.,* 13, 1379, 1981.
14. **Marsh, B. T.,** Preservatives in heparin, *Lancet,* 1, 860, 1977.
15. **Borody, T., Chinwah, P. M., Graham, G. G., Wade D. N., and Williams, K. M.,** Chlorbutol toxicity and dependence, *Med. J. Aust.,* 1, 288, 1979.

CHLOROCRESOL

I. REGULATORY CLASSIFICATION

Chlorocresol is a colorless, dimorphous crystal with a phenolic odor and is classified as an antimicrobial preservative. It has bactericidal activities against Gram-positive and Gram-negative species, spores, molds, and yeasts.

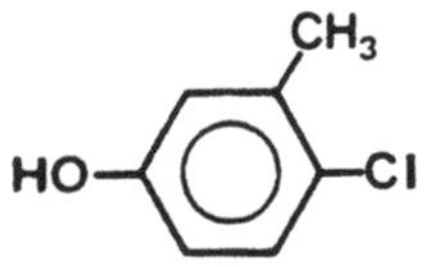

FIGURE 15. Chlorocresol.

II. SYNONYMS

Parachlorometacresol
PCMC
4-Chloro-3-methylphenol
2-Chloro-5-hydroxytoluene

III. AVAILABLE FORMULATIONS

A. Drugs
Chlorocresol is found as a preservative in injectable products in concentrations of 0.1 to 0.15%. It is used in topical creams and lotions in concentrations of 0.075 to 0.12% and is present in at least 22 topical cream NDA-approved formulations.[1]

B. Cosmetics
Chlorocresol is a biocide and preservative found in bath oils and bath salts.[2] It has a distinctive pungent odor which renders it unsuitable for use in perfumed cosmetics.

IV. TABLE OF COMMON PRODUCTS

A. Topical Drug Products

Trade name	Manufacturer
Aclovate cream	Glaxo
Alphatrex cream	Savage
Betatrex cream	Savage
Dermolate cream	Schering
Diprolene AF cream	Schering
Drithocreme	American Dermal
Dritho-scalp	American Dermal
Maxivate cream	Westwood-Squibb
Temovate cream	Glaxo
Tinactin Jock Itch cream	Schering

V. HUMAN TOXICITY DATA

A. Contact Dermatitis

Chlorocresol is a weak primary irritant in a concentration of 5% in petrolatum, and a moderate irritant in a concentration of 10%.[3] The recommended patch test concentration is 2% in petrolatum;[4] however, irritant reactions may occur with this concentration.[3] A 1% concentration may be optimal, but can occasionally be irritating.[5]

Chlorocresol hypersensitivity was documented in 8 of 1000 patients with contact dermatitis in Australia. All were positive to the related compound chloroxylenol, which was thought to be the primary sensitizer.[6] Cross-reactivity between these two compounds was demonstrated in three patients reported by Hjorth and Trolle-Lassen,[7] and in two cases reported by Burry et al.[5] An ECG paste containing chloroxylenol was also speculated to have been the primary sensitizer in a patient with contact dermatitis from a chlorocresol-containing corticosteroid cream; however, patch testing to chloroxylenol was not done.[8]

The incidence of hypersensitivity to chlorocresol in contact dermatitis patients was 3% in a Belgian series of 167 patients.[8]

B. Contact Urticaria

One case of contact urticaria has been described following open and prick tests with 10% chlorocresol. Urticaria, followed by superficial necrosis, was described in the patient, while tests in ten controls elicited occasional slight erythema without necrosis.[9]

The simultaneous occurrence of immediate and delayed hypersensitivity was reported in a 35-year-old woman exposed to chlorocresol-containing disinfectants in a pathology laboratory. Symptoms of rhinitis, conjunctivitis, and lip and eyelid edema occurred within 15 to 30 min after entering the laboratory. Open and prick testing with 1 and 5% chlorocresol in an alcohol/water mixture resulted in both an immediate wheal and a delayed eczematous reaction.[10]

C. Toxic Reactions

Occupational exposure to chlorocresol in the sterilizing department of a pharmacy was associated with repeated episodes of left-sided facial palsy, lasting 15 min to 3 h, in a 42-

year-old woman. These episodes were traced to preparation of ampules containing heparin and chlorocresol. Open inhalation provocation with a very dilute chlorocresol solution, 0.1% diluted in water and aerosolized, reproduced the symptoms after 3 min. EMG results showed severe reduction of motor unit potentials of the left orbicularis oris.[11]

A systemic reaction attributed to irritant or allergic phlebitis was described in a 21-year-old woman following administration of heparin preserved with chlorocresol via an indwelling venous canula. Symptoms consisted of severe burning pain at the injection site, which radiated along the veins, followed by nausea, light-headedness, drowsiness, pallor, and sweating. A red papule developed 2 d later at the injection site. Intradermal testing with the preserved heparin was positive, and the patient tolerated subsequent treatment with chlorocresol-free heparin.[12]

D. Anaphylactoid Reactions

One documented case of an anaphylactoid reaction was described in a 55-year-old man receiving chlorocresol-preserved heparin injections. Intravenous injection of 10,000 U of preserved heparin was followed by nasal congestion, sweating, and generalized urticaria within 1 h. Intradermal testing with nonpreserved heparin was negative.[13] In another immediate hypersensitivity reaction described to the same product, intradermal tests suggested hypersensitivity to heparin rather than to the preservative.[13]

E. Local Reactions

Subcutaneous heparin preserved with chlorocresol was associated with indurated erythematous pruritic reactions within hours of injection at the injection site in seven patients; all had positive intradermal tests to chlorocresol-preserved heparin, while four tested with preservative-free heparin had a negative response.[13]

VI. CLINICAL RELEVANCE

A. Contact Dermatitis

The significance of positive patch tests to chlorocresol 2% in petrolatum has been questioned, based on data indicating a lack of reproducibility during retesting and negative usage tests in 11 patients with positive initial tests; six of these tests were judged to be irritant reactions.[3]

Hypersensitivity to chlorocresol has almost always been associated with primary hypersensitivity to chloroxylenol when this relationship has been explored.[5] One chlorocresol-sensitive patient was reported to have a negative patch test to chloroxylenol.[14]

B. Systemic Reactions

Although many cases of systemic anaphylactoid reactions have been reported following chlorocresol-containing heparin, only one has been definitely proven to be related to the preservative.

REFERENCES

1. **Food and Drug Administration,** *Inactive Ingredients Guide,* FDA, Washington, D.C., March 1990.
2. **Nikitakis, J. M.,** *CTFA Cosmetic Ingredient Handbook,* 1st ed., The Cosmetic, Toiletry and Fragrance Association, Washington, D.C., 1988.

3. **Andersen, K. E. and Hamann, K.,** How sensitizing is chlorocresol? Allergy tests in guinea pigs versus the clinical experience, *Contact Dermatitis,* 11, 11, 1984.

4. **Fisher, A. A.,** *Contact Dermatitis,* 3rd ed., Lea & Febiger, Philadelphia, 1986.

5. **Burry, J. N., Kirk, J., Reid, J. G., and Turner, T.,** Chlorocresol sensitivity, *Contact Dermatitis,* 1, 41, 1975.

6. **Burry, J. N., Kirk, J., Reid, J. G., and Turner, T.,** Environmental dermatitis: patch tests in 1000 cases of allergic contact dermatitis, *Med. J. Aust.,* 2, 681, 1973.

7. **Hjorth, N. and Trolle-Lassen, C.,** Skin reaction to ointment bases, *Trans. St. John's Hosp. Soc.,* 49, 127, 1963.

8. **Oleffe, J. A., Blondeel, A., and de Coninck, A.,** Allergy to chlorocresol and propylene glycol in a steroid cream, *Contact Dermatitis,* 5, 1979.

9. **Freitas, J. P. and Brandao, F. M.,** Contact urticaria to chlorocresol, *Contact Dermatitis,* 15, 252, 1986.

10. **Goncalo, M., Concalo, S., and Moreno, A.,** Immediate and delayed sensitivity to chlorocresol, *Contact Dermatitis,* 17, 46, 1987.

11. **Dossing, M., Wulff, C. H., and Olsen, P. Z.,** Repeated facial palsies after chlorocresol inhalation, *J. Neurol. Neurosurg. Psychiatry,* 49, 1452, 1986.

12. **Ainley, E. J., Mackie, I. G., and Macarthur, D.,** Adverse reaction to chlorocresol-preserved heparin, *Lancet,* 1, 705, 1977.

13. **Hancock, B. W. and Naysmith, A.,** Hypersensitivity to chlorocresol-preserved heparin, *Br. Med. J.,* 3, 746, 1975.

14. **Dooms-Goossens, A., Degreef, H., Vanhee, J., Kerkhofs, L., and Chrispeels, M. T.,** Chlorocresol and chloracetamide: allergens in medications, glues, and cosmetics, *Contact Dermatitis,* 7, 51, 1981.

CHLOROFLUOROCARBONS

I. REGULATORY CLASSIFICATION

Dichlorodifluoromethane, dichlorotetrafluoromethane, and trichloromonofluoromethane are classified as aerosol propellants. These substances are prohibited for nonessential uses in self-pressurized containers, such as cosmetic products, but are allowed in pharmaceuticals. Applications for new pharmaceuticals must demonstrate that there are no technically feasible alternatives to the use of chlorofluorocarbons, that the product provides a substantial health benefit, and that usage would not involve a significant release of chlorofluorocarbons (CFCs) into the atmosphere.

II. SYNONYMS

A. Dichlorodifluromethane

CFC 12
Difluorodichloromethane (CFC 12)
Freon® 12

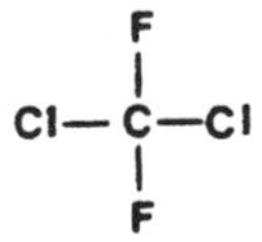

FIGURE 16. CFC 12.

B. Trichlorofluoromethane

CFC 11
Fluorotrichloromethane
Freon® 11
Trichloromonofluoromethane

91

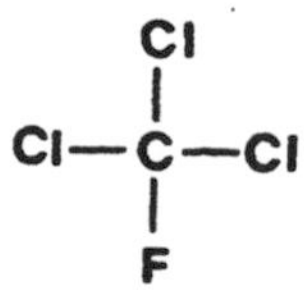

FIGURE 17. CFC 11.

C. Dichlorotetrafluoromethane

CFC 114
Cryfluorane
Freon® 114
Tetrafluorodichloroethane

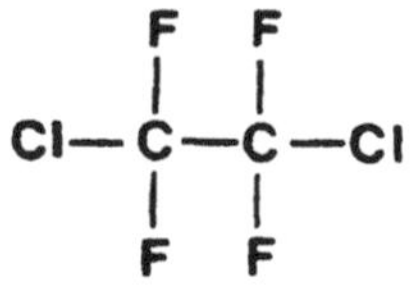

FIGURE 18. CFC 114.

III. AVAILABLE FORMULATIONS

Chlorofluorocarbons are used as propellants in drug products for oral inhalation, nasal inhalation, and topical application. CFC 12 is the most commonly used propellant; CFC 11 and CFC 114 may be mixed with CFC 12 to modify the vapor pressure and as diluents and solvents.

CFC 12 is listed as an inactive ingredient in 30 FDA-approved pharmaceuticals, including metered inhalation aerosols, nasal aerosols, rectal foams, topical aerosols and foams, and vaginal foams. Concentrations range from 11 to 68%.[1]

CFC 114 is listed in 22 products, including 1 nasal aerosol with a concentration of 90% and 2 vaginal aerosol foams containing 80 to 85%. CFC 11 is listed in 18 products, mostly metered inhalation aerosols.[1]

The amount of chlorofluorocarbon inhaled with each metered dose of an inhaled bronchodilator has been estimated to range from 6.5 to 20 ml.[2,3]

IV. TABLE OF COMMON PRODUCTS AND ALTERNATIVES

A. Inhalation Aerosol Drug Products

Trade name	Manufacturer	Alternative
AeroBid	Forest	None
Alupent aerosol	Boehringer Ingelheim	Alupent solution
Asthmahaler	Norcliff Thayer	AsthmaNefrin
Atropine sulfate	U.S. Army	None
Atrovent	Boehringer Ingelheim	None

Trade name (cont'd)	Manufacturer (cont'd)	Alternative (cont'd)
Azmacort	Rorer	None
Beclovent	Allen & Hanburys	None
Beconase	Allen & Hanburys	Beconase AQ
Brethaire	Geigy	None
Bronitin mist	Whitehall	AsthmaNefrin
Bronkaid mist	Winthrop	AsthmaNefrin
Bronkaid mist suspension	Winthrop	AsthmaNefrin
Bronkometer	Winthrop	Bronkosol solution
Decadron phosphate respihaler	Merck Sharp & Dohme	None
Decadron phosphate turbinaire	Merck Sharp & Dohme	None
Duo-Medihaler	3M Riker	None
Intal inhaler	Fisons	Intal solution
Isuprel mistometer	Winthrop	Isuprel solution
Maxair	3M Riker	None
Medihaler-Ergotamine	3M Riker	None
Medihaler-Epi	3M Riker	AsthmaNefrin
Medihaler-Iso	3M Riker	Isuprel solution
Metaprel aerosol	Sandoz	Metaprel solution
Nitrolingual	Rorer	None
Norisodrine aerotrol	Abbott	Isuprel solution
Primatene mist	Whitehall	AsthmaNefrin
Primatene mist suspension	Whitehall	AsthmaNefrin
Proventil inhaler	Schering	Ventolin solution
		Ventolin rotacaps
Tornalate	Winthrop	None
Vancenase	Schering	Vancenase AQ
Vanceril	Schering	None
Ventolin aerosol	Allen & Hanburys	Ventolin solution
		Ventolin rotacaps

V. ANIMAL TOXICITY DATA

A. Direct Cardiovascular Toxicity

Pressurized aerosol bronchodilators first became available in the 1950s. Over the next decade, reports emerged of an increasing incidence of sudden unexpected death in asthmatic patients, particularly in children aged 10 to 14 years, often associated with excessive use of the aerosols.[4-6] Reports of deaths in asthmatics who received epinephrine injections soon after taking a metered-dose bronchodilator[7,8] combined with a report of sudden sniffing deaths in adolescents who deliberately abused chlorofluorocarbons, 18 of 110 cases in association with recent activity or stress,[9] suggested a primary role of the propellant in causing the asthmatic deaths. Animal studies were done to attempt to elucidate the mechanism for the sudden deaths.

In a widely criticized study, Taylor and Harris[10] exposed eight anesthetized mice to three doses of an isoproterenol pressurized spray with the liquid particles filtered out, followed by asphyxiation via a tightly fitted plastic bag around the nostril and mouth. Another six mice were exposed to a mixture of 60% CFC 12 and 40% CFC 114 contained in a loosely fitted plastic bag around the head, again followed by asphyxiation. The mice treated with the propellants developed profound sinus bradycardia and AV block in response to asphyxiation, which stopped when the asphyxia was terminated. Administration of the propellant without asphyxia produced minimal cardiac effects, limited to marked T-wave depression in 7 of 12 animals. No ventricular extrasystoles were observed in any of the mice.

This study was criticized because the CFC-treated mice were preexposed to an oxygen-deprived gas prior to asphyxiation and thus were under hypoxic conditions for a longer time than control animals. A study attempting to replicate their findings found that control animals developed the same cardiac changes as exposed mice, but with a delayed onset; thus the cardiac effects were related to the duration of hypoxia in both groups.[11]

A study designed to investigate the contribution of hypoxia in chlorofluorocarbon arrhythmias used 16 anesthetized, artificially ventilated dogs. Arterial oxygen saturation, PO_2, PCO_2, and pH were maintained in the normal range during aerosol administration of high concentrations of a commercial antiseptic spray containing 97% of a fluorocarbon mixture, mimicking the inhalant abuse situation. Lethal arrhythmias occurred both in oxygenated dogs and in animals exposed to room air, consisting of initial sudden sinus node slowing, followed by asystole and ventricular escape, and ultimately electrical asystole or ventricular fibrillation.[12]

These authors compiled the results of 55 experiments using Freon® 11 and 25 experiments using Freon® 12 in the same dog model. Administration of Freon® 11 at concentrations below 15% for 10 min were nonlethal, and concentrations in excess of 21% (215,000 ppm) were uniformly lethal. Lethal concentrations were comparable to those achieved by filling plastic bags to simulate the deliberate sniffer, which reached up to 350,000 ppm.[13]

B. Sensitization to Epinephrine

Nonanesthetized dogs exposed to 1.25% CFC 11 in inspired air by mask for 5 min, followed immediately by epinephrine 5 mcg/kg, developed serious cardiac arrhythmias 10 to 20 s after injection, consisting of a short run of multifocal ventricular ectopic beats, then ventricular fibrillation, which responded to defibrillation. Administration of epinephrine 10 min after cessation of exposure to CFC 11 had no adverse effect. The human exposure from one puff of an aerosolized bronchodilator was estimated to be a maximum of 0.3% of CFC 11. This dose of epinephrine was large enough to cause ventricular ectopic beats in some control injections in animals not exposed to the fluorocarbon.[14]

A similar experiment designed to mimic the clinical situation of excessive aerosol administration in asthmatic patients failed to produce cardiac arrhythmias in dogs given the equivalent of 25 times the recommended dose of isoproterenol within 5 min, 20 times the concentration of fluorocarbon associated with propellant administration in humans, and a degree of hypoxia comparable with that found in a severe asthma attack.[14]

An investigation of anesthetized monkeys reported a threshold inhaled concentration for signs of cardiotoxicity of 2.5% for myocardial contractility depression and 5% for cardiac arrhythmias. When epinephrine was given concurrently in doses of 0.5 to 1 mcg/min for 5 min, proarrhythmic activity occurred at a concentration of 2.5%, with abolishment of the contractility depressant effects. The minimal proarrhythmic concentration was further decreased to 1.25% in animals with coronary artery occlusion and to 0.5% in animals with a combination of coronary artery occlusion and epinephrine administration. There were marked differences among individual chlorofluorocarbons. CFC 12 and CFC 114 produced cardiac toxicity only at twice the concentrations reported for CFC 11. Sensitization to epinephrine could not be produced with these agents in the monkey.[15]

VI. HUMAN TOXICITY DATA

A. Allergic Contact Dermatitis

Three patients with axillary contact eczema secondary to use of deodorant sprays were found to have positive patch tests to Freon® 11; one of these patients also had a positive reaction to Freon® 12. Histological examination showed evidence of acute allergic eczema.[16]

Contact dermatitis to an aerosol deodorant and an antimycotic aerosol in a 36-year-old man prompted investigation of the role of the propellants. Closed patch testing was positive to Freon® 12, but not to Freon® 11 and Freon® 114. A skin biopsy at the site of the reaction showed histologic findings compatible with the diagnosis of allergic contact dermatitis.[17]

B. Bronchoconstriction

Chlorofluorocarbons have been shown to reduce the ventilatory capacity in healthy adults, asthmatic adults, and asthmatic children, possibly due to a local irritant effect. The decrease in FEV_1 was only observed during the first 2 h after inhalation.[18-20] Combining a bronchodilator with a chlorofluorocarbon propellant mixture had no effect on efficacy of the bronchodilator in a double-blind, crossover study of eight asthmatics and eight patients with chronic bronchitis.[2]

Significant maximal airway resistance changes and bronchoconstriction were observed in 5 of 13 subjects after inhalation of a placebo aerosol containing sorbitan trioleate and chlorofluorocarbons.[21]

Paradoxical bronchoconstriction was observed in 52 of 900 asthmatics in an evaluation of a metered-dose metaproterenol product, with an incidence of 4.4% in the drug group and 6.9% in the placebo group (chlorofluorocarbons and oleic acid). Patients with bronchoconstriction from the placebo system did not improve after receiving the active metered-dose inhaler, but did improve after nebulized metaproterenol, implicating one or more of the excipients as causing the adverse response. The investigators did not determine which of the excipients was responsible.[22]

In another study of 12 asthmatic subjects with a history of severe cough and wheezing after inhalation of beclomethasone aerosol, patients were enrolled in a double-blind crossover trial comparing a single dose of either beclomethasone aerosol or a placebo metered-dose inhaler containing oleic acid and fluorocarbon propellants. The incidence of coughing averaged 31 times after the active inhaler and 19 times after the placebo inhaler. The forced expiratory volume in 1 s (FEV_1) decreased to a similar degree with both treatments, a mean of 22.6% after the active inhaler and 22% after the placebo. Pretreatment with an inhaled bronchodilator enabled 7 of the 12 subjects to tolerate beclomethasone. The overall incidence of this adverse reaction was 20% of 70 patients prescribed the drug.[23]

A case report described paradoxical bronchoconstriction following the use of a metered-dose isoetharine aerosol containing the inactive ingredients chlorofluorocarbons, ethanol, ascorbic acid, menthol, and saccharin. Testing with a complete placebo system showed a decline in FEV_1 1 h after administration of the aerosol to as much as 40% below baseline. Testing with the individual components was not done. Based on a lack of reported reactions to other aerosol bronchodilators, the chlorofluorocarbons were not implicated.[24]

C. Aerosol Abuse

A review of 16 cases of aerosol addiction in asthmatics over an 18-year period found albuterol to be the agent involved in all of the cases, with terbutaline and beclomethasone inhalers as additional agents in 3 cases. Ten of the cases were teenagers, aged 11 to 17 years; one case occurred in a 3.5-year-old child. Because of the virtual inability of albuterol to distribute to brain tissue, the fluorinated hydrocarbons were considered to be responsible for the addiction. A nonfluorinated hydrocarbon-containing albuterol nebulizer solution was implicated in one case.[25]

D. Cardiovascular Toxicity

Continuous electrocardiogram monitoring in 24 subjects receiving nebulized albuterol, followed by inhalation of chlorofluorocarbons or oxygen, demonstrated ventricular extrasystoles in 3 subjects, 2 occurring in the CFC-treated group.[2]

VII. CLINICAL RELEVANCE

A. Allergic Contact Dermatitis

Chlorofluorocarbons are volatile, chemically stable compounds, which have been responsible for only four published cases of contact dermatitis, despite use in 75% of topical aerosol drugs and cosmetics in some countries. The use of CFCs has been banned in the U.S. for all uses except pharmaceutical; therefore, the potential for sensitization is very low.

B. Bronchoconstriction

The incidence of paradoxical bronchoconstriction potentially related to the chlorofluorocarbon excipient in some bronchodilator or inhaled corticosteroid products is confounded by the inherent irritability of the corticosteroid and partial or complete reversal of the adverse effect when combined with a bronchodilator. Testing with the CFC alone has not been frequently investigated in this population, but three studies implicate a transient primary irritant effect. Further studies are needed to determine the role of chlorofluorocarbons, oleic acid, sorbitan trioleate, and other excipients in exacerbation of asthma following use of inhalation aerosol products.

C. Aerosol Abuse

Patients with excessive use or abuse of metered-dose bronchodilators may benefit from switching to a dry powder inhalation system. If excessive use continues, oral theophylline therapy may be considered. Although sudden death in asthmatics was originally attributed to excessive use of aerosols containing chlorofluorocarbon propellants, it is currently believed that most deaths can be attributed to inadequate assessment and treatment of the asthmatic condition.[26,27]

D. Cardiac Toxicity

Based primarily on animal data, cardiac arrhythmias and sensitization to the effects of epinephrine are extremely unlikely to occur in patients exposed to the amounts of chlorofluorocarbons present in antiasthmatic aerosol medications. Sudden death is a well-documented sequela of deliberate inhalation of high concentrations of chlorofluorocarbons in the context of abuse.

REFERENCES

1. **Food and Drug Administration,** *Inactive Ingredients Guide,* FDA, Washington, D.C., March 1990.
2. **Thiessen, B. and Pedersen, O. F.,** Effect of freon inhalation on maximal expiratory flows and heart rhythm after treatment with salbutamol and ipratropium bromide, *Eur. J. Respir. Dis.,* 61, 156, 1980.
3. **Dollery, C. T., Draffan, G. H., Davies, D. S., Williams, F. M., and Conolly, M. E.,** Blood concentrations in man of fluorinated hydrocarbons after inhalation of pressurized aerosols, *Lancet,* 2, 1164, 1970.
4. **Speizer, F. E., Doll, R., and Heaf, P.,** Observations on recent increase in mortality from asthma, *Br. Med. J.,* 1, 335, 1968.
5. **Speizer, F. E., Doll, R., Heaf, P., and Strang, L. B.,** Investigation into use of drugs preceding death from asthma, *Br. Med. J.,* 1, 339, 1968.
6. **Inman, W. H. W. and Adelstein, A. M.,** Rise and fall of asthma mortality in England and Wales in relation to use of pressurised aerosols, *Lancet,* 2, 279, 1969.
7. **McManis, A. J.,** Adrenaline and isoprenaline: a warning, *Med. J. Aust.,* 3, 76, 1964.
8. **Greenberg, M. J.,** Isoprenaline in myocardial failure, *Lancet,* 2, 442, 1965.
9. **Bass, M.,** Sudden sniffing death, *JAMA,* 212, 2075, 1970.
10. **Taylor, G. J. and Harris, W. S.,** Cardiac toxicity of aerosol propellants, *JAMA,* 214, 81, 1970.
11. **Azar, A., Zapp, J. A., Reinhardt, C. F., and Stopps, G. J.,** Cardiac toxicity of aerosol propellants, *JAMA,* 215, 1501.
12. **Flowers, N. C. and Horan, L. G.,** Nonanoxic aerosol arrhythmias, *JAMA,* 219, 33, 1972.
13. **Flowers, N. C., Hand, R. C., and Horan, L. G.,** Concentrations of fluoroalkanes associated with cardiac conduction system toxicity, *Arch. Environ. Health,* 30, 353, 1975.
14. **Clark, D. G. and Tinston, D. J.,** Cardiac effects of isoproterenol, hypoxia, hypercapnia and fluorocarbon propellants and their use in asthma inhalers, *Ann. Allergy,* 30, 536, 1972.
15. **Belej, M. A., Smith, D. G., and Aviado, D. M.,** Toxicity of aerosol propellants in the respiratory and circulatory systems. IV. Cardiotoxicity in the monkey, *Toxicology,* 2, 381, 1974.
16. **van Ketel, W. G.,** Allergic contact dermatitis from propellants in deodorant sprays in combination with allergy to ethyl chloride, *Contact Dermatitis,* 2, 115, 1976.
17. **Valdivieso, R., Pola, J., Zapata, C., Cuesta, J., Puyana, J., Martin, C., and Losada, E.,** Contact allergic dermatitis caused by Freon 12 in deodorants. *Contact Dermatitis,* 17, 243, 1987.
18. **Sterling, G. M. and Batten, J. C.,** Effect of aerosol propellants and surfactants on airway resistance, *Thorax,* 24, 228, 1969.
19. **Valie, F., Skuvie, Z., Bantic, Z., Rudar, M., and Hecej, M.,** Effects of fluorocarbon propellants on respiratory flow and ECG, *Br. J. Ind. Med.,* 34, 130, 1977.
20. **Graff-Lonnevig, V.,** Diurnal expiratory flow after inhalation of Freons and fenoterol in childhood asthma, *J. Allergy Clin. Immunol.,* 64, 534, 1979.
21. **Brooks, S. M., Mintz, S., and Weiss, E.,** Changes occurring after Freon inhalation, *Am. Rev. Respir. Dis.,* 105, 640, 1972.
22. **Yarborough, L., Mansfield, L., and Ting, S.,** Metered dose inhaler induced bronchospasm in asthmatic patients, *Ann. Allergy,* 55, 25, 1985.
23. **Shim, C. and Williams, M. H.,** Cough and wheezing from beclomethasone aerosol, *Chest,* 91, 207, 1987.
24. **Witek, R. J., Schachter, E. N., and Zuskin, E.,** Paradoxical bronchoconstriction following inhalation of isoetharine aerosol: a case report, *Respir. Care,* 32, 29, 1987.
25. **Prasher, V. P. and Corbett, J. A.,** Aerosol addiction, *Br. J. Psychiatry,* 157, 922, 1990.
26. **Benatar, S. R.,** Fatal asthma, *N. Engl. J. Med.,* 314, 423, 1986.
27. **Lanes, S. F. and Walker, A. M.,** Do pressurized bronchodilator aerosols cause death among asthmatics?, *Am. J. Epidemiol.,* 125, 755, 1987.

CINNAMON OIL

I. REGULATORY CLASSIFICATION

Cinnamon and cinnamon oil are classified as flavoring agents. Cinnamon and cinnamaldehyde are on the GRAS list.

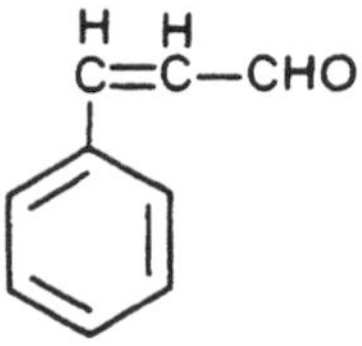

FIGURE 19. Cinnamic aldehyde. 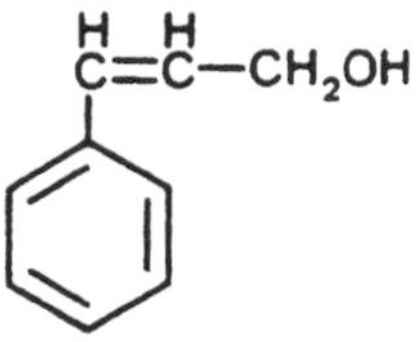　　FIGURE 20. Cinnamic alcohol.

II. SYNONYMS

Cassia oil
Saigon cinnamon
Oil of Chinese cinnamon

III. AVAILABLE FORMULATIONS

A. Constituents

Most cinnamon oil used in the U.S. is derived from the leaves, inner bark, and twigs of *Cinnamomum cassia* and contains cinnamic aldehyde (cinnamaldehyde) (minimum 80%) and smaller amounts of cinnamyl alcohol, cinnamyl acetate, eugenol, and 2-methoxy-cinnamaldehyde.[1]

B. Foods

Cinnamon oil is used as a flavoring in baked goods, candies, chewing gum, ice cream, soft drinks, cola beverages, alcoholic beverages, vermouths, bitters, and processed meats.[2]

C. Cosmetics

Cosmetics containing cinnamon oil include dentifrices, toilet soaps, mouthwashes, and

99

lipsticks. There are over 180 naturally occurring cinnamic acid esters. Methylcinnamate is widely used as a perfume in cosmetics. Other cinnamates may be present in sunscreen products.[3]

D. Drugs

Cinnamon oil is a component of official preparations of Compound Vanillin Elixir and Aromatic Cascara sagrada fluidextract.

IV. TABLE OF COMMON PRODUCTS

A. Oral Drug Products

Trade name	Manufacturer
Mycostatin pastilles	Bristol-Myers
PBZ elixir	Geigy
Phos-Flur	Colgate-Hoyt
Ryna C liquid	Wallace

V. ANIMAL TOXICITY DATA

As a result of a 16-week rat study, the conditional acceptable daily intake of cinnamaldehyde was recommended to be 1.25 mg/kg.[4] Dietary levels of 1% in the rat model resulted in slight hepatic cell swelling and forestomach hyperkeratosis.[5]

VI. HUMAN TOXICITY DATA

A. Primary Irritation

Irritant patch test reactions were reported in 5 of 58 control subjects (8.6%) tested with 5% cinnamic aldehyde in petrolatum.[6] Irritancy was also reported in 5 of 18 (27.8%) control patients tested with cinnamic aldehyde 2% in petrolatum.[7] Patch testing with 1% in petrolatum is currently recommended.[8]

B. Allergic Contact Dermatitis

Testing of 713 patients with cosmetic dermatitis revealed the causative agent to be most frequently related to a fragrance ingredient. When the specific fragrance could be identified, the most frequently implicated was cinnamic alcohol. Cinnamic aldehyde was responsible in 6 cases (0.8%) and cinnamic alcohol in 17 cases (2.4%).[9]

Occupational exposure to cinnamon in food handlers resulted in allergic contact dermatitis in 4 of 20 cases of food handler dermatitis reported in Canada.[10]

A deodorant sanitary napkin perfumed with cinnamaldehyde and cinnamic alcohol was responsible for development of perineal dermatitis in a 38-year-old woman.[11]

A bullous eruption resembling bullous pemphigoid, but without the immunological markers, was reported in a patient with contact allergy to cinnamic aldehyde.[12]

Other immunologic reactions to cinnamon derivatives include chronic urticaria.[13]

C. Oral Mucosal Reactions

There are numerous cases of cheilitis, gingivitis, and stomatitis secondary to the presence of cinnamon in chewing gum, lipstick, or toothpaste. Symptoms and lesions include lip swelling, lip fissuring, a burning sensation, erythema, gingivitis, vesiculation, and ulceration. The burning sensation is the most common finding. Lesions are usually confined to the buccal mucosa and lateral border of the tongue, appearing as erythematous patches with superimposed keratosis or ulceration.[2,3,14-23]

Chronic cheilitis of 1-year duration was described in an 82-year-old woman following the use of a sunscreen lipstick and a toothpaste containing cinnamon. Patch testing was positive only to 1% cinnamic aldehyde in petrolatum. Discontinuation of these products resulted in clearing over 3 weeks.[24]

While the reaction was confined to the lip in this patient, presumably due to the low concentration in the lipstick, Thyne et al. reported a case of contact stomatitis with ulceration of the lower labial mucosa and lower lip swelling, due to a toothpaste containing cinnamon. Patch testing was strongly positive to 2% cinnamic aldehyde and weakly positive to 0.1%.[2]

Orofacial granulomatosis, a triad of lip or facial swelling, seventh nerve palsy, and a fissured tongue, has been associated with the presence of cinnamon in foods with histological findings of plasma cell infiltration, changes in stratified squamous epithelium, and capillary dilation.[25-27]

Perioral leukoderma with complete skin depigmentation at the oral commissures has been described following the use of a cinnamon-flavored toothpaste in a 25-year-old woman with delayed-type hypersensitivity to cinnamic aldehyde. Leukodermic areas were sharply marginated by a thin border of hyperpigmentation.[28]

D. Nonimmunological Contact Urticaria

Cinnamon, cinnamic acid, cinnamates, and cinnamaldehyde readily produce contact urticaria when applied to the skin.[13] Sixteen control subjects without a history of cinnamon intolerance were tested with 0.2% cinnamic aldehyde in 10% ethyl alcohol applied to the anticubital fossa, mimicking concentrations in a commercial mouthwash. All subjects had a perceptible response, which was 2+ or greater in 12 subjects. Changing the vehicle to petrolatum reduced the incidence of positive responses to 50%. When applied in the alcohol vehicle, the response began in 5 to 15 min and disappeared within 30 min. The onset was delayed with the petrolatum vehicle to 45 to 60 min. The minimum concentration producing urticaria ranged from 0.01 to 0.1% in alcohol. None reacted to amyl or methyl cinnamates.[29]

Of 40 children tested with cinnamaldehyde, 12 developed contact urticaria. Positive reactions were seen in 2 of the 12 children with cinnamon, and in 1 of the 12 with cinnamyl alcohol.[27] Of 18 nonallergic control subjects tested with cinnamic aldehyde 2% in petrolatum, immediate urticarial reactions were observed in 8 (44.4%) and significant erythema in 4. If patients with proven perfume allergy were included, this incidence increased to 93%.[30]

E. Photosensitivity

Positive photopatch testing was reported to cinnamaldehyde in 4 of 35 patients with a diagnosis of either polymorphic light eruption or contact dermatitis.[31]

VII. CLINICAL RELEVANCE

Unlike benzoic acid and sorbic acid, cinnamic aldehyde may produce delayed hypersensitivity contact dermatitis as well as nonimmunologic contact urticaria. Oral mucosa reactions have been most frequently reported, including cheilitis, gingivitis, stomatitis, and orofacial granulomatosis. Common sources are lipsticks, toothpastes, and mouthwashes. Flare-ups of contact dermatitis may occur after ingestion of cinnamon oil.[32]

REFERENCES

1. **Archer, A. W.,** Determination of cinnamaldehyde, coumarin and cinnamyl alcohol in cinnamon and cassia by high-performance liquid chromatography, *J. Chromatogr.,* 447, 272, 1988.
2. **Thyne, G., Young, D. W., and Ferguson, M. M.,** Contact stomatitis caused by toothpaste, *N.Z. Dent. J.,* 85, 124, 1989.
3. **Drake, T. E. and Maibach, H. I.,** Allergic contact dermatitis and stomatitis caused by a cinnamic aldehyde flavoured toothpaste, *Arch. Dermatol.,* 112, 202, 1976.
4. **World Health Organization,** Toxicologic evaluation of some flavouring substances and non-nutritive sweetening agents, Tech. Rep. Ser. No. 44A, WHO, Geneva, 1967.
5. **Hagan, E. C., Hansen, W. H., Fitzhugh, O. G., Jenner, P. M., Jones, W. I., Taylor, J. M., Long, E. L., Nelson, A. A., and Brouwer, J. B.,** Food flavourings and compounds of related structure. II. Subacute and chronic toxicity, *Food Cosmet. Toxicol.,* 5, 141, 1967.
6. **Hjorth, N.,** Eczematous allergy to balsams, allied perfumes and flavouring agents, *Acta Derm. Venereol.,* 41 (Suppl. 46), 1, 1961.
7. **Speight, E. L. and Lawrence, C. M.,** Cinnamic aldehyde 2% pet. is irritant on patch testing, *Contact Dermatitis,* 23, 379, 1990.
8. **Fisher, A. A.,** *Contact Dermatitis,* 3rd ed., Lea & Febiger, Philadelphia, 1986.
9. **Adams, R. M. and Maibach, H. I.,** A five-year-study of cosmetic reactions, *J. Am. Acad. Dermatol.,* 13, 1062, 1985.
10. **Nethercott, J. R. and Holness, D. L.,** Occupational dermatitis in food handlers and bakers, *J. Am. Acad. Dermatol.,* 21, 485, 1989.
11. **Larsen, W. G.,** Sanitary napkin dermatitis due to the perfume, *Arch. Dermatol.,* 115, 363, 1979.
12. **Goh, C. L. and Ng, S. K.,** Bullous contact allergy from cinnamon, *Derm. Beruf. Umwelt.,* 36, 186, 1988.
13. **Hannuksela, M. and Haahtela, T.,** Hypersensitivity reactions to food additives, *Allergy,* 42, 561, 1987.
14. **Silverman, S., Jr. and Lozada, F.,** An epilogue to plasma-cell gingivostomatitis (allergic gingivostomatitis), *Oral Surg. Oral Med. Oral Pathol.,* 43, 211, 1977.
15. **Laubach, J. L., Malkinson, F. D., and Ringrose, E. J.,** Cheilitis caused by cinnamon (cassia) oil in toothpaste, *JAMA,* 152, 404, 1953.
16. **Millard, L. G.,** Contact sensitivity to toothpaste, *Br. Med. J.,* 1, 676, 1973.
17. **Millard, L.,** Acute contact sensitivity to a new toothpaste, *J. Dent.,* 1, 168, 1973.
18. **Kirton, V. and Wilkinson, D. S.,** Sensitivity to cinnamic aldehyde in a toothpaste, *Contact Dermatitis,* 1, 77, 1975.
19. **Zakon, S. J., Goldberg, A. L., and Kahn, J. B.,** Lipstick cheilitis. A common dermatitis: report of 32 cases, *Arch. Dermatol. Syph.,* 56, 499, 1947.
20. **Miller, J.,** Cheilitis from sensitivity to oil of cinnamon present in bubble gum, *JAMA,* 116, 131, 1941.
21. **Kerr, D. A., McClatchey, K. D., and Regezi, J. A.,** Allergic gingivostomatitis (due to chewing gum), *J. Periodontol.,* 42, 709, 1971.
22. **Allen, C. M. and Blozis, G. G.,** Oral mucosal reactions to cinnamon-flavored chewing gum, *J. Am. Dent. Assoc.,* 116, 664, 1988.
23. **Lamey, P. J., Lewis, M. A., Rees, T. D., Fowler, C., Binnie, W. H., and Forsyth, A.,** Sensitivity reaction to the cinnamonaldehyde component of toothpaste, *Br. Dent. J.,* 168, 115, 1990.
24. **Maibach, H. I.,** Cheilitis: occult allergy to cinnamic aldehyde, *Contact Dermatitis,* 15, 106, 1986.
25. **Patton, D. W., Ferguson, M. M., Forsyth, A., et al.,** Oro-facial granulomatosis: a possible allergic basis, *Br. J. Oral Maxillofac. Surg.,* 23, 235, 1985.

26. **Ferguson, M. M. and MacFadyen, E. E.,** Orofacial granulomatosis — a 10-year review, *Ann. Acad. Med.,* 15, 370, 1986.

27. **Rademaker, M. and Forsyth, A.,** Contact dermatitis in children, *Contact Dermatitis,* 20, 104, 1989.

28. **Mathias, C. G., Maibach, H. I., and Conant, M. A.,** Perioral leukoderma simulating vitiligo from use of a toothpaste containing cinnamic aldehyde, *Arch. Dermatol.,* 116, 1172, 1980.

29. **Mathias, C. G. T., Chappler, R. R., and Maibach, H. I.,** Contact urticaria from cinnamic aldehyde, *Arch. Dermatol.,* 116, 74, 1980.

30. **Safford, R. J., Basketter, D. A., Allenby, C. F., and Goodwin, B. F. J.,** Immediate contact reactions to chemicals in the fragrance mix and a study of the quenching action of eugenol, *Br. J. Dermatol.,* 123, 595, 1990.

31. **Addo, H. A., Ferguson, J., Johnson, B. E., and Frain-Bell, W.,** The relationship between exposure to fragrance materials and persistent light reaction in the photosensitivity dermatitis with actinic reticuloid syndrome, *Br. J. Dermatol.,* 107, 261, 1982.

32. **Fisher, A. A.,** Systemic eczematous "contact-type" dermatitis medicamentosa, *Ann. Allergy,* 24, 415, 1966.

Corn Starch

I. REGULATORY CLASSIFICATION

Pregelatinized starch is classified as a tablet binder, tablet disintegrant, and tablet and/or capsule diluent. Starch NF may be corn, wheat, or potato starch.

$$n = 300-1000$$

FIGURE 21. Starch.

II. SYNONYMS

Maize starch

III. AVAILABLE FORMULATIONS

A. Constituents
Corn starch is derived from the caryopsis of maize (*Zea mays*) and contains the polysaccharides amylose 18 to 27% and amylopectin. Corn starch or other corn sweeteners may contain residual byproducts of corn substances other than starch or sugars. One allergenic contaminant was identified to be D-psicose.

Pregelatinized starch is processed to rupture all or part of the granules in an aqueous medium and then dried.

B. Foods
Corn sweeteners, such as high fructose corn syrups, are increasing in usage as sucrose

105

is progressively declining. In 1985, they accounted for 47% of total sweeteners in the food supply. This increase is primarily due to the use of high fructose corn syrup in soft drinks.[1] Other food products include piecrusts, shortbread, cookies, popcorn, corn chips, tortillas, and cold breakfast cereals.[2]

Modified corn starches include distarch phosphate (treated with phosphorus oxychloride or sodium trimetaphosphate), acetylated distarch adipate (treated with a mixed anhydride of adipic and acetic anhydride), and acetylated distarch phosphate. The raw starch is cross-linked or substituted with various chemical processes to produce a product with less gel-forming properties.[3] They are used in concentrations of 5 to 6.5% in baby foods as a stabilizer to suspend the fine food particles and contribute to consistency, texture, appearance, nutrient distribution, and storage.[4]

C. Drugs

Corn starch paste, containing 5 to 25% starch, is widely used as a tablet binder. It is also commonly used as a tablet disintegrant in concentrations of 3 to 15%.

IV. HUMAN TOXICITY DATA

A. Gastrointestinal Intolerance

Clinical intolerance to corn products was found in 152 of 514 healthy French infants. Manifestations were either generalized, with urticaria, asthma, or rhinitis, or confined to the gastrointestinal tract, with diarrhea. One case of painless diarrhea in a 66-year-old atopic French woman was attributed to the ingestion of about 500 mg of corn starch per day as excipients in four medications (diazepam, chlordiazepoxide, clidinium bromide, and amineptine). The symptoms resolved after discontinuing the medications; provocative challenge was not done, nor were attempts made to obtain lists of other excipients from the manufacturer.[5]

B. Immediate Hypersensitivity Reactions

Contact urticaria and life-threatening anaphylactoid reactions have been reported from occupational exposure of physicians and nurse to surgical gloves containing corn starch powder. Practically all such gloves in the U.S. contain a mixture of corn starch, magnesium oxide, and tricalcium phosphate, with the exception of Pristine® gloves, available from World Medical Supply, San Jose, CA.

The pathogenic mechanism for these reactions is unclear. In one case, open patch testing to both an unwashed rubber glove and a small amount of the powder produced urticaria in 20 min.[6] In another case, testing with the glove was positive, with no reaction to powder supplied by the manufactuer and a negative RAST test.[7] A positive usage test and inhalation challenge was demonstrated in a third patient, but patch testing only showed urticaria for the glove and cornstarch powder after 48 h, and the RAST test was negative.[8]

Patients with hemagglutinating antibodies to corn starch also had antibodies to dextrins derived from corn syrup.[2]

C. Delayed Hypersensitivity Reactions

Delayed onset of food allergy to corn starch and corn sugar-dextrose, 2 h to several days after ingestion, was reported in seven children and five adults. Six patients had positive leukocyte inhibition factor tests, and four had positive lymphocyte transformation tests, indicating a cell-mediated hypersensitivity reaction.[9]

D. Granulomas

Granulomatous peritonitis is a well-recognized complication of wound contamination with surgical glove powder with an incidence of about 0.1% of abdominal surgical procedures.[10] Onset of peritonitis is evident in the third or forth postoperative week, commonly with the presenting symptom of pain. Surgical exploration reveals a straw-colored ascitic fluid, with nodules and seeding of all organs, along with localized granulomas that may cause obstruction or appendicitis. Eosinophilia may be present. Starch granulomas appear as Maltese crosses under polarized light and stain pink-purple with PAS and dark blue with iodine. Earlier cases were attributed to talc, talc-contaminated corn starch, or rice starch, but cases solely attributed to corn starch are sporadically reported.

Patients may respond to conservative treatment with prednisone and/or indomethacin and display positive lymphocyte transformation and intradermal tests, suggesting that the reaction may be a cell-mediated delayed hypersensitivity response.[11,12] Another explanation is a dose-related phenomenon with local clumping of corn starch powder.[13]

Starch granulomata has also been reported following dusting of a surgical wound with a corn starch-containing antibiotic powder[14] and following intravenous drug abuse with corn starch-containing crushed tablets.[15,16] Sites of granuloma formation are not confined to the peritoneal cavity, but have been reported in the pulmonary parenchyma, endocardium,[17] myocardium,[18] pleura, paranasal sinuses, mastoid cavities, testes,[14] biopsy scars, and synovial space.[16]

V. CLINICAL RELEVANCE

A. Immediate Hypersensitivity

Contact urticaria is a potential occupational hazard in health professionals using surgical gloves. Patients with negative testing to latex allergens should be patch tested with corn starch powder and undergo usage tests.

B. Food Allergy

Corn allergy may occur to corn starch, although less frequent than milk allergy. Scratch tests may be positive, but some patients may exhibit delayed hypersensitivity responses. Avoidance of all corn-related products, including corn starch, corn syrup, and corn dextrins, is prudent in patients with documented corn allergy.

C. Granulomatous Peritonitis

Starch granulomas are uncommonly reported. Wiping and/or washing surgical gloves preoperatively is essential in preventing this potentially fatal complication.

REFERENCES

1. **Glinsmann, Irausquin, and Park,** II. Consumer exposure, *J. Nutr.,* 116 (Suppl.), S20, 1986.
2. **Lietze, A.,** Laboratory research in food allergy. I. Food allergens, *J. Asthma Res.,* 7, 25, 1969.
3. **Wurzburg, O. B.,** Nutritional aspects and safety of modified food starches, *Nutr. Rev.,* 44, 74, 1986.
4. **Lebenthal, E.,** Use of modified food starches in infant nutrition, *Am. J. Dis. Child.,* 132, 850, 1978.
5. **Lagier, G., Pontal, P., and Gervais, P.,** Intolerance to cornstarch in many medications as an excipient?, *Therapie,* 36, 365, 1981.
6. **Fisher, A. A.,** Contact urticaria and anaphylactoid reaction due to corn starch surgical glove powder, *Contact Dermatitis,* 16, 224, 1987.

7. **Van der Meeren, H. L. M. and Van Erp, P. E. J.,** Life-threatening contact urticaria from glove powder, *Contact Dermatitis,* 14, 190, 1986.

8. **Assalve, D., Cicioni, C., Perno, P., and Lisi, P.,** Contact urticaria and anaphylactoid reaction from cornstarch surgical glove powder, *Contact Dermatitis,* 19, 61, 1988.

9. **Minor, J. D., Tolber, S. G., and Frick, O. L.,** Leukocyte inhibition factor in delayed-onset food allergy, *J. Allergy Clin. Immunol.,* 66, 314, 1980.

10. **Ignaties, J. A. and Hartman, W. H.,** The glove starch peritonitis syndrome, *Ann. Surg.,* 175, 388, 1972.

11. **Goodacre, R. L., Clancy, R. L., Davidson, R. A., and Mullens, J. E.,** Cell mediated immunity to corn starch in starch-induced granulomatous peritonitis, *Gut,* 17, 202, 1976.

12. **Grant, J. B. F., Davies, J. D., Espiner, H. J., and Eltringham, W. K.,** Diagnosis of granulomatous starch peritonitis by delayed hypersensitivity skin reactions, *Br. J. Surg.,* 69, 197, 1982.

13. **Klink, B. and Boynton, C. J.,** Starch peritonitis: a case report and clinicopathologic review, *Am. Surg.,* 56, 672, 1990.

14. **Pemberton, M. and Johnson, M.,** Dangers of corn starch powder, *Br. Med. J.,* 3, 1973.

15. **Michelson, J. B., Whitcher, J. P., Wilson, S., and O'Connor, G. R.,** Possible foreign body granuloma of the retina associated with intravenous cocaine addiction, *Am. J. Ophthalmol.,* 87, 278, 1979.

16. **Freemont, A. J., Porter, M. L., Tomlinson, I., Clague, R. B., and Jayson, M.,** Starch synovitis, *J. Clin. Pathol.,* 37, 990, 1984.

17. **McKee, P. H. and McKeown, E. F.,** Starch granulomata of the endocardium, *J. Pathol.,* 126, 103, 1978.

18. **Brynjolfsson, G., Eshaghy, B., Talano, J. V., and Gunnar, R.,** *Am. Heart J.,* 94, 353, 1977.

COTTONSEED OIL

I. REGULATORY CLASSIFICATION

Cottonseed oil is classified as an oleaginous vehicle and solvent.

II. SYNONYMS

Aceite de Algodon
Cotton oil
Oleum Gossypii Seminis

III. AVAILABLE FORMULATIONS

A. Constituents
Cottonseed oil is the fixed oil obtained from the seeds of *Gossypium hirsutum* and other cultivated species.

B. Drugs
Cottonseed oil is listed as an ingredient in 11 intramuscular preparations registered with the FDA. Concentrations range from 56 to 92% in these products. It is also present in at least 72 solid oral dosage forms, including capsules, tablets, coated tablets, sustained-release tablets, and sublingual tablets in amounts of 0.002 to 402 mg.[1]

C. Cosmetics
Cosmetics containing cottonseed oil include bath oils, bath tablets, bath salts, eyebrow pencils, and skin-conditioning products.[2]

IV. HUMAN TOXICITY DATA

A. Immediate Hypersensitivity Reactions
Cottonseed protein flour in a whole-grain bread was responsible for an episode of anaphylaxis with angioedema and generalized urticaria in a 29-year-old woman with a history of childhood atopic dermatitis. Skin prick testing with the moistened bread produced a large wheal. Passive

109

cutaneous transfer was documented. Subsequent skin testing and RAST assays were positive for cottonseed protein.[3]

A 51-year-old man experienced anaphylaxis on a commercial airline flight 30 s after ingesting two pieces of candy. Skin prick tests were strongly positive to cottonseed extract. Skin test with the candy extract was negative; this was attributed to the small amount available for testing. Although the candy could not be definitely shown to contain cottonseed, it had five protein bands that contained identical bands found in cottonseed, which bound labeled anti-IGE, leading the investigators to conclude that cottonseed was the putative agent.[4]

Double-blind oral challenge with cottonseed flour produced immediate hypersensitivity reactions with 100 and 500 mg, respectively, in two patients with reported anaphylactic reactions after ingesting a new fiber supplement bar. Skin prick tests to cottonseed oil were negative in these two patients and in six others with similar reactions to cottonseed flour.[5]

V. CLINICAL RELEVANCE

Although cottonseed protein is a potent allergen, causing severe anaphylactic reactions, the oil has not been implicated. Cottonseed allergens have been demonstrated to be present in the kernel that remains after the oil is extracted. Patients allergic to cottonseed flour can tolerate cottonseed oil.[5-7]

REFERENCES

1. **Food and Drug Administration,** *Inactive Ingredients Guide,* FDA, Washington, D.C., March 1990.
2. **Nikitakis, J. M.,** *CTFA Cosmetic Ingredient Handbook,* 1st ed., The Cosmetic, Toiletry and Fragrance Association, Washington, D.C., 1988.
3. **Malanin, G. and Kalimo, K.,** Angioedema and urticaria caused by cottonseed protein in whole-grain bread, *J. Allergy Clin. Immunol.,* 82, 261, 1988.
4. **O'Neil, C. E., Lehrer, S. B., and Gutman, A. A.,** Anaphylaxis apparently caused by a cottonseed-containing candy ingested on a commercial airliner, *J. Allergy Clin. Immunol.,* 84, 407, 1989.
5. **Atkins, F. M., Wilson, M., and Bock, S. A.,** Cottonseed hypersensitivity: new concerns over an old problem, *J. Allergy Clin. Immunol.,* 82, 242, 1988.
6. **Bernton, H. S., Spies, J. R., and Stevens, H.,** Significance of cottonseed sensitiveness, *J. Allergy,* 11, 138, 1940.
7. **Figley, K. D.,** Sensitivity to edible vegetable oils, *J. Allergy,* 20, 198, 1949.

D&C Red No. 22

I. REGULATORY CLASSIFICATION

D&C Red No. 22 is an indelible xanthene dye used as a color additive in drugs and cosmetics. It is the disodium salt of eosin (D&C Red No. 21).

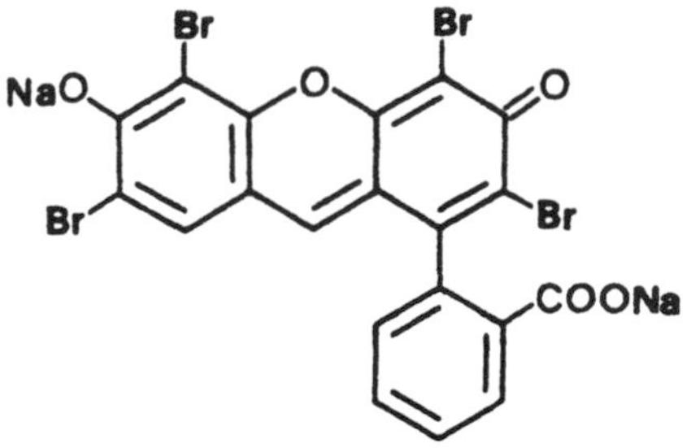

FIGURE 22. D&C Red No. 22.

II. SYNONYMS

C.I. Acid Red 87
C.I. 45380
Eosin disodium
Bromofluroesceic acid
Tetrabromofluorescein

III. AVAILABLE FORMULATIONS

A. Cosmetics
Eosin may be present in lipsticks in concentrations of 0.05 to 0.5%. Concentrations of 2% or more may be found in dark red lipstick shades.[1] Other types of products include permanent waves and moisturizing creams and lotions.[2]

B. Drugs

Eosin is infrequently used in pharmaceuticals. There are six oral capsules and two oral tablets listed with the FDA as containing eosin.[2]

IV. TABLE OF COMMON PRODUCTS

A. Oral Drug Products

Trade name	Manufacturer
Dristan Ultra Cold formula capsule	Whitehall
Pepto-Bismol liquid	Proctor & Gamble
Maximum Strength Pepto-Bismol liquid	Proctor & Gamble
Sectral capsule 200, 400 mg	Wyeth-Ayerst
Serax capsule 10 mg	Wyeth-Ayerst

V. ANIMAL TOXICITY DATA

Eosin (D&C Red No. 21) was moderately comedogenic in the rabbit ear model, with a rating of 3 on a scale of 0 to 5. This and other red pigments may explain the severity of cosmetic acne on the upper cheekbones where blush is applied.[4]

VI. HUMAN TOXICITY DATA

Eosin-induced lipstick cheilitis was commonly described in the literature between 1925 and 1959.[5-8] Since that time, only sporadic cases have been reported. Four patients who applied a 2% aqueous eosin solution to leg ulcers or dermatitis developed allergic contact dermatitis associated with positive patch tests to eosin 1 or 2%.[1]

Sensitization to eosin is believed to be related to an unidentified impurity in commercial formulations, since repeated precipitation purification eliminated or ameliorated positive patch test reactions in patients with lipstick cheilitis.[9] Strongly positive patch tests were obtained following testing with residues and impurities remaining after a purification procedure. Only weak reactions were demonstrated to the purified eosin.[8-10]

VII. CLINICAL RELEVANCE

Eosin has been an uncommon source of allergic contact dermatitis since the 1960s. In a prospective study of 713 cases of cosmetic-related dermatitis identified between 1977 and 1983 in the U.S., no cases were attributed to eosin.[11] The scarcity of hypersensitivity probably reflects improved purity of the product and lower concentrations used in present-day cosmetics.

REFERENCES

1. **Tomb, R. R.,** Allergic contact dermatitis from eosin, *Contact Dermatitis,* 24, 27, 1991.
2. **Nikitakis, J. M.,** *CTFA Cosmetic Ingredient Handbook,* 1st ed., The Cosmetic, Toiletry and Fragrance Association, Washington. D.C., 1988.
3. **Food and Drug Administration,** *Inactive Ingredients Guide,* FDA, Washington, D.C., March 1990.

4. **Fulton, J. E., Jr., Pay, S. R., and Fulton, J. E., III,** Comedogenicity of current therapeutic products, cosmetics, and ingredients in the rabbit ear, *J. Am. Acad. Dermatol.,* 10, 96, 1984.

5. **Miller, H. E. and Taussig, L. R.,** Lipstick dermatitis due to carthamin and eosin, *JAMA,* 84, 1999, 1925.

6. **Sulzberger, M. B., Goodman, J., Byrne, L. A., and Mallozzi, E. D.,** Acquired specific hypersensitivity to simple chemicals: cheilitis with special reference to sensitivity to lipsticks, *Arch. Dermatol.,* 37, 597, 1938.

7. **Calnan, C. D. and Sarkany, I.,** Studies in contact dermatitis (II): lipstick cheilitis, *Trans. St. John's Dermatol. Soc.,* 39, 28, 1957.

8. **Calnan, C. D.,** Allergic sensitivity to eosin, *Acta Allergol.,* 13, 493, 1959.

9. **Sulzberger, M. B. and Hecht, R.,** Acquired specific hypersensitivity to simple chemicals: further studies on the purification of dyes in relation to allergic reactions, *J. Allergy,* 12, 129, 1941.

10. **Hecht, R., Schwarzschild, L., and Sulzberger, M. B.,** Sensitisation to simple chemicals: comparison between reactions to commercial and to purified dyes, *N.Y. State J. Med.,* 39, 2170, 1939.

11. **Adams, R. M. and Maibach, H. I.,** A five-year-study of cosmetic reactions, *J. Am. Acad. Dermatol.,* 13, 1062, 1985.

D&C YELLOW No. 10

I. REGULATORY CLASSIFICATION

D&C Yellow No. 10 is a greenish-yellow quinoline compound classified as a color additive and approved for use in drugs and cosmetics. It is a mixture of the disodium salt of the mono and disulfonic acids of 2-(2-quinolyl)-1,3-indandione.

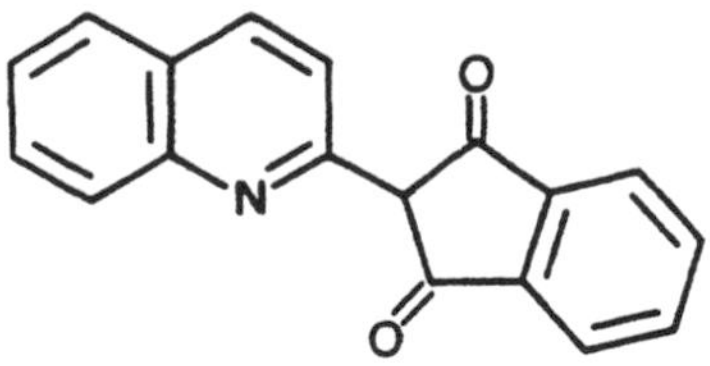

FIGURE 23. Quinophthalone.

II. AVAILABLE FORMULATIONS

A. Cosmetics
D&C Yellow No. 10 and its aluminum lake is used in cosmetic products such as lipsticks, shampoos, hair conditioners and dyes, creams and lotions, colognes, toilet waters, permanent waves, bubble baths, and aftershave lotions.[1]

III. TABLE OF COMMON DRUG PRODUCTS

A. Oral Drug Products

Trade name	Manufacturer
Achromycin V capsule	Lederle
Accutane capsule 40 mg	Roche
Aldoclor tablet 150, 250	Merck Sharp & Dohme
Aldomet tablet 125, 250, 500 mg	Merck Sharp & Dohme
Allergy-Sinus Comtrex	Bristol-Myers

115

Trade name (cont'd)	Manufacturer (cont'd)
Alu-tab tablet	3M Riker
Alu-cap capsule	3M Riker
Ancoban capsule	Roche
Antivert tablet 25, 50 mg	Roerig
Anturane capsule	Ciba
Apresazide capsule 100/50	Ciba
Apresoline tablet 10 mg	Ciba
Aristocort tablet	Lederle
Asbron G elixir, tablet	Sandoz
Atarax tablet 10 mg	Roerig
Atromid S capsule	Wyeth-Ayerst
Augmentin oral suspension	Beecham
Bactrim tablet	Roche
Basaljel	Wyeth-Ayerst
Beelith tablet	Beach
Benemid tablet 500 mg	Merck Sharp & Dohme
Berocca tablet	Roche
Bromfed capsule	Muro
Bronkodyl capsule 200 mg	Winthrop
Bumex tablet 0.5, 1 mg	Roche
Butazolidin capsule, tablet 100 mg	Geigy
Cafergot PB tablet	Sandoz
Calan SR caplet	Searle
Cardene capsule 30 mg	Syntex
Cardizem tablet 60, 120 mg	Marion
Celontin capsule 150, 300 mg	Parke-Davis
Centrax capsule 5, 20 mg	Parke-Davis
Chloraseptic menthol liquid	Richardson-Vicks
Chloraseptic menthol aerosol spray	Richardson-Vicks
Chloraseptic menthol lozenge	Richardson-Vicks
Choledyl elixir	Parke-Davis
Choledyl tablet 200 mg	Parke-Davis
Cinobac capsule	Dista
Comhist LA capsule	Norwich Eaton
Compazine tablet, spansule	Smith Kline & French
Comtrex caplet	Bristol-Myers
Comtrex liqui-gel capsule	Bristol-Myers
Comtrex liquid	Bristol-Myers
Comtrex tablet	Bristol-Myers
Congespirin chewable cold tablet	Bristol-Myers
Coumadin tablet 2.5, 7.5 mg	DuPont
Creon capsule	Reid-Rowell
Cuprimine capsule	Merck Sharp & Dohme
Cycrin tablet	Wyeth-Ayerst

Trade name (cont'd)	Manufacturer (cont'd)
Doryx capsule	Parke-Davis
Dristan tablet, caplet	Whitehall
Dristan maximum strength tablet	Whitehall
Dulcolax tablet	Boehringer Ingelheim
Ecotrin tablet	Smith Kline Beecham
Edecrin tablet 50 mg	Merck Sharp & Dohme
E.E.S. 400 liquid	Abbott
E.E.S. 400 filmtab	Abbott
Elavil tablet 25, 50 mg	Stuart
Endep tablet 10, 25, 50, 75 mg	Roche
Enduronyl tablet	Abbott
Erythromycin delayed-release capsule	Abbott
Eskalith capsule	Smith Kline &French
Estratest tablet	Reid-Rowell
Estratest H.S. tablet	Reid-Rowell
Ethatab	Allen & Hanburys
Etrafon tablet 2-10	Schering
Eulexin capsule	Schering
Excedrin PM caplet, tablet	Bristol-Myers
Extra Strength Tylenol gelcaps	McNeil Consumer
Extra Strength Tylenol adult liquid	McNeil Consumer
Fedahist gyrocap	Schwarz Pharma
Feosol capsule, tablet	Smith Kline Beecham
Ferro-sequels	Lederle
Fioricet capsule	Sandoz
Fiorinal with codeine No. 3 capsule	Sandoz
Gantanol tablet	Roche
Gaviscon liquid antacid	Marion
Geocillin tablet	Roerig
Grisactin capsule 125, 250 mg	Wyeth-Ayerst
Hycomine pediatric syrup	DuPont
Hydromox R tablet	Lederle
Hydropres tablet	Merck Sharp & Dohme
Hytrin tablet 10 mg	Abbott
Ilosone tablet 500 mg	Dista
Inderal tablet 10, 40, 80 mg	Wyeth-Ayerst
Inversine tablet	Merck Sharp & Dohme
Ismelin tablet 10 mg	Ciba
Isoptin tablet 80 mg	Knoll
Isoptin SR tablet	Knoll
Keflet tablet 250, 500, 1000 mg	Dista
Keftab 250, 500 mg	Dista
K-tab	Abbott
Klor-Con 10 tablet	Upsher-Smith

Trade name (cont'd)	Manufacturer (cont'd)
K-lyte	Bristol
K-lyte DS lime	Bristol
K-lyte/Cl	Bristol
K-lyte/Cl 50 citrus	Bristol
Kutrase capsule	Schwarz Pharma
Kuzyme capsule	Schwarz Pharma
Lanoxicaps 0.1, 9.2 mg	Burroughs Wellcome
Lanoxin tablet 0.125, 0.5 mg	Burroughs Wellcome
Lanoxin elixir pediatric	Burroughs Wellcome
Larobec tablet	Roche
Lasix oral solution	Hoechst-Roussel
Levatol tablet 20 mg	Reed & Carnrick
Librax capsule	Roche
Librium capsule 5, 10, 25 mg	Roche
Loestrin 1.5/30 green tablet	Parke-Davis
Loestrin Fe 1.5/30 green tablet	Parke-Davis
Lopressor HCT tablet 100/50	Geigy
Loxitane capsule 5, 10, 25, 50 mg	Lederle
Lufyllin GG tablet	Wallace
Macrodantin capsule	Norwich Eaton
Mandelamine granules	Parke-Davis
Matulane capsule	Roche
Maxzide tablet	Lederle
Maxzide-25 tablet	Lederle
Mellaril tablet 10, 100, 150 mg	Sandoz
Mellaril suspension 100 mg	Sandoz
Menrium tablet 5-2, 5-4	Roche
Mevacor tablet 40 mg	Merck Sharp & Dohme
Mexitil capsule 250 mg	Boehringer Ingelheim
Micronor tablet	Ortho
Midamor tablet 5 mg	Merck Sharp & Dohme
Milontin capsule	Parke-Davis
Mitrolan tablet	Robins
Modicon 28 green tablet	Ortho
MS Contin tablet 60 mg	Purdue Frederick
Mysoline suspension	Wyeth-Ayerst
Naldecon children's syrup	Bristol
Naldecon EX pediatric drops	Bristol
Nalfon capsule 300 mg	Dista
Naturetin tablet	Princeton
Navane capsule	Roerig
Norgesic tablet	3M Riker
Norgesic forte tablet	3M Riker
Norlestrin 1/50 yellow tablet	Parke-Davis

Trade name (cont'd)	Manufacturer (cont'd)
Norpace CR capsule 100, 150 mg	Searle
Norpramin tablet 25, 50, 75, 100 mg	Merrell-Dow
Novahistine expectorant	Lakeside
Nucofed syrup	Beecham
Nucofed capsule	Beecham
Nucofed expectorant	Beecham
Nuprin caplet, tablet	Bristol-Myers
Ogen tablet 0.625, 5 mg	Abbott
Ortho-Novum 7/7/7 green tablet	Ortho
Ortho-Novum 10/11 green tablet	Ortho
Ortho-Novum 1/35 green tablet	Ortho
Ortho-Novum 1/50 yellow, green tablet	Ortho
Orudis capsule	Wyeth-Ayerst
Os-cal 250 + D tablet	Marion
Os-cal 500 tablet	Marion
Ovcon 50-21 day, 28-day	Mead Johnson
Ovcon 35-28-day	Mead Johnson
Pamelor capsule 10, 25, 75 mg	Sandoz
Parafon forte DSC caplet	McNeil
Periactin syrup	Merck Sharp & Dohme
Peritrate tablet 10, 20, 40 mg	Parke-Davis
Peritrate SA tablet 80 mg	Parke-Davis
Phenaphen with codeine No. 2, 3, 4	Robins
Phenergan syrup plain	Wyeth-Ayerst
Phenergan with DM syrup	Wyeth-Ayerst
Plegine tablet	Wyeth-Ayerst
PMB tablet 200 mg	Winthrop
Ponstel capsule 250 mg	Parke-Davis
Pramet FA tablet	Ross
Prelu-2	Boehringer Ingelheim
Premarin tablet 0.3, 1.25 mg	Wyeth-Ayerst
Premarin with methyltestosterone 1.25	Wyeth-Ayerst
Primatene tablet	Whitehall
Primatene M tablet	Whitehall
Primatene P tablet	Whitehall
Principen with probenecid capsule	Squibb
Procan SR 250 mg	Parke-Davis
Procan SR tablet 500 mg	Parke-Davis
Prolixin tablet 5, 10 mg	Princeton
Proloprim tablet 200 mg	Burroughs Wellcome
Pronestyl capsule 250, 500 mg	Princeton
Pronestyl SR tablet	Princeton
Quarzan capsule 2.5, 5 mg	Roche
Reglan tablet	Robins

Trade name (cont'd)	Manufacturer (cont'd)
Renese tablet 2 mg	Pfizer
Ritalin tablet 5, 20 mg	Ciba
Robaxin-750 tablet	Robins
Salutensin	Bristol
Salutensin demi	Bristol
Seromycin capsule	Lilly
Sinemet tablet 25-100	Merck Sharp & Dohme
Sinequan capsules	Roerig
Sinus Excedrin	Bristol-Myers
Solatene capsule	Roche
Soma compound with codeine tablet	Wallace
Sorbitrate SL tablet 10 mg	ICI
Sorbitrate chewable tablet 5, 10 mg	ICI
Sorbitrate oral tablet 5, 10 mg	ICI
Sorbitrate SA tablet	ICI
Stelazine concentrate	Smith Kline & French
Sumycin syrup	Squibb
Surmontil capsule 25 mg	Wyeth-Ayerst
Tagamet tablets	Smith Kline & French
Tavist D tablet	Sandoz
Tedral SA tablet	Parke-Davis
Tenex tablet 2 mg	Robins
Terramycin capsule	Pfizer
Tessalon	Forest
Thorazine tablets	Smith Kline & French
Tolectin 600 tablet	McNeil
Trancopal caplet 200 mg	Winthrop
Triavil tablet 4-25	Merck Sharp & Dohme
Triaminic expectorant DH	Sandoz
Triaminic expectorant with codeine	Sandoz
Triaminic TR tablet	Sandoz
Trimox capsule	Squibb
Tussionex suspension	Fisons
Tussi-Organidin DM liquid	Wallace
Tylenol cold caplet, tablet	McNeil Consumer
Tylenol cold no drowsiness caplet	McNeil Consumer
Tylenol maximum strength allergy sinus	McNeil Consumer
Tylenol maximum strength sinus caplet,tab	McNeil Consumer
Unipen capsule 250 mg	Wyeth-Ayerst
Urecholine tablet 25, 50 mg	Merck Sharp & Dohme
Urobiotic 250	Roerig
Valrelease capsule	Roche
Valium tablet 5 mg	Roche
Velosef capsule	Squibb

Trade name (cont'd)	Manufacturer (cont'd)
Vi-Daylin F chewable tablet	Ross
Vistaril capsule	Pfizer
Vivactil tablet 10 mg	Merck Sharp & Dohme
Wygesic tablet	Wyeth-Ayerst
Wymox capsules	Wyeth-Ayerst
Zantac tablet 300 mg	Glaxo
Zarontin capsule	Parke-Davis

IV. HUMAN TOXICITY DATA

A. Hypersensitivity Reactions

D&C Yellow No. 10 is a quinoline compound, chemically related to D&C Yellow No. 11, a known potent sensitizer. Although occasional positive patch tests to D&C Yellow No. 10 have been reported in patients with a positive response to D&C Yellow No. 11,[2] a study of 15 subjects sensitized to D&C Yellow No. 11 did not demonstrate cross-reactivity.[3]

V. CLINICAL RELEVANCE

D&C Yellow No. 10 is an extremely rare sensitizer and does not appear to cross-react with other yellow dyes.

REFERENCES

1. **Nikitakis, J. M.,** *CTFA Cosmetic Ingredient Handbook,* 1st ed., The Cosmetic, Toiletry and Fragrance Association, Washington, D.C., 1988.
2. **Bjorkner, B. and Magnusson, B.,** Patch test sensitization to D and C Yellow II and simultaneours reaction to quinoline yellow, *Contact Dermatitis,* 7, 1, 1981.
3. **Rapaport, M.,** Allergy to yellow dyes, *Arch. Dermatol.,* 120, 535, 1984.

DIAZOLIDINYL UREA

I. REGULATORY CLASSIFICATION

Diazolidinyl urea is a heterocyclic substituted urea used as a preservative in topical pharmaceutical and cosmetic products.

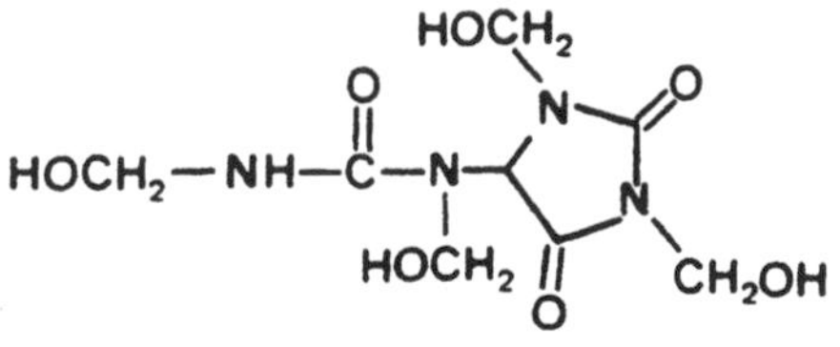

FIGURE 24. Diazolidinyl urea.

II. SYNONYMS

Germall II®

III. AVAILABLE FORMULATIONS

A. Drugs
Diazolidinyl urea is not commonly used in topical pharmaceutical products. In 1990, there was one cream and one lotion formulation listed with the FDA containing this preservative.[1]

B. Cosmetics
Diazolidinyl urea is a broad spectrum cosmetic preservative introduced in 1982, which is structurally related to imidazolidinyl urea. Concentrations in both wash-off and leave-on cosmetics generally range from 0.1 to 0.3%. Similar to imidazolidinyl urea, it is synergistic with other antimicrobials and is often used in combination with parabens.[2]

The antimicrobial spectrum is wider than that of the older compound, imidazolidinyl urea. Activity against *Pseudomonas*, most Gram-negative bacteria, Gram-positive bacteria, yeast, and molds has been demonstrated.[3]

123

Because diazolidinyl urea is a fairly recently available preservative, the prevalence in cosmetic products is low. In 1985, there were 52 cosmetic products listed with the FDA which included makeup, eye makeup, nail products, personal cleanliness products, skin care products, and hair products.[4]

IV. TABLE OF COMMON PRODUCTS

Trade name	Manufacturer
Aqua-A cream	Baker-Cummins
Aquaderm cream, lotion	Baker-Cummins
Ben-Gay Warming Ice	Leeming
Burn cream with lidocaine	Watkins
Eurax cream	Westwood-Squibb
Flex-all 451	Chattem
Massengill medicated soft cloth towelette	Beecham
Moisturel cream	Westwood-Squibb
Oxy 10 wash	Smith Kline Beecham
Pen Kera creme	Ascher

V. ANIMAL TOXICITY DATA

Maximization sensitization studies in guinea pigs have demonstrated a low sensitization potential. Only mild reactions, limited to well-defined, confluent erythema, were seen in 20% of the animals following challenges with 50% aqueous solution in one study. Five of eight animals with positive challenges to diazolidinyl urea also had positive challenges to imidazolidinyl urea, and six had positive challenges to formaldehyde.[2]

VI. HUMAN TOXICITY DATA

Three years after the introduction of diazolidinyl urea as a cosmetic preservative, the first case report of allergic contact dermatitis was published. This 42-year-old male had been using a new hair curl activator gel for 2 weeks and presented with marked facial edema and acute eczema on the neck, face, and ears. He required admission and a brief course of systemic corticosteroids. Patch testing showed positive reactions to the gel and to 1% diazolidinyl urea. Cross-reactions to imidazolidinyl urea, but not to formaldehyde, were noted after 1 week.[4]

A patient with primary sensitivity to formaldehyde acquired sensitivity to diazolidinyl urea after treatment of her dermatitis with a liquid soap containing this preservative.[5]

Initial attempts to define the incidence of contact allergy to diazolidinyl urea in dermatitis patients were negative.[6] This ingredient then began to appear in "hypoallergenic" cosmetics. Case reports of contact dermatitis secondary to the use of these cosmetics became evident.[7]

Subsequent patch test series involving 2400 Dutch patients revealed an incidence of 0.54% positive reactions to diazolidinyl urea. Approximately one half (6 of 13) had concomitant positive tests to formaldehyde and other formaldehyde releasers.[8] A patch test series in 270 Italian dermatitis patients showed a higher incidence (1.1%). Formaldehyde tests were negative in all of these cases.[9]

Limited data suggest, but do not prove, that diazolidinyl urea may be more sensitizing than imidazolidinyl urea. Attempts to sensitize normal volunteers were successful in 19 of 150 with diazolidinyl urea and in 2 of 150 with the same concentration of imidazolidinyl urea.[10]

VII. CLINICAL RELEVANCE

At this time, diazolidinyl urea carries a low risk of sensitization. This risk is likely to increase with wider use of the compound. Although it is not clear whether patients with concomitant reactions to diazolidinyl urea and formaldehyde are reacting to the intact molecule or formaldehyde released during decomposition, patients with a documented hypersensitivity to formaldehyde should perform a ROAT (repeated open application test) or be patch tested with diazolidinyl urea before using products containing this preservative.

Cross-reactions to imidazolidinyl urea may occur regardless of whether formaldehyde sensitivity is present. As with imidazolidinyl urea, patch tests using aqueous solutions may be more sensitive.[7]

REFERENCES

1. **Food and Drug Administration,** *Inactive Ingredients Guide.* FDA, Washington, D.C., March 1990.
2. **Stephens, T. J., Drake, K. D., and Drotman, R. B.,** Experimental delayed contact sensitization to diazolidinyl urea (Germall II) in guinea pigs, *Contact Dermatitis,* 16, 164, 1987.
3. **Berke, P. A. and Rosen, W. E.,** Germall II: new broad-spectrum cosmetic preservative, *Cosmet. Toilet.,* 97, 49, 1982.
4. **Kantor, G. R., Taylor, J. S., Ratz, J. L., and Evey, P. L.,** Acute allergic contact dermatitis from diazolidinyl urea (Germall II) in a hair gel, *J. Am. Acad. Dermatol.,* 13, 116, 1985.
5. **Zaugg, T. and Hunziker, T.,** Germall II and triclosan, *Contact Dermatitis,* 17, 262, 1987.
6. **De Groot, A. C., Bos, J. D., Gagtman, B. A., et al.,** Contact allergy to preservatives(II), *Contact Dermatitis.* 15, 218, 1986.
7. **De Groot, A. C., Bruynzeel, D. P., Jagtman, B. A., and Weyland, J. W.,** Contact allergy to diazolidinyl urea (Germall II), *Contact Dermatitis.* 18, 202, 1988.
8. **Perret, C. M. and Happle, R.,** Contact sensitivity to diazolidinyl urea (Germall II), *Arch. Dermatol. Res.,* 281, 57, 1989.
9. **Tosti, A., Restani, S., and Lanzarini, M.,** Contact sensitization to diazolidinyl urea: report of 3 cases, *Contact Dermatitis,* 22, 127, 1990.
10. **Jordan, W. P.,** Human studies that determine the sensitizing potential of haptens. Experimental allergic contact dermatitis, *Dermatologic Clin.,* 2, 533, 1984.

ETHANOL

I. REGULATORY CLASSIFICATION

Undenatured ethanol is used as a component of flavored vehicles or as a solvent in pharmaceuticals.

Specially denatured alcohol, consisting of ethanol with various adulterants added, is used in cosmetics. The specific adulterant depends on the type of product.

$$CH_3CH_2OH$$

FIGURE 25. Ethanol.

II. AVAILABLE FORMULATIONS

A. Drugs

Ethanol is commonly used in liquid pharmaceutical preparations. In 1978, ethanol was present in over 700 products.[1] Since the American Academy of Pediatrics recommendation in 1984 that manufacturers limit the amount of ethanol to a maximum of 5% v/v in pharmaceuticals intended for use in children, many pediatric products have been reformulated to remove or reduce the ethanol content.[2] Currently, there are 240 FDA-approved oral pharmaceuticals listing ethanol as an excipient.[3]

B. Cosmetics

Ethanol is present in high concentrations in some personal care products, such as hair sprays, mouthwashes, perfumes, colognes, and aftershaves. These products may be deliberately abused by chronic alcoholics seeking alcoholic beverage substitutes.[4]

Children with a history of accidental ingestion of household cosmetics may present with ethanol intoxication. In general, products with a pleasant flavor (due to the methyl salicylate or wintergreen flavor used to denature the alcohol), such as mouthwashes, are ingested in larger amounts with more severe consequences.[5,6]

Products containing alcohol denatured with bitter substances, such as perfumes and colognes, have been shown to be less likely to result in a significant ethanol intoxication in children, presumably due to the lesser amount ingested.[7]

Other products that may contain ethanol include skin fresheners; hair mousses; face, body, and hand creams or lotions; facial masks; breath fresheners; moisturizing creams or lotions; and fragrances.[8]

III. TABLE OF COMMON PRODUCTS AND ALTERNATIVES

A. Liquid Oral Drug Products

Trade name	Ethanol (%)	Manufacturer
Actifed with codeine cough syrup	4.3	Burroughs-Wellcome
Alurate elixir	20	Roche
Ambenyl D syrup	9.5	Forest
Artane elixir	5	Lederle
Asbron G elixir	15	Sandoz
Atarax syrup	Unknown	Roerig
Bayer Children's Cough syrup	5	Glenbrook
Calcidrine syrup	6	Abbott
Cerose DM	2.4	Wyeth-Ayerst
Choledyl elixir	20	Parke-Davis
Colace syrup	0.6	Mead Johnson
Comtrex cough formula	20	Bristol
Decadron elixir	5	Merck Sharp & Dohme
Dexedrine elixir	10	Smith Kline & French
Donnagel elixir	23	Robins
Donnagel PG	5	Robins
Dramamine syrup	5	Searle
Eldertonic	13.5	Mayrand
Elixophyllin elixir	20	Forest
Elixophyllin KI elixir	10	Forest
Feosol elixir	5	Smith Kline Consumer
Iberet liquid	1	Abbott
Imodium A-D	5.25	McNeil Consumer
Indocin oral suspension	1	Merck Sharp & Dohme
Kaochlor SF 10% liquid	5	Adria
Kaochlor 10% liquid	5	Adria
Kaon elixir	5	Adria
Kaon Cl 20% liquid	5	Adria
Lanoxin elixir Pediatric	10	Burroughs-Wellcome
Lasix oral solution	11.5	Hoechst-Roussel
Levsin elixir	20	Schwarz-Pharma
Levsin oral drops	5	Schwarz-Pharma
Lortab liquid	7	Rugby
Lufyllin elixir	20	Wallace
Lufyllin GG elixir	17	Wallace
Marax syrup	5	Roerig

Trade name (cont'd)	Ethanol (%) (cont'd)	Manufacturer (cont'd)
May-Vita elixir	13	Mayrand
Mellaril concentrate	3	Sandoz
Methadone oral solution	8	Roxane
Minocin oral suspension	5	Lederle
Naldecon DX Children's syrup	5	Bristol
Navane concentrate	7	Roerig
Niferex elixir	10	Central
Niferex forte elixir	10	Central
Norisodrine with calcium iodide	6	Abbott
Novahistine DH	5	Lakeside
Novahistine expectorant	7.5	Lakeside
Nucofed expectorant	12.5	Beecham
Nucofed pediatric expectorant	6	Beecham
Nu-Iron elixir	10	Mayrand
Organidin elixir	21.75	Wallace
Paradione oral solution	65	Abbott
PBZ elixir	12	Geigy
Pediacof	5	Winthrop
Peri-Colace syrup	10	Mead Johnson
Permitil oral concentrate	1	Schering
Phenergan syrup plain	7	Wyeth-Ayerst
Phenergan syrup fortis	1.5	Wyeth-Ayerst
Phenergan with codeine syrup	7	Wyeth-Ayerst
Phenergan with dextromethorphan	7	Wyeth-Ayerst
Phenergan VC syrup	7	Wyeth-Ayerst
Phenergan VC with codeine syrup	7	Wyeth-Ayerst
Polaramine syrup	6	Schering
Prelone syrup 15mg/5ml	5	Muro
Prolixin elixir	14	Princeton
Prolixin oral concentrate	14	Princeton
Quelidrine syrup	2	Abbott
Robitussin AC	3.5	Robins
Robitussin DAC	1.9	Robins
S-T Forte syrup	5	Scot-Tussin
S-T Forte Sugar-Free syrup	5	Scot-Tussin
Tagamet liquid	2.8	Smith Kline & French
Tavist syrup	5.5	Sandoz
Temaril syrup	5.7	Herbert
Theo-Organidin elixir	15	Wallace
Trilafon concentrate	3.4	Schering
Triaminic expectorant	5	Sandoz
Triaminic expectorant with codeine	5	Sandoz

Trade name (cont'd)	Ethanol (%) (cont'd)	Manufacturer (cont'd)
Tussar 2	6	Rorer
Tussar SF	12	Rorer
Tuss-Ornade liquid	5	Smith Kline & French
Tylenol Liquid Cold Medication	7	McNeil Consumer

B. Parenteral Drug Products

Trade name	Amount (v/v) (%)	Manufacturer	Alternative
Bactrim IV	10	Roche	None
Brevibloc 2.5 g ampule	25	DuPont	100 mg ampule
Cedilanid-D	9.8	Sandoz	None
D.H.E. 45	6.1	Sandoz	None
Dilantin	10	Parke-Davis	None
Lanoxin	10	Burroughs Wellcome	None
Lanoxin pediatric	10	Burroughs Wellcome	None
Luminol sodium	10 Winthrop	Breon	None
Nembutol sodium	10	Abbott	None
Nitro-BID IV 5 mg/ml	70	Marion	Nitropress
Nitrostat IV 0.8 mg/ml	5	Parke-Davis	Nitropress
Nitrostat IV 5 mg/ml	28.5	Parke-Davis	Nitropress
Nitrostat IV 10 mg/ml	50	Parke-Davis	Nitropress
Prostin VR Pediatric	100 (diluent)	Upjohn	None
Septra IV	10	Burroughs Wellcome	None
Syntocinon	0.61	Sandoz	None
Toradol injection	10	Syntex	None
Tridil 5 mg/ml	10	DuPont	Nitropress
Tridil 25, 50, or 100 mg/ml	30	DuPont	Nitropress
Valium	10	Roche	None
Vepesid	30.5	Bristol-Myers	None

Note: Most of these drugs require dilution prior to use, thus final ethanol concentrations will generally be less than those listed here.

IV. HUMAN TOXICITY DATA

A. Acute Ethanol Intoxication

Inadvertent intoxication with ethanol has been reported after intravenous infusion of nitroglycerin products containing ethanol as a component of the solvent. This has only been reported in elderly patients receiving high-dose therapy, ranging from 1250 to 2470 μg of nitroglycerin per minute, resulting in serum ethanol levels of 21 to 267 mg/dl.[9-12]

Children may develop CNS signs of ethanol intoxication after blood ethanol concentrations of 25 mg/dl or greater. Computer simulation studies have predicted that the administration of a medication containing a 10% ethanol vehicle, given at a dose of 40 ml every 4 h, would maintain this blood level in a 6-year-old child.[2]

B. Drug Interactions

Patients maintained on chronic disulfiram therapy or treated with agents reported to cause "disulfiram-like" reactions must avoid ethanol-containing formulations.

Consequences of inadvertent ethanol exposure have ranged from mild transient reactions (nausea, headache, feeling of warmth, perspiration) after topical exposure [13] to severe, life-threatening reactions (hypotension, cardiac arrhthymias) after oral exposure to an elixir.[14] Therapeutic agents given by other routes of exposure, such as nebulizer solutions and otic solutions, have also produced reactions.[15]

Cosmetics containing ethanol may also present a risk to the patient receiving these medications. A disulfiram reaction has been described in a patient using a beer-containing shampoo[16] and following application of an aftershave lotion containing 50% ethanol.[17]

C. Total Parenteral Nutrition

Patients with underlying cardiac disease have been reported to develop deterioration in cardiac function following administration of total parenteral nutrition solutions containing ethanol.[18]

D. Venous Phlebitis

Intravenous administration of absolute ethanol (99.5% v/v) is used therapeutically as a venous sclerosant. There is a dose-related venous irritant effect with an incidence of phlebitis of 8% in patients receiving rapid infusions of 5% w/v ethanol and 30% in patients receiving 10% w/v solutions.[19]

E. Precipitation of Gout

Four cases of precipitation of acute gouty arthritis were reported within 12 h of receiving intravenous nitroglycerin therapy. The alcohol vehicle was implicated; however, blood alcohol levels were not documented during therapy.[20]

F. Immediate Hypersensitivity Reactions

Acute urticaria has been reported following ingestion of alcoholic beverages. It occurs within 5 or 10 min after ingestion and lasts for up to 1 h. Urticaria can be either localized to the face or generalized.[21-25] This IgE-mediated reaction must be distinguished between a nonspecific cutaneous flushing reaction observed in some ethnic groups and preexisting ethanol-exacerbated urticaria.

Anaphylaxis, manifested by generalized pruritus, dizziness, facial flushing, abdominal pain, dyspnea, and collapse, was documented in a patient with a prior history of vinegar hypersensitivity. Scratch tests were positive to ethanol and to two of its metabolites, acetaldehyde and acetic acid. Dyspnea was provoked by oral challenge.[26]

Allergic contact urticaria, occasionally accompanied by systemic symptoms such as pruritus and transient hypotension, has been rarely reported. Passive transfer can confirm these reactions.[27]

G. Delayed Hypersensitivity Reactions

A fine vesicular eruption on the fingers and toes, beginning 12 h after ethanol ingestion, was reported in one case.[22] Delayed contact hypersensitivity was induced in 6 of 93 normal volunteers exposed to 50% aqueous ethanol.[28] Occupational exposure to ethanol-containing disinfectants has also resulted in delayed contact dermatitis.[29]

H. Pseudoallergic Reactions

An idiosyncratic ethnic response to ethanol is demonstrated by the frequent cutaneous flushing observed in Asian people. The reaction progressed to severe bronchospasm in several cases.[30,31] The physiologic basis for this reaction is unclear; it is unrelated to histamine and is not prevented by cromolyn sodium.[32]

V. CLINICAL RELEVANCE

A. High-Dose Nitroglycerin Therapy

Clinically evident ethanol intoxication has resulted from infusions of ethanol-containing nitroglycerin given in doses of greater than 1200μ g/min. Although this has only been reported in elderly patients, this population may be more likely to display signs of intoxication than a population of younger patients receiving the same doses. Gouty arthritis may also be precipitated in these patients.

Most of the ethanol-containing preparations also contain propylene glycol and present a risk of adverse reactions from this solvent as well. Thus, avoidance of solvent-containing nitroglycerin products seems prudent when high-dose therapy is indicated.

B. Disulfiram Therapy

Patients receiving disulfiram therapy are at risk of serious sequelae from ingestion of ethanol as an excipient in medications. Although small amounts of ethanol may be tolerated by some individuals, it is advisable to avoid any amount of ethanol in this population. Other agents reported to produce a disulfiram-like reaction when combined with ethanol include metronidazole, moxalactam, cefoperazone, griseofulvin, chloramphenicol, cefamandole, nitrofurantoin, chlorpropamide, sulfonamides, ketoconazole, procarbazine, and moxalactam.

C. Hypersensitivity

Ethanol is rarely implicated as a cause of hypersensitivity reactions when used as a vehicle excipient. Careful double-blind challenge testing is essential to rule out more common nonallergic types of skin reactions.

REFERENCES

1. **Petroni, N. C. and Cardoni, A. A.,** Alcohol content of liquid medicinals, *Drug Ther.*, 8, 72, 1978.
2. **Pruitt, A. W., Anyan, W. R., Hill, R. M., et al.,** American Academy of Pediatrics Committee on Drugs: ethanol in liquid preparations intended for children, *Pediatrics*, 73, 405, 1984.
3. **Weiner, M. and Bernstein, I. L.,** *Adverse Reactions to Drug Formulation Agents,* Marcel Dekker, New York, 1989.
4. **Sperry, K. and Pfalzgraf, R.,** Fatal ethanol intoxication from household products not intended for ingestion, *J. Forensic Sci.*, 35, 1138, 1990.
5. **Weller-Fahy, E. R., Berger, L. R., and Troutman, W. G.,** Mouthwash: a source of acute ethanol intoxication, *Pediatrics,* 66, 302, 1980.
6. **Varma, B. K. and Cincotta, J.,** Mouthwash-induced hypoglycemia, *Am. J. Dis. Child.*, 132, 930, 1978.

7. **Scherger, D. L., Wruk, K. M., Kulig, K. W., and Rumack, B. H.,** Ethyl alcohol (ethanol)-containing cologne, perfume, and after-shave ingestions in children, *Am. J. Dis. Child.*, 142, 630, 1988.

8. **Nikitakis, J. M.,** *CTFA Cosmetic Ingredient Handbook,* 1st ed., The Cosmetic, Toiletry and Fragrance Association, Washington, D.C., 1988.

9. **Andrien, P. and Lemberg, L.,** An unusual complication of intravenous nitroglycerin, *Heart Lung,* 15, 534, 1986.

10. **Korn, S. H. and Comer, J. B.,** Intravenous nitroglycerin and ethanol intoxication, *Ann. Intern. Med.,* 102, 274, 1985.

11. **Shook, T. L., Kirshenbaum, J. M., Hundley, R. F., Shorey, J. M., and Lamas, G. A.,** Ethanol intoxication complicating intravenous nitroglycerin therapy, *Ann. Intern. Med.,* 101, 498, 1984.

12. **Shorey, J., Bhardwaj, N., and Loscalzo, J.,** Acute Wernicke's encephalopathy after intravenous infusion of high-dose nitroglycerin, *Ann. Intern. Med.,* 101, 500, 1984.

13. **Ellis, C. N., Mitchell, A. J., and Beardsley, G. R.,** Tar gel interaction with disulfiram, *Arch. Dermatol.,* 115, 1367, 1979.

14. **Brown, K. R., Guglielmo, B. J., Pons, V. G., and Jacobs, R. A.,** Theophylline elixir, moxalactam, and a disulfiram reaction, *Ann. Intern. Med.,* 97, 621, 1982.

15. **Talbott, G. D. and Gander, O.,** Antabuse, *Md. State Med. J.,* 22, 60, 1973.

16. **Stoll, D. and King, L. E.,** Disulfiram-alcohol skin reaction to beer-containing shampoo, *JAMA,* 244, 2045, 1980.

17. **Mercurio, F.,** Antabuse-alcohol reaction following use of after-shave lotion, *JAMA,* 149, 82, 1952.

18. **Klein, H. and Harmjanz, D.,** Effect of ethanol infusion on the ultrastructure of human myocardium, *Postgrad. Med. J.,* 51, 325, 1975.

19. **Isaac, M. and Dundee, J. W.,** Clinical studies of induction agents XXX: venous sequelae following ethanol anaesthesis, *Br. J. Anaesth.,* 41, 1070, 1969.

20. **Shergy, W. J., Gilkeson, G. S., and German, D. C.,** Acute gouty arthritis and intravenous nitroglycerin, *Arch. Intern. Med.,* 148, 2505, 1988.

21. **Dees, S. C.,** An experimental study of the effect of alcohol and alcoholic beverages on allergic reactions, *Ann. Allergy,* 7, 185, 1949.

22. **Drevets, C. C. and Seebohm, P. M.,** Dermatitis from alcohol, *J. Allergy,* 32, 277, 1961.

23. **Karvonen, J. and Hannuksela, M.,** Urticaria from alcoholic beverages, *Acta Allergol.,* 31, 167, 1976.

24. **Oermerod, A. D. and Holt, P. J. A.,** Acute urticaria due to alcohol, *Br. J. Dermatol.,* 108, 723, 1983.

25. **Elphinstone, P. E., Black, A. K., and Greaves, M. W.,** Alcohol-induced urticaria, *J. R. Soc. Med.,* 78, 340, 1985.

26. **Przybilla, B. and Ring, J.,** Anaphylaxis to ethanol and sensitisation to acetic acid, *Lancet,* 1, 483, 1983.

27. **Rilliet, A., Hunziker, N., and Brun, R.,** Alcohol contact urticaria syndrome (immediate-type hypersensitivity). Case report, *Dermatologica,* 161, 361, 1980.

28. **Stotts, J. and Ely, W. J.,** Induction of human skin sensitization to ethanol, *Ann. Invest. Dermatol.,* 69, 219, 1977.

29. **Melli, M. C., Giorgini, S., and Sertoli, A.,** Sensitization from contact with ethyl alcohol, *Contact Dermatitis,* 14, 1986.

30. **Gong, H., Jr., Tashkin, D. P., and Calvarese, B. M.,** Alcohol-induced bronchospasm in an asthmatic patient, *Chest,* 80, 167, 1981.

31. **Geppert, E. F. and Boushey, H. A.,** An investigation of the mechanism of ethanol-induced bronchoconstriction, *Am. Rev. Respir. Dis.,* 118, 135, 1978.

32. **Seta, A., Tricomi, S., Goodwin, D. W., et al.,** Biochemical correlates of ethanol-induced flushing in orientals, *J. Stud. Alcohol,* 39, 1, 1978.

ETHYLENEDIAMINE

I. REGULATORY CLASSIFICATION

Ethylenediamine forms a stable soluble complex with theophylline and neomycin and is used to stabilize formulations containing these ingredients. It is an alkaline compound that is also used as a buffering agent.

$$H_2NCH_2CH_2NH_2$$

FIGURE 26. Ethylenediamine.

II. SYNONYMS

EDA
1,2-Ethanediamine

III. AVAILABLE FORMULATIONS

A. Drugs

Ethylenediamine hydrochloride is added to some nystatin preparations as a stabilizer in the presence of neomycin. Although Mycolog® was reformulated to remove the neomycin and ethylenediamine, some generic "equivalents" to the new neomycin-free formulation (Mycolog® II) inexplicably contain ethylenediamine. Aminophylline consists of theophylline and ethylenediamine hydrochloride.[1]

Ethylenediamine base is added to tincture of Merthiolate®, thimerosal topical solution USP, and thimerosal topical aerosol USP to act as a buffer and prevent deterioration of the mercury compound; it is not added to thimerosal-containing vaccines.

IV. TABLE OF COMMON PRODUCTS

A. Topical Drug Products

Trade name	Manufacturer
Myco-Triacet II cream	Lemmon
Mytrex cream	Savage
Nystex cream	Savage
Nystatin cream	Pharmaderm-Fougera

Note: Ethylenediamine-free nystatin creams include Nilstat® (Lederle), Mycostatin® (Squibb), and Mycolog II® (Squibb).[2]

V. HUMAN TOXICITY DATA

A. Delayed Hypersensitivity Reactions

1. Incidence

Delayed hypersensitivity reactions to systemically administrated aminophylline are usually attributed to the ethylenediamine component. Most reactions are reported in patients previously sensitized to EDA from topical medication or occupational exposure. Of 18 case reports of EDA sensitivity secondary to aminophylline exposure, the drug was given orally in 7, intravenously in 9, and rectally in 2 patients.[3]

Total avoidance of ethylenediamine for 10 years was associated with disappearance of contact hypersensitivity to EDA in 25% of patients with original 3+ positive patch tests. Twelve of 16 patients had unchanged results, with one patient experiencing an outbreak at the original site of eczema.[4]

Allergic contact dermatitis was reported in three infants, aged 6 weeks to 1 year, following the use of Mycolog® cream to treat diaper rash. All had positive patch tests to ethylenediamine 1%.[5]

Occupational contact dermatitis has developed in pharmacists preparing aminophylline suppositories[6,7] and in a nurse who prepared and administered injectable theophylline preparations.[8]

The incidence of positive patch tests to ethylenediamine is summarized in the following:

No. patients	No. positive (%)	Diagnosis	Country	Ref.
1007	46(4.6)	Dermatitis	Italy	9
8230	252(3)	Dermatitis	Italy	10
159	20(12.6)	Dermatitis	Denmark	11
100	18(18)	Medication dermatitis	U.S.	12
150	(13.2)	Dermatitis	U.S.	13
1200	85(7)	Dermatitis	U.S.	14
127	(3.1)	Dermatitis	U.S.	15
1158	(0.43)	Healthy volunteers	U.S.	15

2. Manifestations

Typically, generalized pruritus, burning, and maculopapular erythroderma occurs 12 to 24 h after systemic exposure to aminophylline, given IV, orally, or rectally, followed by exfoliative erythroderma. Patch tests to ethylenediamine are positive. Bronchospasm may also be evident.[16-20] The onset of dermatitis may occur 6 to 8 h after subsequent exposures, suggesting a combined cell-mediated and humoral mechanism.[21]

Topical application of ethylenediamine-containing preparations may cause a typical delayed contact eczematous dermatitis. Nummular eruptions, mostly on the limbs, have been reported in children and persisted for 20 to 25 d despite treatment.[9]

Prolonged daily contact with ethylenediamine for 12 months, which had transferred from a corticosteroid cream to a chewing gum wrapper and trouser pocket lining, resulted in a local lymphoblastic reaction, simulating mycosis fungoides, in a 44-year-old man.[22]

3. Mechanism

The delayed hypersensitivity reactions to ethylenediamine are primarily typical Type IV cell-mediated responses. The accelerated response could be a humoral response or lymphokine-mediated memory response.[23]

4. Cross-Reactivity

Ethylenediamine is structurally related to antihistamines such as antazoline, promethazine, hydroxyzine, tripelennamine, methapyrilene, and pyrilamine. Other related chemicals include triethanolamine, piperazine, triethylenetetramine, and EDTA. Cross-sensitivity to ethylenediamine has been documented for hydroxyzine,[24-26] piperazine,[26-29] antazoline,[30] and dimethindene.[29] Despite earlier reports of cross-reactivity to promethazine,[31] an attempt to confirm these findings by oral challenge with promethazine in 12 EDA-sensitive patients revealed no exacerbation of dermatitis.[32]

Despite the sporadic case reports of cross-sensitivity to these antihistamines, the incidence is relatively low. In extensive testing of 32 ethylenediamine-sensitive patients to 23 topical antihistamines, only 2 cross-reactions occurred to antihistamines, both to dimethindene. A high incidence of cross-reactions were observed to related industrial chemicals, such as diethylenetriamine (53%), triethylenetetramine (37.5%), tetraethylenepentamine (37.5%), piperazine (6.25%), hexamethylenetetramine (3%), and diethanolamine (3%).[29]

Cross-sensitivity to EDTA is controversial. While two patients with ocular sensitivity reactions to EDTA-preserved ophthalmic products were confirmed to have positive patch tests to both EDA and EDTA,[33] other clinicians have been unable to confirm cross-reactions.[29,34]

B. Immediate Hypersensitivity Reactions

While most cases of dermatitis following systemic aminophylline administration have been attributed to a delayed hypersensitivity response, several cases of acute urticaria indicative of a Type I hypersensitivity response have been reported. One case can be attributed to the theophylline component since the patient developed a similar rash following ingestion of anhydrous theophylline.[35]

C. Photocontact Dermatitis

Two patients with classic delayed-type hypersensitivity to ethylenediamine displayed exacerbation or a decreased latency of the reaction in the presence of sunlight.[36,37]

D. Erythema Multiforme

Two cases of erythema-multiforme-like eruptions have been attributed to ethylenediamine. These lesions are thought to be secondary disseminations of the original eczematous contact dermatitis, characterized by pinhead-sized erythematovesicular lesions, beginning at the site of initial lesions and becoming bilateral and often symmetrical.[38,39]

VI. CLINICAL RELEVANCE

A. Delayed Hypersensitivity Reactions

Ethylenediamine is one of the most common causes of allergic contact dermatitis with a prevalence of less than 1% in the general population, 3 to 7% in unselected patients with contact dermatitis, and up to 18% in patients with medication-related dermatitis. Ethylenediamine sensitivity persists for 10 years or longer in 75% of patients with significant reactions despite avoidance of the chemical. Reactions can be exacerbated by systemic administration of ethylenediamine-containing medications.

Ethylenediamine is responsible for most cases of aminophylline-related hypersensitivity reactions, which may have an accelerated onset on subsequent exposures.

Patients with ethylenediamine sensitivity can be given EDTA and probably the phenothiazine group of antihistamines (i.e., promethazine). Asthmatics sensitive to ethylenediamine may be given anhydrous theophylline. An EDA-free injectable preparation is available from Travenol Laboratories.

The presence of corticosteroids in some topical EDA-containing medications may mask hypersensitivity. A tenfold increase in concentration has been found to be nonirritating, thus a 1% concentration is recommended for patch testing.[39]

REFERENCES

1. **Fisher, A. A.,** The significance of ethylenediamine hydrochloride dermatitis caused by a "generic" nystatin-triamcinolone II cream, *Cutis,* 41, 241, 1988.
2. **Fisher, A. A.,** Unnecessary addition of ethylenediamine hydrochloride to "generic" nystatin creams, *J. Am. Acad. Dermatol.,* 20, 129, 1989.
3. **Anon.,** Allergy to aminophylline, *Lancet,* 1, 290, 1985.
4. **Nielsen, M. and Jorgensen, J.,** Persistence of contact sensitivity to ethylenediamine, *Contact Dermatitis,* 16, 275, 1987.
5. **Fisher, A. A.,** Allergic contact dermatitis in early infancy, *Cutis,* 35, 315, 1985.
6. **Baer, R., Cohen, H., and Neidorff, A.,** Allergic eczematous sensitivity to aminophylline, *Arch. Dermatol.,* 79, 647, 1959.
7. **Tas, J. and Weisberg, D.,** Allergy to aminophylline, *Acta Allergol.,* 12, 39, 1958.
8. **Dal Monte, A., De Benedictis, E., and Laffi, G.,** Occupational dermatitis from ethylenediamine hydrochloride, *Contact Dermatitis,* 17, 254, 1987.
9. **Caraffini, S. and Lisi, P.,** Nummular dermatitis-like eruption from ethylenediamine hydrochloride in 2 children, *Contact Dermatitis,* 17, 313, 1987.
10. **Angelini, G., Vena, G. A., and Meneghini, C. L.,** Allergic contact dermatitis to some medicaments, *Contact Dermatitis,* 12, 263, 1985.
11. **White, M. I., Douglas, W. S., and Main, R. A.,** Contact dermatitis attributed to ethylenediamine, *Br. Med. J.,* 1, 415, 1978.
12. **Fisher, A. A., Pascher, F., and Kanol, N. B.,** Allergic contact dermatitis due to ingredients of vehicles, *Arch. Dermatol.,* 104, 286, 1971.
13. **Baer, R. L., Ramsey, D. L., and Biondi, E.,** The most common contact allergens, *Arch. Dermatol.,* 106, 74, 1973.

14. **North American Contact Dermatitis Group,** Epidemiology of contact dermatitis in North America: 1972, *Arch. Dermatol.,* 108, 537, 1973.

15. **Prystowsky, S. D., Allen, A. M., Smith, R. W., Nonomura, J. H., Odom, R. B., and Akers, W. A.,** Allergic contact hypersensitivity to nickel, neomycin, ethylenediamine, and benzocaine, *Arch. Dermatol.,* 115, 959, 1979.

16. **Bernstein, J. E. and Lorincz, A. L.,** Ethylenediamine-induced exfoliative erythroderma, *Arch. Dermatol.,* 115, 360, 1979.

17. **Mohsenifar, Z., Lehrman, S., Carson, S. A., and Tashkin, D.,** Two cases of allergy to aminophylline, *Ann. Allergy,* 49, 281, 1982.

18. **Elias, J. A. and Levinson, A. I.,** Hypersensitivity reactions to ethylenediamine in aminophylline, *Am. Rev. Respir. Dis.,* 123, 560, 1981.

19. **Hardy, C., Schofield, O., and George, C. F.,** Allergy to aminophylline, *Br. Med. J.,* 286, 2051, 1983.

20. **Thompson, P. J., Gibb, W. R. G., Cole, P., and Citron, K. M.,** Generalised allergic reactions to aminophylline, *Thorax,* 39, 600, 1984.

21. **Gibb, W. and Thompson, P. J.,** Allergy to aminophylline, *Br. Med. J.,* 287, 508, 1983.

22. **Wall, L. M.,** Lymphomatoid contact dermatitis due to ethylenediamine dihydrochloride, *Contact Dermatitis,* 8, 51, 1982.

23. **deShazo, R. D. and Stevenson, H. C.,** Generalized dermatitis to aminophylline, *Ann. Allergy,* 46, 152, 1981.

24. **Pascher, F.,** Systemic reactions to topically applied drugs, *Bull. N.Y. Acad. Med.,* 49, 613, 1973.

25. **Lawyer, C. H., Bardana, E. J., Rodgers, R., et al.,** Utilization of intravenous dihydroxypropyl theophylline (dyphylline) in an aminophylline-sensitive patient, and its pharmacokinetic comparison with theophylline, *J. Allergy Clin. Immunol.,* 65, 353, 1980.

26. **Calnan, C. D.,** Sensitivity to ethylenediamine, piperazine and hydroxyzine, *Contact Dermatitis,* 1, 176, 1975.

27. **Wright, S. and Harman, R. R. M.,** Ethylenediamine and piperazine sensitivity, *Br. Med. J.,* 287, 463, 1983.

28. **Price, M. L. and Hall-Smith, S. P.,** Allergy to piperazine in a patient sensitive to ethylenediamine, *Contact Dermatitis,* 10, 120, 1984.

29. **Balato, N., Cusano, F., Lembo, G., and Ayala, F.,** Ethylenediamine dermatitis, *Contact Dermatitis,* 15, 263, 1986.

30. **Berman, B. A. and Ross, R. N.,** Ethylenediamine: systemic eczematous contact-type dermatitis, *Cutis,* 31, 594, 1983.

31. **Eriksen, K. E.,** Allergy to ethylenediamine, *Arch. Dermatol.,* 111, 791, 1975.

32. **King, C. M. and Beck, M.,** Oral promethazine hydrochloride in ethylenediamine-sensitive patients, *Contact Dermatitis,* 9, 444, 1983.

33. **Raymond, J. Z. and Gross, P. R.,** EDTA: preservative dermatitis, *Arch. Dermatol.,* 100, 436, 1969.

34. **Fisher, A. A.,** Does ethylenediamine hydrochloride crossreact with ethylenediamine tetracetate?, *Contact Dermatitis,* 1, 267, 1975.

35. **Wong, D., Lopapa, A. F., and Haddad, Z. H.,** Immediate hypersensitivity reaction to aminophylline, *J. Allergy Clin. Immunol.,* 48, 165, 1971.

36. **Burry, J. N.,** Photocontact dermatitis from ethylenediamine, *Contact Dermatitis,* 15, 305, 1986.

37. **Romaguera, C. Grimalt, F., and Lecha, M.,** Photoallergic dermatitis from ethylenediamine, *Contact Dermatitis,* 14, 130, 1986.

38. **Meneghini, C. L. and Angelini, G.,** Contact dermatitis from pyrrolnitrin, *Contact Dermatitis,* 8, 55, 1982.

39. **Fisher, A. A.,** Erythema multiforme-like eruptions due to topical medications: Part II, *Cutis,* 37, 158, 1986.

FD&C BLUE DYE NO. 1

I. REGULATORY CLASSIFICATION

FD&C Blue No. 1 is a triphenylmethane dye permanently listed by the FDA for use in oral and topical drug products and cosmetics.

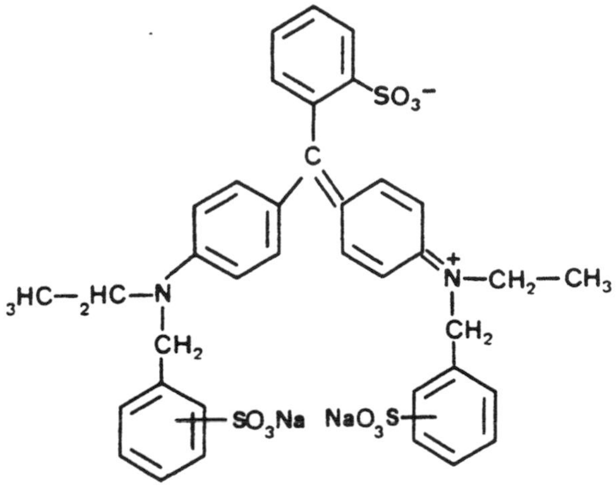

FIGURE 27. FD&C Blue No. 1.

II. SYNONYMS

Brilliant Blue FCF
C.I. Acid Blue 9
C.I. Food Blue 2
C.I. 42090
Patent Blue AC

III. AVAILABLE FORMULATIONS

A. Cosmetics
FD&C Blue No. 1 is also available as the aluminum lake, a water-insoluble form. The

141

ammonium salt, D&C Blue No. 4, is used in drugs and cosmetics. These dyes are used in shampoos; colognes; toilet water; perfumes; bubble baths; creams and lotions; aftershave lotions; skin fresheners; cleansing products; hair conditioners; wave sets; and hair tonics, dressing, and grooming aids.[1]

IV. TABLE OF COMMON PRODUCTS (INCLUDES LAKES)

A. Oral Drug Products

Trade name	Manufacturer
Achromycin V capsule	Lederle
Achromycin V oral suspension	Lederle
Accutane capsule 20 mg	Robins
Adeflor M tablet	Upjohn
Allergy-Sinus Comtrex	Bristol-Myers
Alu-tab tablet	3M Riker
Alu-cap capsule	3M Riker
Ambenyl D decongestant cough formula	Forest
Anacin-3 tablet	Whitehall
Anacin-3 maximum strength tablet, caplet	Whitehall
Ancoban capsule	Roche
Antivert 12.5, 50 mg	Roerig
Anturane capsule	Ciba
Apresazide capsule 25/25, 50/50	Ciba
Apresoline tablet 25, 50 mg	Ciba
Artane sequels	Lederle
Arthritis Strength Tri-Buffered Bufferin	Bristol-Myers
Asbron G elixir, tablet	Sandoz
Atarax tablet 100 mg	Roerig
Atromid-S capsule	Wyeth-Ayerst
Bactrim tablet	Roche
Basaljel	Wyeth-Ayerst
Benadryl capsule 25, 50 mg	Parke-Davis
Bentyl capsule 10 mg	Lakeside
Bentyl tablet 20 mg	Lakeside
Bentyl syrup	Lakeside
Berocca tablet	Roche
Brexin LA	Savage
Bromfed capsule	Muro
Bronkodyl capsule 100, 200 mg	Winthrop
Bumex tablet 0.5 mg	Roche
Butisol sodium tablet 15, 30 mg	Wallace
Cafergot PB tablet	Sandoz
Carafate tablet	Marion
Cardene capsule 20, 30 mg	Syntex
Cardizem tablet 30, 90 mg	Marion

Trade name (cont'd)	Manufacturer (cont'd)
Catapres tablet 0.1, 0.2 mg	Boehringer Ingelheim
Ceclor pulvule 250, 500 mg	Lilly
Cefol filmtab	Abbott
Ceftin tablet 250, 500 mg	Allen & Hanburys
Celontin capsule 150 mg	Parke-Davis
Centrax capsule 5, 10 mg	Parke-Davis
Cephulac syrup	Merrell Dow
Children's Chloraseptic lozenges	Richardson Vicks
Children's Tylenol chewable cold tablet	McNeil consumer
Children's Tylenol liquid cold formula	McNeil consumer
Chloraseptic cool mint liquid	Richardson Vicks
Chloraseptic lozenge menthol, cherry, mint	Richardson Vicks
Choloxin tablet 6 mg	Boots-Flint
Cholybar	Parke-Davis
Chronulac syrup	Merrell Dow
Cinobac capsule	Dista
Cleocin capsule 75, 150, 300 mg	Upjohn
Combipres tablet 0.2 mg	Boehringer Ingelheim
Compazine tablet, spansule	Smith Kline & French
Constant-T tablet 300 mg	Geigy
Coumadin tablet 2.5 mg	DuPont
Dalmane capsule 30 mg	Roche
Datril extra strength	Bristol-Myers
Decadron tablet 0.75 mg	Merck Sharp & Dohme
Declomycin capsule	Lederle
Demser capsule	Merck Sharp & Dohme
Demulen 1/35 blue tablet	Searle
Depakote tablet 125 mg	Abbott
Dexedrine spansule	Smith Kline & French
DiaBeta tablet 5 mg	Hoechst-Roussel
Diabinese tablet	Pfizer
Dilantin capsule 30 mg	Parke-Davis
Dilaudid tablet 1 mg	Key
Dimetane DC cough syrup	Robins
Disalcid capsule, tablet	3M Riker
Ditropan tablet	Marion
Donnatal capsule	Robins
Donnazyme tablet	Robins
Doral tablet	Wallace
Doryx capsule	Parke-Davis
Duricef capsule	Mead Johnson
Edecrin tablet 50 mg	Merck Sharp & Dohme
Elavil tablet 10, 25 mg	Stuart
Ery-Tab 333 mg	Abbott

Trade name (cont'd)	Manufacturer (cont'd)
Esidrix tablet 100 mg	Ciba
Estinyl tablet 0.05 mg	Schering
Estrace tablet 1, 2 mg	Mead Johnson
Estratest tablet	Reid-Rowell
Estratest H.S. tablet	Reid-Rowell
Estrovis tablet	Parke-Davis
Eulexin capsule	Schering
Excedrin extra strength caplet	Bristol-Myers
Excedrin PM tablet, caplet	Bristol-Myers
Extra Strength Tri-Buffered Bufferin	Bristol-Myers
Extra Strength Tylenol gelcaps	McNeil consumer
Estra Strength Tylenol adult liquid	McNeil consumer
Fastin capsule	Beecham
Fedahist timecap, gyrocap	Schwarz Pharma
Feldene capsule	Pfizer
Fero-Folic-500 filmtab	Abbott
Fero-Grad-500 filmtab	Abbott
Feosol capsule	Smith Kline Beecham
Ferro-sequels	Lederle
Fioricet capsule, tablet	Sandoz
Fioricet with codeine #3 capsule	Sandoz
Gantanol tablet	Roche
Gaviscon liquid antacid	Marion
Guaifed-PD capsule	Muro
Haldol tablet 5, 10 mg	McNeil
Heptuna plus capsule	Roerig
Hydropres tablet	Merck Sharp & Dohme
Inderal tablet 20, 40 mg	Wyeth-Ayerst
Inderal LA capsule 60, 80, 120, 160 mg	Wyeth-Ayerst
Inderide LA capsule 120/50, 160/50	Wyeth-Ayerst
Indocin capsule 25, 50 mg	Merck Sharp & Dohme
Indocin SR capsule	Merck Sharp & Dohme
Keflet tablet 250, 500, 1000 mg	Dista
Keftab 250, 500 mg	Dista
Klonopin tablet 1 mg	Roche
Klor-Con 8	Upsher-Smith
Lanoxicaps 0.2 mg	Burroughs Wellcome
Larodopa capsule 100, 250 mg	Roche
Levsinex timecap	Schwarz Pharma
Librax capsule	Roche
Librium capsule 5, 10, 25 mg	Roche
Limbitrol tablet	Roche
Lincocin capsules	Upjohn
Lithane tablet	Miles

Trade name (cont'd)	Manufacturer (cont'd)
Lithobid tablet 300 mg	Ciba
Loestrin 1.5/30 green tablet	Parke-Davis
Loestrin FE 1.5/30 green tablet	Parke-Davis
Lopid capsule 300 mg	Parke-Davis
Lopressor HCT tablet 50/25 mg	Geigy
Lortab tablet	Russ
Lortab 7.5/500	Russ
Loxitane capsule 5, 10, 25, 50 mg	Lederle
Maximum Strength Sine-Aid caplet, tablet	McNeil consumer
Maxzide 25 mg	Lederle
Mediatric capsule	Wyeth-Ayerst
Megace tablet	Bristol-Myers
Mellaril tablet 10, 100 mg	Sandoz
Menrium tablet 5-2, 5-4, 10-4	Roche
Meprospan 400 capsule	Wallace
Mestinon syrup	ICN
Methergine tablet	Sandoz
Mexitil capsule 150, 200, 250 mg	Boehringer Ingelheim
Micronase tablet 5 mg	Upjohn
Minipress capsules	Pfizer
Minizide capsule	Pfizer
Minocin capsule 100 mg	Lederle
Modane soft capsule	Adria
MS Contin tablet 30 mg Purdue	Frederick
Naldecon CX liquid adult	Bristol
Naldecon Senior DX cough/cold liquid	Bristol
Naldecon Senior EX cough/cold liquid	Bristol
Naturetin tablet 5 mg	Princeton
Navane capsules	Roerig
Nembutal sodium capsule 50, 100 mg	Abbott
Niferex PN tablet	Central
Noludar capsule 300 mg	Roche
Norcept-E 1/35-28 green tablet	GynoPharm
Norethin 1/35, 1/50 blue tablet	Schiapparelli Searle
Norgesic tablet	3M Riker
Norgesic forte tablet	3M Riker
Norpace capsule 150 mg	Searle
Norpace CR capsule 100, 150 mg	Searle
Norpramin tablet 10, 25, 50, 75, 100 mg	Merrell Dow
Novafed capsule	Lakeside
Novafed A capsule	Lakeside
Novahistine DH	Lakeside
Novahistine expectorant	Lakeside
Nucofed capsule	Beecham

Trade name (cont'd)	Manufacturer (cont'd)
Nucofed syrup	Beecham
Omnipen capsules	Wyeth-Ayerst
Omnipen oral suspension 125	Wyeth-Ayerst
Ornade spansule	Smith Kline & French
Orudis capsules	Wyeth-Ayerst
Os-cal 250 + D tablet	Marion
Os-cal 500 tablet	Marion
Ovcon 35 green tablet	Mead Johnson
Ovcon 50 green tablet	Mead Johnson
Panwarfin tablet 2 mg	Abbott
Paradione capsule 300 mg	Abbott
Parafon forte DSC caplet	McNeil
Pathocil capsule	Wyeth-Ayerst
Pentids 800 tablet	Squibb
Peritrate tablet 10, 20 mg	Parke-Davis
Peritrate SA tablet 80 mg	Parke-Davis
Permax tablet 0.25 mg	Lilly
Persantine tablet 25, 50, 75 mg	Boehringer Ingelheim
Phenaphen with codeine	Robins
Phenergan syrup plain	Wyeth-Ayerst
Phenergan with codeine syrup	Wyeth-Ayerst
Phillips laxcaps	Glenbrook
Placidyl capsule 750 mg	Abbott
PMB 200 tablet	Wyeth-Ayerst
Ponstel capsule 250 mg	Parke-Davis
Pramet FA tablet	Ross
Pramilet FA tablet	Ross
Prelu-2	Boehringer Ingelheim
Premarin tablet 0.3 mg	Wyeth-Ayerst
Principen oral suspension 125	Squibb
Principen with probenecid capsule	Squibb
Procan SR tablet 250 mg	Parke-Davis
Prolixin tablet 5, 10 mg	Princeton
Prozac capsule	Dista
Quarzan capsule 2.5 mg	Roche
Rauzide tablet	Princeton
Reglan tablet	Robins
Renese R tablet	Pfizer
Restoril capsule	Sandoz
Ridaura capsule	Smith Kline & French
Rifadin capsule 150, 300 mg	Merrell Dow
Rifamate capsule	Merrell Dow
Rimactane capsule 300 mg	Ciba
Rondec TR tablet	Ross

Trade name (cont'd)	Manufacturer (cont'd)
Rondec DM syrup, oral drops	Ross
Rynatuss tablet, pediatric suspension	Wallace
Salutensin tablet	Bristol
Sansert tablet	Sandoz
Sectral capsule	Wyeth-Ayerst
Septra suspension grape	Burroughs Wellcome
Ser-as-es tablet	Ciba
Serax capsule 10, 30 mg	Wyeth-Ayerst
Seromycin capsule	Lilly
Sinequan capsules	Roerig
Solatene capsule	Roche
Sorbitrate chewable tablet 5 mg	ICI
Sorbitrate oral tablet 5, 20, 40 mg	ICI
Sumycin capsules	Squibb
Surmontil capsules	Wyeth-Ayerst
Synalgos DC capsule	Wyeth-Ayerst
Synthroid tablet 125, 175, 300 mcg	Boots-Flint
Tace capsule	Merrell Dow
Talacen caplet	Winthrop
Taractan concentrate	Roche
Theo-24 capsule 300 mg	Searle
Theobid	Allen & Hanburys
Theobid Jr.	Allen & Hanburys
Tofranil PM capsule 75, 100, 125, 150 mg	Geigy
Trancopal caplet 200 mg	Winthrop
Triavil tablet 2-10	Merck Sharp & Dohme
Triaminic expectorant DH	Sandoz
Triaminic expectorant with codeine	Sandoz
Tri-Buffered Bufferin	Bristol-Myers
Tridione capsule 300 mg	Abbott
Tri-levlen 28 tablet	Berlex
Trimox capsules	Squibb
Trinsicon	Russ
Triphasil 28 green tablet	Wyeth-Ayerst
Tuss-Ornade spansule	Smith Kline & French
Tylenol liquid	McNeil consumer
Tylenol cold caplet	McNeil consumer
Tylenol no drowsiness caplet	McNeil consumer
Tylenol maximum strenth allergy sinus caplet	McNeil consumer
Tylenol maximum strength sinus caplet, tablet	McNeil consumer
Tylox capsule	McNeil
Unipen capsules	Wyeth-Ayerst
Unisom dual relief tablet	Pfizer Consumer
Unisom tablet	Pfizer Consumer

Trade name (cont'd)	Manufacturer (cont'd)
Valmid capsule	Dista
Valium tablet 10 mg	Roche
Valrelease capsule	Roche
Velosef capsules	Squibb
Verelan capsule 240 mg	Wyeth-Ayerst
Vibramycin capsule	Pfizer
Vibramycin oral suspension	Pfizer
Vicon forte	Russ
Wygesic tablet	Wyeth-Ayerst
Wymox capsules	Wyeth-Ayerst
Zenate tablet	Reid-Rowell

V. ANIMAL TOXICITY DATA

Chronic feeding studies in rats given dietary concentrations of 0.1, 1, or 2% demonstrated no adverse effects in male rats and decreased body weight and survival in females at the highest dose (2%). Similar studies in mice given 0.5, 1.5, or 5% dietary concentrations did not demonstrate adverse effects. The highest concentration (5%) corresponds to a dose of 7354 mg/kg/d for male mice and 8966 mg/kg/d for female mice.[2]

Repeated subcutaneous injections in animals established the WHO acceptable daily intake of 12.5 mg/kg/d as a food additive.[3]

VI. HUMAN TOXICITY DATA

A. Hypersensitivity Reactions

In general, the triphenylmethane dyes are not considered to be potent sensitizers, however, these are rarely tested for in routine screening or in patients with suspected food or drug allergy. In a series of 45 patients with moderately severe perennial asthma, none of the patients developed bronchoconstriction after double-blind challenge with 20 mg orally of FD&C Blue No. 1.[4]

Brilliant blue has been included in food dye patch test series in a 2% concentration in petrolatum[5] and was listed as a cause of unspecified adverse reactions in a survey of British drug formulations.[6]

VII. CLINICAL RELEVANCE

Due to the paucity of reported adverse reactions to this dye, there is no need for routine testing in cases of suspected color additive allergy. If other, more allergenic dyes have been excluded as the cause, such as azo dyes, testing with this dye, as well as other non-azo dyes, may be appropriate.

REFERENCES

1. **Nikitakis, J. M.,** *CTFA Cosmetic Ingredient Handbook,* 1st ed., The Cosmetic, Toiletry and Fragrance Association, Washington, D.C., 1988.
2. **Borzelleca, J. F., Depukat, K., and Hallagan, J. B.,** Lifetime toxicity/carconogenicity studies of FD&C Blue No. 1 (brilliant blue FCF) in rats and mice, *Food Chem. Toxicol.,* 28, 221, 1990.
3. **WHO,** Thirteenth Report of an FAO/WHO Expert Committee on Food Additives, Tech. Rep. Ser. No. 445, WHO, Geneva, 1970.
4. **Weber, R. W., Hoffman, M., Raine, D. A., Jr., and Nelson, H. S.,** Incidence of bronchoconstriction due to aspirin, azo dyes, non-azo dyes, and preservatives in a population of perennial asthmatics, *J. Allergy Clin. Immunol.,* 64, 32, 1979.
5. **Mitchell, J. C.,** Allergic contact dermatitis from a certified food dye presenting as "sock dermatitis", *Contact Dermatitis Newsletter,* 11, 247, 1972.
6. **Pollock, I., Young, E., Stoneham, M., Slater, N., Wilkinson, J. D., and Warner, J. O.,** Survey of colourings and preservatives in drugs, *Br. Med. J.,* 299, 649, 1989.

FD&C Blue No. 2

I. REGULATORY CLASSIFICATION

FD&C Blue No. 2 is an indigoid dye producing a grape shade. It was approved for use in foods and ingested drugs in 1987 and for use in sutures in 1971.

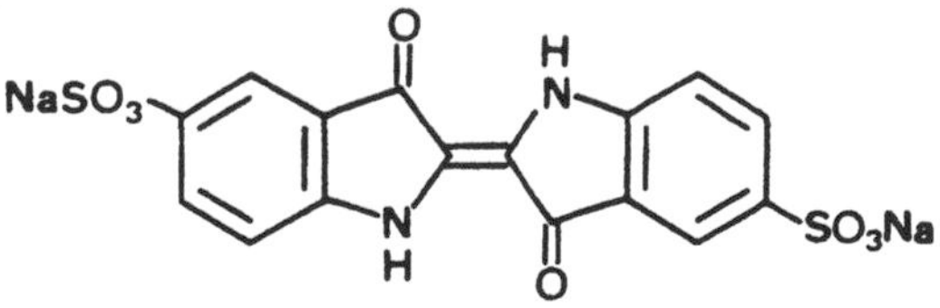

FIGURE 28. FD&C Blue No. 2.

II. SYNONYMS

Indigo carmine
Indigotin
C.I. Food Blue 1
C.I. 73015
E132
Indigotindisulfonate sodium
Ceruleinum
Blue X

III. AVAILABLE FORMULATIONS

A. Drugs

Indigo carmine is a common color additive in oral tablets and capsules and is also used in some nylon sutures. An injectable preparation, containing 0.8% indigo carmine, is available in 5 ml ampules for diagnostic use.

IV. TABLE OF COMMON DRUG PRODUCTS

A. Oral Drug Products

Trade name	Manufacturer
Aldoclor 250 tablet	Merck Sharp & Dohme
Ansaid tablet	Upjohn
Asendin tablet 100 mg	Lederle
Beelith tablet	Beach
Blocadren tablet	Merck Sharp & Dohme
Butisol sodium tablet 100 mg	Wallace
Calan SR caplet	Searle
Caltrate 600 + vitamin D	Lederle
Cardene capsule 30 mg	Syntex
Centrax tablet 10 mg	Parke-Davis
Cesamet pulvule	Lilly
Comhist LA	Norwich Eaton
Compazine tablets	Smith Kline & French
Contac allergy capsule	Smith Kline Beecham
Coumadin tablet 2 mg	DuPont
Cystospaz tablet	Webcon
Dantrium capsule 25, 50, 100 mg	Norwich Eaton
Demser capsule	Merck Sharp & Dohme
Depakote tablet 500 mg	Abbott
Diethylstilbestrol enseal	Lilly
Diulo tablet 5 mg	Schiapparelli Searle
Donnatal capsule	Robins
Elavil tablet 100, 150 mg	Stuart
Entozyme tablet	Robins
Estinyl tablet 0.02 mg	Schering
Estratest H.S. tablet	Reid-Rowell
Euthroid tablet 1, 2, 3	Parke-Davis
Extra Strength Tylenol gelcaps	McNeil Consumer
Fedahist gyrocap	Schwarz Pharma
Feosol tablet	Smith Kline Beecham
Flagyl tablet	Searle
Furoxone tablet	Norwich Eaton
Geocillin tablet	Roerig
Grisactin capsule 125, 250 mg	Wyeth-Ayerst
Halcion tablet 0.25 mg	Upjohn
Haldol tablet 2 mg	McNeil
Halotestin tablet 5, 10 mg	Upjohn
Hytrin tablet 10 mg	Abbott
Indocin SR capsule	Merck Sharp & Dohme
Isoptin tablet 40 mg	Knoll
Isoptin SR tablet	Knoll

Trade name (cont'd)	Manufacturer (cont'd)
Keflet tablet 1 g	Dista
Klonopin tablet 1 mg	Roche
Levsinex timecaps	Schwarz Pharma
Lithium carbonate capsule	Roxane
Lopressor tablet 100 mg	Geigy
Mellaril tablet 100 mg	Sandoz
Mevacor tablet 20, 40 mg	Merck Sharp & Dohme
Moban tablet 10, 25, 50, 100 mg	DuPont
Modane soft capsule	Adria
Modicon 28 green tablet	Ortho
MS Contin tablet 15 mg	Purdue Frederick
Mycostatin tablets	Squibb
Normodyne tablet 100, 300 mg	Schering
Normozide tablet 100/25, 300/25	Schering
Ogen tablet 2.5, 5 mg	Abbott
Ortho-Novum 7/7/7 green tablet	Ortho
Ortho-Novum 10/11 green tablet	Ortho
Ortho-Novum 1/35 green tablet	Ortho
Ortho-Novum 1/50 green tablet	Ortho
PBZ-SR tablet	Geigy
Percocet demi tablet	DuPont
Peritinic tablet	Lederle
PMB 400 tablet	Wyeth-Ayerst
Preludin tablet	Boehringer Ingelheim
Premarin tablet 0.625, 2.5 mg	Wyeth-Ayerst
Premarin with methyltestosterone tablets	Wyeth-Ayerst
Procan SR tablet 500 mg	Parke-Davis
Prolixin tablet 2.5 mg	Princeton
Pronestyl SR tablet	Princeton
Pyridium tablet	Parke-Davis
Pyridium plus tablet	Parke-Davis
Quinidex extentabs	Robins
Retrovir capsule	Burroughs Wellcome
Sinemet tablet 10-100, 25-250	Merck Sharp & Dohme
Stelazine tablets	Smith Kline & French
Synthroid tablet 75, 112, 150 mcg	Boots-Flint
Tace capsule 25 mg	Merrell Dow
Tagamet tablets	Smith Kline &French
Taractan tablet 10, 25, 50, 100 mg	Roche
Tegison capsule	Roche
Teldrin capsule	Smith Kline Beecham
Thorazine tablets	Smith Kline & French
Timolide tablet	Merck Sharp & Dohme
Tranxene tablet 3.75 mg	Abbott

Trade name (cont'd)	Manufacturer (cont'd)
Tranxene SD 11.25 mg	Abbott
Trilisate tablet 1000 mg	Purdue Frederick
Tri-vi-flor chewable tablet 1 mg	Mead Johnson
Unisom Dual Relief tablet	Pfizer Consumer
Valpin 50 tablet	DuPont
Vancomycin pulvule 125, 250 mg	Lilly
Wyamycin S tablet	Wyeth-Ayerst
Wytensin tablet 8 mg	Wyeth-Ayerst
Xanax tablet 1 mg	Upjohn
Zovirax capsule	Burroughs Wellcome

V. ANIMAL TOXICITY DATA

The WHO has estimated the acceptable daily dietary intake of indigo carmine to be up to 2.5 mg/kg/d.[1]

VI. HUMAN TOXICITY DATA

A. Intravenous Injection

Indigo carmine has been used diagnostically by urologists as a intravenously injected dye during cystoscopy. The usual dose is 5 ml (20 mg) intravenously.[2] Adverse reactions during these procedures have been attributed to alpha-adrenergic stimulation, resulting in increased total peripheral resistance, and include severe headache, acute pulmonary edema, cardiac arrest, and hypertension. Cerebral artery spasm has also been described.[3] Although in most cases the hemodynamic changes are transient and not severe, patients with underlying cardiovascular disease may experience severe reactions.[4-8] These reactions have not been described following other routes of exposure.

B. Hypersensitivity Reactions

Three of 19 children with recurrent urticaria challenged with an oral dose of indigo carmine 0.1 mg in a double-blind fashion reacted with urticaria or angioedema.[9] In a series of 45 patients with moderately severe perennial asthma, none of the patients developed bronchoconstriction after double-blind challenge with 20 mg orally of FD&C Blue No. 2.[10]

Allergic contact dermatitis of the hands was related to occupational exposure to indigo carmine, as documented by positive patch testing, in 3 of 204 animal feed mill workers tested for various feed additive substances; an additional 2 workers had positive patch tests without a history of dermatitis.[11]

VII. CLINICAL RELEVANCE

A. Hypersensitivity

Both immediate allergic reactions, such as urticaria, and delayed allergic contact dermatitis have been documented following exposure to indigo carmine. Patients with a history of food-related urticaria or occupational dermatitis from contact with animal feeds should be be tested with indigo carmine as part of the evaluation for potential hypersensitivity.

REFERENCES

1. **WHO,** Thirteenth Report of FAO/WHO Expert Committee on Food Additives, Tech. Rep. Ser. No. 445, WHO. Geneva, 1971.

2. **Lin, B. and Iwata, Y.,** Modified cystoscopy to evaluate unilateral traumatic injury of the ureter during pelvic surgery, *Am. J. Obstet. Gynecol.,* 162, 1343, 1990.

3. **Hammann, B.,** The adverse effects of indigo carmine, *Anaesthesist,* 38, 136, 1989.

4. **Kennedy, W. F., Jr., Wirjoatmdja, K., Akamatsu, T. J., and Bonica, J. J.,** Cardiovascular and respiratory effects of indigo carmine, *J. Urol.,* 100, 775, 1968.

5. **Erickson, J. C. and Widmer, B. A.,** The vasopressor effect of indigo carmine, *Anesthesiology,* 29, 188, 1968.

6. **Ng, T. Y., Datta, T. D., and Kirimli, B. I.,** Reaction to indigo carmine, *J. Urol.,* 116, 132, 1976.

7. **Jeffords, D. L., Lange, P. H., and DeWolf, W. C.,** Severe hypertensive reaction to indigo carmine, *Urology,* 9, 180, 1977.

8. **Harioka, T., Mori, H., and Mori, K.,** Hypertensive reaction to indigo carmine during transurethral resection of a bladder tumor, *Anesth. Analg.,* 66, 1049, 1987.

9. **Supramaniam, G. and Warner, J. O.,** Artificial food additive intolerance in patients with angio-oedema and urticaria, *Lancet,* 2, 907, 1986.

10. **Weber, R. W., Hoffman, M., Raine, D. A., Jr., and Nelson, H. S.,** Incidence of bronchoconstriction due to aspirin, azo dyes, non-azo dyes, and preservatives in a population of perennial asthmatics, *J. Allergy Clin. Immunol.,* 64, 32, 1979.

11. **Mancuso, G., Staffa, M., Errani, A., Berdondini, R. M., and Fabbri, P.,** Occupational dermatitis in animal feed mill workers, *Contact Dermatitis,* 22, 37, 1990.

FD&C RED No. 3

I. REGULATORY CLASSIFICATION

FD&C Red No. 3 is classified as a dye, imparting a bluish-red (watermelon) color. Chemically, it is the disodium salt of 2',4',5',7'-tetraiodofluorescein. Aluminum or calcium lakes, water-insoluble forms, were also formerly used.

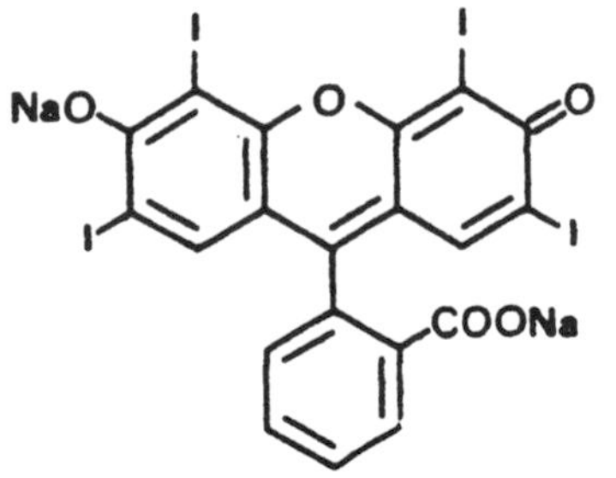

FIGURE 29. FD&C Red No. 3.

Erythrosine was permanently listed for use in ingested foods in 1969. External pharmaceutical uses of erythrosine were listed as provisional in 1960, pending conduction of studies on skin exposure. These studies showed clear evidence of an association with thyroid cancer in male rats exposed dermally to large amounts of the dye, therefore, the provisional use of erythrosine in externally applied drugs was terminated in January 1990 as part of a FDA review.

Even though the small amounts of the dye used in consumer products was unlikely to pose a significant hazard, the Delaney Clause of the 1960 Color Additive Amendments prevented the FDA from approving a color additive shown to induce cancer in humans or animals in any amount. All use of the lakes was also terminated. Erythrosine was permanently listed for use in ingested drugs in 1969, thus the new ruling only affects the lakes in oral drug products and all external drug products. The procedure for banning permanent listed uses of a color additive are more complex, allowing for public comment.[1]

The provisional use of erythrosine or its lakes in cosmetics was also terminated in January 1990. Because the ban was primarily based on legal provisions of the Delaney Clause and not on perceived risk of acute, imminent hazard, existing products were not recalled and will be allowed to remain available to consumers until current supplies are exhausted.[1]

II. SYNONYMS

Erythrosine
C.I. Acid Red 51
C.I. No. 45430
C.I. Food Red 14.

III. AVAILABLE FORMULATIONS

A. Foods

Available sources in foods include maraschino cherries, baked goods, chewing gum, soft candy, dairy products, desserts, dietary supplements (including vitamin and mineral supplements), food seasonings, jellies, jams, cake decorations, fruit cocktail, toaster pastries, and pistachio nuts.[1,2]

B. Drugs

Tablet, capsule, and liquid oral formulations are still in use. Erythrosine is commonly used as a disclosing agent for plaque on teeth.

C. Cosmetics

Former uses in cosmetics included creams and lotions, face and body powder, colognes, toilet waters, lipsticks, bubble baths, hair grooming aids, and rouges.[3,4]

IV. TABLE OF COMMON DRUG PRODUCTS

A. Oral Drug Products

Note: The use of erythrosine lakes was banned in January 1990. The products listed as containing the lakes are marked with an * and will be required to be reformulated. Manufacturers should be contacted if necessary to determine which dye will replace erythrosine.

Trade name	Manufacturer
Accutane capsule 20 mg	Roche
Aldoril tablet 15 mg	Merck Sharp & Dohme
Aldoril D30 tablet	Merck Sharp & Dohme
Amoxil capsule, oral suspension	Beecham
Apresazide capsule 25/25, 50/50	Ciba
Atarax tablet 100 mg	Roerig
Atromid S capsule	Wyeth-Ayerst
Azo-Gantanol tablet	Roche
Azo-Gantrisin tablet	Roche
Bayer children's cold tablet	Glenbrook
Benadryl capsule 25, 50 mg	Parke-Davis
Brexin LA	Savage
Bronkodyl capsule 100 mg	Winthrop
Butazolidin capsule 100 mg	Geigy
Butisol sodium tablet	Wallace

Trade name (cont'd)	Manufacturer (cont'd)
Catapres tablet 0.1 mg	Boehringer Ingelheim
Ceclor pulvule 250, 500 mg	Lilly
Celontin capsule 150, 300 mg	Parke-Davis
Cinobac capsule	Dista
Combipres tablet 0.2 mg	Boehringer Ingelheim
Constant-T tablet 200 mg	Geigy
Dalmane capsule 15 mg	Roche
Damason-P tablet	Mason
Danocrine capsule 50, 200 mg	Winthrop
Dantrium capsule 25, 50, 100 mg	Norwich Eaton
Darvon compound 65 pulvule	Lilly
Darvon pulvule with A.S.A.	Lilly
Demulen 1/50 pink tablet	Searle
Didrex tablet 50 mg	Upjohn
Diethylstilbestrol enseal	Lilly
Dilantin capsule 30, 100 mg	Parke-Davis
Diupres tablet 250, 500 mg	Merck Sharp & Dohme
E.E.S. Granules	Abbott
Enduron tablet 5 mg	Abbott
Enovid tablet 5 mg	Searle
Entex capsule	Norwich Eaton
Erythrocin stearate tablet 500 mg	Abbott
Erythromycin delayed-release capsule	Abbott
Esidrix tablet 25 mg	Ciba
Estinyl tablet 0.05 mg	Schering
Eulexin capsule	Schering
Feldene capsule	Pfizer
Fiorinal with codeine #3 capsule	Sandoz
Grisactin capsule 125, 250 mg	Wyeth-Ayerst
Heptuna Plus	Roerig
Ilosone pulvule 250 mg	Dista
Ilosone tablet 500 mg	Dista
Inderal LA 60, 80, 120 mg	Wyeth-Ayerst
Indocin capsule 25, 50 mg	Merck Sharp & Dohme
Indocin SR capsule	Merck Sharp & Dohme
Larodopa capsule 100 mg	Roche
Levlen 28 tablet	Berlex
Librium capsule 10 mg	Roche
Lincocin capsule	Upjohn
Lithobid tablet 300 mg*	Ciba
LoOvral 28 pink tablet	Wyeth-Ayerst
Lopid capsule 300 mg	Parke-Davis
Lopressor HCT tablet 100/25 mg	Geigy
Marplan tablet	Roche

Trade name (cont'd)	Manufacturer (cont'd)
Menrium tablet 10-4	Roche
Mesantoin tablet	Sandoz
Mexitil 150 mg	Boehringer Ingelheim
Milontin capsule	Parke-Davis
Minipress capsule	Pfizer
Minizide capsule	Pfizer
Mycostatin tablet	Squibb
Naldecon tablet	Bristol
Navane capsules	Roerig
Nembutal sodium capsule 50, 100 mg	Abbott
Noludar capsule 300 mg	Roche
Nordette-28 pink tablet	Wyeth-Ayerst
Norpace capsule 100, 150 mg	Searle
Norpace CR capsule 150 mg	Searle
Norpramin tablet 10, 25, 50, 75, 100 mg	Merrell Dow
Novafed capsule	Lakeside
Novafed A capsule	Lakeside
Ornade spansule	Smith Kline & French
Ovral 28 pink tablet	Wyeth-Ayerst
Paraflex tablet	McNeil
Parnate tablet	Smith Kline & French
Percocet demi	DuPont
Persantine tablet 25 mg*	Boehringer Ingelheim
Phenaphen with codeine No. 2, 3	Robins
Phillips Laxcaps	Glenbrook
Ponstel capsule 250 mg	Parke-Davis
Preludin tablet	Boehringer Ingelheim
Premarin with methyltestosterone 0.625	Wyeth-Ayerst
Principen oral suspension 125	Squibb
Principen with probenecid capsule	Squibb
Procan SR tablet 1000 mg*	Parke-Davis
Prolixin tablet 1, 10 mg	Princeton
Quarzan capsule 2.5 mg	Roche
Raudixin tablet	Princeton
Restoril capsule	Sandoz
Rifadin capsule 150 mg	Merrell Dow
Robaxisal tablet	Robins
Rocaltrol capsule 0.5 mcg	Roche
Rynatan pediatric suspension	Wallace
Rynatuss pediatric suspension	Wallace
Seromycin capsule	Lilly
Sinequan caspule	Roerig
Solatene capsule	Roche
Sumycin capsule	Squibb

Trade name (cont'd)	Manufacturer (cont'd)
Synthroid tablet 112 mcg	Boots-Flint
Taractan tablet 10, 25, 50, 100 mg	Roche
Tegretol chewable tablet 100 mg	Geigy
Terramycin capsule	Pfizer
Theo-24 capsule 200 mg	Searle
Theobid Jr. capsule	Allen & Hanburys
Theragran hematinic	Squibb
Tolectin DS capsule	McNeil
Tranxene tablet 15 mg	Abbott
Trecator SC tablet	Wyeth-Ayerst
Trental tablet	Hoechst-Roussel
Tridione capsule 300 mg	Abbott
Trimox oral suspension	Squibb
Trinsicon capsule	Russ
Tuss Ornade spansule	Smith Kline & French
Tylox capsule	McNeil
Urecholine tablet 10 mg	Merck Sharp & Dohme
Valmid capsule	Dista
Velosef capsule 250 mg	Squibb
Zarontin capsule	Parke-Davis

V. ANIMAL TOXICITY DATA

Two chronic feeding studies, in which rats were given varying amounts of erythrosine in food, 0.1, 0.5, 1, or 4%, demonstrated an increased incidence of combined thyroid follicular cell adenomas and carcinomas in male rats fed at the highest level (4%). Results of histopathology examination showed follicular cell adenomas in 20.6% of male rats in the 4% group compared to 1.5% in controls. The Certified Color Manufacturer's Association maintained that these cancers were a secondary effect caused by iodine released by the color additive, resulting in a thyroid hormone inbalance. A Color Additives Review Panel appointed by the FDA concluded that a secondary mechanism was likely, but requested further studies to determine the mechanism of action. This mechanism was not supported by a subsequent short-term study done by the Primate Research Institute.[4]

The WHO has estimated the acceptable daily dietary intake of erythrosine to be up to 600 mcg/kg/d, pending results of pharmacokinetic studies.[5]

VI. HUMAN TOXICITY DATA

A. Thyroid Effects

Several studies have been conducted in human volunteers in order to confirm a secondary mechanism of thyroid carcinomas found in animals. In studies of 35 healthy volunteers, oral doses of up to 200 mg/d resulted in increased serum total iodine and serum protein-bound iodine (PBI), suggesting that erthrosine is absorbed and slowly deiodinated. Basal TSH was significantly increased in the 200 mg/d group. In another study, using 30 or 75 mg of radiolabeled erythrosine found a terminal elimination half-life of the radiolabeled iodine to be 8.4 d. Total absorption was estimated to be 1% or less of the dose. No effects on thyroid hormone or TSH concentrations were demonstrated.[4]

Elevated PBI and decreased radioactive iodine uptake was found in a patient treated with erythrosine-containing lithium carbonate capsules.[6]

B. Hypersensitivity Reactions

Patients with "iodide-bromide intolerance" may develop chronic urticaria, asthma, or rhinitis following exposure to erythrosine, which contains over 50% iodine.[7] In a study of 45 patients with moderately severe perennial asthma, 1 patient was confirmed to develop bronchoconstriction after three separate double-blind challenges with 20 mg of erythrosine over a 7-week period. Repeat testing 2 years later did not result in bronchoconstriction.[8]

Erythrosine is a fluorescein derivative and has been associated with phototoxic reactions, which occurred with concentrations of less than 1 μ mol/l in an *in vitro* model.[9] Photosensitization reactions reported in humans have involved erythroderma, desquamation, and alopecia.[10]

VII. CLINICAL RELEVANCE

A. Thyroid Cancer

The amount of erythrosine in ingested foods and drugs is unlikely to present a significant risk of cancer in humans. The estimated lifetime risk was estimated to be a maximum of 1 in 100,000.[1]

B. Hypersensitivity Reactions

Patients with known hypersensitivity to iodides and bromides should avoid exposure to FD&C Red No. 3. The incidence of hypersensitivity to this dye in asthmatics is lower than that to aspirin and tartrazine, but may be considered if the history implicates foods or drugs known to contain erythrosine.

REFERENCES

1. **Blumenthal, D.,** Red No. 3 and other colorful controversies, *FDA Consumer,* 21, 18, 1990.
2. **Food and Drug Administration,** Termination of provisional listings of FD&C Red No. 3 for use in cosmetics and externally applied drugs and of lakes of FD&C Red No. 3 for all uses, *Fed. Reg.,* 55, 3516, 1990.
3. **Nikitakis, J. M.,** *CTFA Cosmetic Ingredient Handbook,* 1st ed., The Cosmetic, Toiletry and Fragrance Association, Washington, D.C., 1988.
4. **Food and Drug Administration,** Color additives; denial of petition for listing of FD&C Red No. 3 for use in cosmetics and externally applied drugs; withdrawal of petition for use in cosmetics intended for use in the area of the eye, *Fed. Reg.,* 55, 3520, 1990.
5. **WHO,** Thirtieth Report of the Joint FAO/WHO Expert Committee on Food Additives, Tech. Rep. Ser. No. 751, WHO, Geneva, 1987.
6. **Haas, S.,** Contamination of protein-bound iodine by pink gelatin capsules colored with erythrosine, *Ann. Intern. Med.,* 72, 549, 1970.
7. **Taylor, F.,** Iodine, going from hypo to hyper, HHS Publication No. FDA 81-2153, 1981.
8. **Weber, R. W., Hoffman, M., Raine, D. A., Jr., and Nelson, H. S.,** Incidence of bronchoconstriction due to aspirin, azo dyes, non-azo dyes, and preservatives in a population of perennial asthmatics, *J. Allergy Clin. Immunol.,* 64, 32, 1979.
9. **Valenzeno, D. P. and Pooler, J. P.,** Phototoxicity: the neglected factor, *JAMA,* 242, 453, 1979.
10. **Castelain, P. Y. and Piriou, A.,** Photosensitization eczema with positive erythrosine test, *Contact Dermatitis,* 4, 305, 1978.

FD&C Red No. 40

I. REGULATORY CLASSIFICATION

FD&C Red No. 40 is a monoazo dye used as a color additive in foods, drugs, and cosmetics.

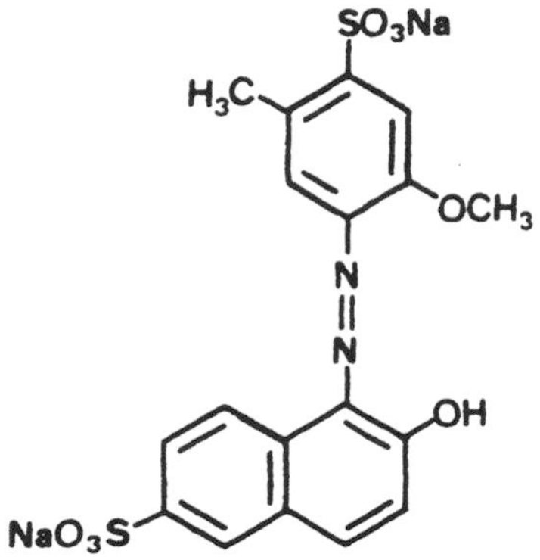

FIGURE 30. FD&C Red No. 40.

II. SYNONYMS

Allura Red AC
C.I. Food Red 17
C.I. 16035

III. AVAILABLE FORMULATIONS

A. Foods

Since decertification of FD&C Red No. 2 in 1976, allura red has become the most widely used certified food coloring. It was petitioned for use and approved in 1971. Types of foods include candy, baked goods, cereals, carbonated beverages, and animal feeds.[1]

B. Drugs

FD&C Red No. 40 was approved for pharmaceutical use in 1971 and is widely used in oral drug products.

C. Cosmetics

FD&C Red No. 40 has been permanently listed for use in cosmetics since 1975. Types of products include shampoos, colognes, toilet waters, creams and lotions, aftershave lotions, bubble baths, mouthwashes, and breath fresheners.[2]

IV. TABLE OF COMMON PRODUCTS

A. Oral Drug Products

Trade name	Manufacturer
Achromycin V oral suspension	Lederle
Allergy-Sinus Comtrex	Bristol-Myers
Alupent syrup	Boehringer Ingelheim
Alu-Cap capsule	3M Riker
Alurate elixir, suspension	Roche
Ambenyl D decongestant cough formula	Forest
Antivert chewable tablet 25 mg	Roerig
Anspor oral suspension 250/5 ml	Smith Kline & French
Aristocort syrup	Lederle
Basaljel	Wyeth-Ayerst
Bayer children's cold tablet	Glenbrook
Benadryl capsule 25 mg	Parke-Davis
Bentyl syrup	Lakeside
Bonine chewable tablet	Pfizer Consumer
Brexin LA	Savage
Caltrate 600 + Vitamin D	Lederle
Caltrate 600 + Iron and Vitamin D	Lederle
Ceclor oral suspension	Lilly
Children's Advil suspension	Whitehall
Children's Chloraseptic lozenges	Richardson-Vicks
Children's Panadol drops, liquid, tablet	Glenbrook
Children's Tylenol liquid cold formula	McNeil Consumer
Chloraseptic cherry liquid	Richardson-Vicks
Chloraseptic cherry aerosol spray	Richardson-Vicks
Chloraseptic cherry lozenge	Richardson-Vicks
Choledyl elixir	Parke-Davis
Cholybar	Parke-Davis
Colace capsule 50, 100 mg	Mead Johnson
Colace syrup	Mead Johnson
Compazine tablet, spansules	Smith Kline & French
Comtrex caplet, tablet, liqui-gel, liquid	Bristol-Myers
Comtrex liquid cough formula	Bristol-Myers
Contac allergy capsule	Smith Kline Beecham
Coumadin tablet 2 mg	DuPont
Creon capsule	Reid-Rowell
Crystodigin tablet 0.1 mg	Lilly

Trade name (cont'd)	Manufacturer (cont'd)
Decadron elixir	Merck Sharp & Dohme
Decadron tablet 1.5 mg	Merck Sharp & Dohme
Declomycin capsule	Lederle
Depakene syrup	Abbott
Depakote tablet 125 mg	Abbott
Dexedrine spansules	Smith Kline & French
DiaBeta tablet 2.5 mg	Hoechst-Roussel
Diflucan tablet	Roerig
Dilantin oral suspension pediatric	Parke-Davis
Dimetane DC cough syrup	Robins
Dimetane DX cough syrup	Robins
Donnatal capsule	Robins
Doral tablet	Wallace
Dramamine liquid	Richardson-Vicks
Duricef capsule	Mead Johnson
Elavil tablet 100 mg	Merck Sharp & Dohme
Elixophyllin elixir	Fisons
Enovid tablet 10 mg	Searle
Entozyme tablet	Robins
Ery-tab 500 mg	Abbott
E.E.S. 200 liquid	Abbott
E.E.S. 400 filmtab	Abbott
Elavil tablet 100 mg	Stuart
Etrafon forte 4-25 tablet	Schering
Eulexin capsule	Schering
Extra Strength Tylenol caplet, gelcap, tab	McNeil Consumer
Feosol capsule	Smith Kline Beecham
Fero-Gradumet	Abbott
Gantanol suspension	Roche
Grifulvin V suspension	Ortho
Grisactin capsule 125, 250 mg	Wyeth-Ayerst
Haldol tablet 20 mg	McNeil
Heptuna Plus	Roerig
Hycodan syrup	DuPont
Hycomine syrup	DuPont
Hycomine compound tablet	DuPont
Hycotuss expectorant	DuPont
Iberet filmtab	Abbott
Iberet-500 filmtab	Abbott
Ilosone chewable tablet 125, 250 mg	Dista
Ilosone suspension	Dista
Inderide capsule 120/50, 160/50	Wyeth-Ayerst
Keflet tablet 250 mg	Dista
Lanoxicaps 0.05 mg	Burroughs Wellcome

Trade name (cont'd)	Manufacturer (cont'd)
Lanoxin tablet 0.5 mg	Burroughs Wellcome
Levsin elixir, drops	Schwarz Pharma
Levsinex timecaps	Schwarz Pharma
Librium capsule 10 mg	Roche
Lithium carbonate capsule	Roxane
Maximum Strength Sine-Aid caplet, tablet	McNeil Consumer
Medipren caplet, tablet	McNeil Consumer
Mestinon syrup	ICN
Metamucil powder strawberry	Procter & Gamble
Methadone oral solution	Roxane
Mexitil capsule 150, 200, 250 mg	Boehringer Ingelheim
Micro-K extencaps, Micro-K 10	Robins
Micronase tablet 2.5 mg	Upjohn
Minipress capsule	Pfizer
Moban tablet 10 mg	DuPont
Modane mild tablet	Adria
Modane tablet	Adria
MSIR oral solution	Purdue Frederick
Naldecon CX liquid adult	Bristol
Naldecon Senior DX cough/cold liquid	Bristol
NegGram suspension	Winthrop
Norlestrin 2.5/50 pink tablet	Parke-Davis
Norlestrin Fe 2.5/50 pink tablet	Parke-Davis
Novahistine DH	Lakeside
Novahistine expectorant	Lakeside
Nucofed expectorant syrup	Beecham
Nucofed pediatric syrup	Beecham
Omnipen oral suspension 125	Wyeth-Ayerst
Organidin tablet	Wallace
Pathocil capsule	Wyeth-Ayerst
Pediacof	Winthrop
Pen Vee K oral solution	Wyeth-Ayerst
Pericolace capsule	Mead Johnson
Peritinic tablet	Lederle
Phenaphen with codeine No. 2, 3	Robins
Phenergan tablet 50 mg	Wyeth-Ayerst
Phillips Laxcaps	Glenbrook
Placidyl capsule 200, 500 mg	Abbott
Polaramine tablet 4, 6 mg	Schering
Poly-vi-flor with iron chewable tablet	Mead Johnson
Premarin tablet 0.625 mg	Wyeth-Ayerst
Premarin with methyltestosterone 1.25	Wyeth-Ayerst
Quinidex extentabs	Robins
Ridaura capsule	Smith Kline & French

Trade name (cont'd)	Manufacturer (cont'd)
Rifadin capsule 150, 300 mg	Merrell Dow
Rifamate capsule	Merrell Dow
Rimactane capsule 300 mg	Ciba
Robitussin A-C syrup	Robins
Robitussin DAC syrup	Robins
Roxicodone oral solution	Roxane
Rynatuss tablet, pediatric suspension	Wallace
Sectral capsule 400 mg	Wyeth-Ayerst
Septra tablet	Burroughs Wellcome
Septra DS tablet	Burroughs Wellcome
Septra suspension, regular and grape	Burroughs Wellcome
Ser-ap-es tablet	Ciba
Serax capsule 15, 30 mg	Wyeth-Ayerst
Serentil tablet 10, 25, 50, 100 mg	Boehringer Ingelheim
Serentil oral concentrate	Boehringer Ingelheim
Sinequan capsules	Roerig
Sinus Excedrin	Bristol-Myers
Soma compound tablet	Wallace
Stelazine tablets	Smith Kline & French
Surmontil capsule 50 mg	Wyeth-Ayerst
Symmetrel capsule	DuPont
Synthroid tablet 75, 125, 200 mcg	Boots-Flint
Tace capsule 25 mg	Merrell Dow
Tagamet tablets	Smith Kline & French
Taractan concentrate	Roche
Tegretol tablet 200 mg	Geigy
Teldrin capsule	Smith Kline Beecham
Temaril syrup	Herbert
Temaril spansule	Herbert
Tenex tablet 1 mg	Robins
Trilisate tablet 1000 mg	Purdue Frederick
Trinsicon capsule	Russ
Tri-vi-flor 1 mg chewable tablet	Mead Johnson
Tussi-Organidin liquid	Wallace
Tussi-Organidin DM liquid	Wallace
Tylenol Maximum Strength Sinus caplet	McNeil Consumer
Tylox capsule	McNeil
Ultracef oral suspension	Bristol
Unipen oral suspension	Wyeth-Ayerst
Urecholine tablet 10 mg	Merck Sharp & Dohme
Veetids oral solution	Squibb
Velosef oral suspension 250 mg/5 ml	Squibb
Verelan capsule	Wyeth-Ayerst
Vicks Children's Nyquil	Richardson-Vicks

Trade name (cont'd)	Manufacturer (cont'd)
Vicks Pediatric Formula 44 syrups	Richardson-Vicks
Vicon forte capsule	Russ
Winstrol tablet	Winthrop
Wymox oral suspension	Wyeth-Ayerst
Zarontin syrup	Parke-Davis

V. ANIMAL TOXICITY DATA

The acceptable daily dietary intake of FD&C Red No. 40 established by the WHO is 7 mg/kg/d, based on animal data.[3] The maximum anticipated daily consumption from food sources is estimated to be 0.18 mg/kg/d.[4]

No adverse teratogenic or developmental effects were observed when FD&C Red No. 40 was given by gavage in doses of up to 1 g/kg/d to pregnant rats.[1]

Lifetime feeding studies in rats, using dietary concentrations of 0.37, 1.39, or 5.19%, resulted in a red tint in fur and feces and decreased body weight. The incidence of neoplasms was not significantly different in the treatment group compared to controls. A no-observable-adverse-effect level of 2829 mg/kg/d for male rats and 901 mg/kg/d for female rats was established.[4]

FD&C Red No. 40 was highly comedogenic in the rabbit ear model with a rating of 4 on a scale of 0 to 5. This and other red pigments may explain the severity of cosmetic acne on the upper cheekbones where blush is applied.[5]

VI. HUMAN TOXICITY DATA

A. Hypersensitivity Reactions

FD&C Red No. 40 was negative in patch tests and photosensitization tests in human volunteers.[4]

VII. CLINICAL RELEVANCE

FD&C Red No. 40 has not been associated directly with any observable adverse reaction in humans.

REFERENCES

1. **Collins, T. F. X., Black, T. N., Welsh, J. J., and Brown, L. H.,** Study of the teratogenic potential of FD & C Red No. 40 when given by gavage to rats, *Food Chem. Toxicol.,* 27, 707, 1989.
2. **Nikitakis, J. M.,** *CTFA Cosmetic Ingredient Handbook,* 1st ed., The Cosmetic, Toiletry and Fragrance Association, Washington, D.C., 1988.
3. **WHO,** Twenty-fifth Report of Joint FAO/WHO Expert Committee on Food Additives, Tech. Rep. Ser. No. 669, WHO, Geneva, 1981.
4. **Borzelleca, J. F., Olson, J. W., and Reno, F. E.,** Lifetime toxicity/carcinogenicity study of FD & C Red No. 40 (allura red) in sprague-dawley rats, *Food Chem. Toxicol.,* 27, 701, 1989.
5. **Fulton, J. E., Jr., Pay, S. R., and Fulton, J. E., III,** Comedogenicity of current therapeutic products, cosmetics, and ingredients in the rabbit ear, *J. Am. Acad. Dermatol.,* 10, 96, 1984.

FD&C YELLOW NO. 5

I. REGULATORY CLASSIFICATION

FD&C Yellow No. 5 is an orange-yellow monoazo dye used as a color additive in drugs. Labeling of tartrazine has been required on all prescription and over-the-counter oral, rectal, vaginal, and nasal pharmaceuticals since 1980.

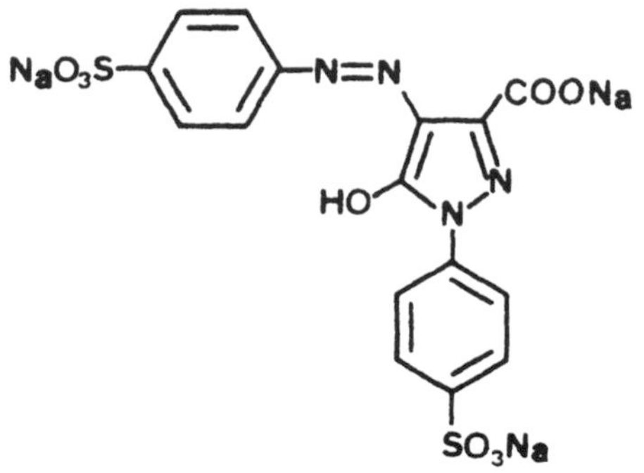

FIGURE 31. FD&C Yellow No. 5.

II. SYNONYMS

Tartrazine
E102
C.I. Food Yellow 4
C.I. 19140

III. AVAILABLE FORMULATIONS

A. Drugs

The prevalence of FD&C Yellow No. 5 in pharmaceuticals has decreased markedly since requirement of labeling with a warning statement. In 1990, there were 244 NDA-approved oral pharmaceuticals listing tartrazine as an ingredient, compared to 932 oral products containing the related dye FD&C Yellow No. 6.[1] Tartrazine is infrequently used in topical preparations; only three topical products are listed with the FDA.

169

B. Foods that May Contain Azo Dyes[2,3]
1. Candy

Butterscotch candy
Candy corn (Brach's)
Caramels
Life Savers
Fruit drops (lemon)
Fruit chews (Skittles)
Filled chocolates

2. Beverages

Soft drinks
Fruit drinks
Lemonade
Orange drinks (Tang, Daybreak, Awake)
Gatorade (lime flavored)

3. Desserts

Butterscotch puddings
Chocolate puddings
Caramel custard
Dessert whips and sauces
Flavored gelatin
Ice cream
Sherbert

4. Baked Goods

Crackers
Cheese puffs
Cheez curls, balls (Planters)
Chips
Cookie and cake mixes (Duncan Hines, Pillsbury)
Cake icing (Cake Mate)
Waffle and pancake mixes
Macaroni and spaghetti
Macaronic and cheese dinner (Kraft)
Cereals

5. Condiments

Mayonnaise
Salad dressings
Golden Blend Italian dressing (Kraft)

Zesty Italian reduced-calorie dressing (Kraft)
Catsup
Mustard
Bearnaise sauce
Hollandaise sauces
Other sauces (curry, fish, onion, tomato, white cream)
Imitation butter flavoring (McCormick)
Imitation banana extract (McCormick)
Imitation pineapple extract (McCormick)
Seasoning salt (French's)

6. Seafood

Canned anchovies
Canned herring
Canned sardines
Canned fish balls
Canned caviar

7. Soups

Packaged soups
Canned soups

III. TABLE OF COMMON PRODUCTS

A. Oral Drug Products

Trade name	Manufacturer
Apresoline tablet 100 mg	Ciba
Bucladin S softab	Stuart
Butisol sodium tablet 30, 50 mg	Wallace
Butisol sodium elixir	Wallace
Choloxin tablet 2, 6 mg	Boots-Flint
Cleocin capsule 75, 150 mg	Upjohn
Deprol tablet	Wallace
Desoxyn gradumet tablet 15 mg	Abbott
Dexedrine spansule, tablet, elixir	Smith Kline & French
Didrex tablet 25 mg	Upjohn
Estinyl tablet 0.02 mg	Schering
Estrace tablet 2 mg	Mead Johnson
Euthroid tablet 1/2, 1, 3	Parke-Davis
Exna tablet	Robins
Haldol tablet 1, 5, 10 mg	McNeil
Halotestin tablets	Upjohn
Hiprex tablet	Merrell Dow

Trade name (cont'd)	Manufacturer (cont'd)
Janimine tablet 10, 25 mg	Abbott
Kaochlor 10% liquid	Adria
Kaon-Cl tablet	Adria
Lithane tablet	Miles
Metandren linguet 10 mg	Ciba
Nembutal sodium capsule 100 mg	Abbott
Nicolar tablet	Rhone-Poulenc Rorer
Norzine tablet 10 mg	Purdue Frederick
Ovrette tablet	Wyeth-Ayerst
Panwarfin tablet 7.5 mg	Abbott
Paradione capsule 300 mg	Abbott
Pentids syrup	Squibb
Pentids 800 tablet	Squibb
Placidyl capsule 750 mg	Abbott
Preludin tablet	Boehringer Ingelheim
Prolixin tablet 2.5, 5, 10 mg	Princeton
Pronestyl tablet	Princeton
Raudixin tablet	Princeton
Rauzide tablet	Princeton
Rynatuss pediatric suspension	Wallace
Sansert tablet	Roxane
Serax tablet	Wyeth-Ayerst
Serpasil-Apresoline tablet	Ciba
Serpasil-Esidrix tablet	Ciba
Synthroid tablet 100, 300 mcg	Boots-Flint
Tace capsule 12, 25, 72 mg	Merrell Dow
Taractan tablet 10, 25, 50, 100 mg	Robins
Theragran hematinic	Squibb
Torecan tablet 10 mg	Roxane
Trisoralen tablet	ICN
Vari-Flavors (lemon flavor)	Ross
Veetids oral solution 125	Squibb
Vontrol tablet	Smith Kline & French

V. ANIMAL TOXICITY DATA

Chronic dietary toxicity studies in rats did not demonstrate any adverse effects of tartrazine with a no-effect level of 5% in the diet, providing 1641 mg/kg/d for males and 3348 mg/kg/d for females.[4] A similar study in mice demonstrated a no-effect level of 8103 mg/kg/d in males and 9735 mg/kg/d in females.[5]

VI. HUMAN TOXICITY DATA

A. Asthma

Since the report in 1967 of 50% cross-idiosyncrasy between aspirin and tartrazine intolerance in 80 patients,[6] most studies investigating the potential for tartrazine to exacerbate asthma have selected patients based on this presumed relationship. The incidence of tartrazine intolerance in the absence of aspirin intolerance has been less frequently studied.

In unselected asthmatic populations, the incidence of tartrazine intolerance, documented by a 20% or greater fall in forced expiratory volume in 1 s, ranged from 4 to 22% in two studies.[7,8] Both of these study protocols included withholding of bronchodilator medication prior to testing, which has been demonstrated to influence the response to tartrazine challenge.

In a study of 44 asthmatics, 16% had positive tartrazine challenges when morning bronchodilators were withheld, compared to none when the same patients were challenged 1 week later without interrruption of therapy.[9] A subsequent study performed in 26 aspirin-tolerant patients tested while continuing bronchodilator therapy showed a borderline decrease in FEV_1 (20.4%) with the highest dose of tartrazine (15 mg) in 1 patient; the asthma worsened during an elimination diet for an overall incidence of 3.8%.[10]

B. Aspirin-Sensitive Patients

Early estimates of the incidence of cross-intolerance to tartrazine in asthmatics with aspirin intolerance were in the range of 8 to 50%.[6,11,12] Studies showing the highest incidence were generally nonplacebo controlled with subjective criteria for positive responses.[13] More recent placebo-controlled double-blind studies involving a total of 490 patients with aspirin-induced asthma found incidences of tartrazine intolerance of 0 to 2.4%.[13-16]

Tartrazine was originally thought to have a similar mechanism as aspirin, the inhibition of cyclooxygenase, leading to a shunt favoring leukotriene production. This mechanism has been demonstrated not to occur with tartrazine in studies using supraphysiological levels of arachinoic acid.[17,18] These studies have been questioned by additional work showing significant inhibition of thromboxane B2 formation by 10 mM of tartrazine in platelets activated by physiologic levels of arachidonic acid generated via noradrenaline stimulation.[19]

C. Urticaria

The first reported cases of tartrazine-induced urticaria in 1959 has also been credited with being the first published adverse reactions associated with a food additive. Three patients were described to develop generalized urticaria, with angioedema in one case, following ingestion of tartrazine-containing corticosteroids. Unblinded challenges with a sublingual tartrazine solution reproduced the reactions in the two patients tested.[20]

Tartrazine is clearly a cause of nonimmunologic urticaria in some patients. While not confirmed by double-blind challenge, a 50-year-old woman developed urticaria following ingestion of tartrazine-containing estradiol tablets, which resolved when substituted for a product that did not contain this dye. The reaction recurred upon deliberate open rechallenge and again after inadvertent ingestion of a relish containing tartrazine.[21]

Studies in series of adults and children given tartrazine in double-blind challenges have confirmed a relationship to this dye in some cases of chronic urticaria or angioedema. In a study of 43 children with chronic urticaria and angioedema that resolved following a 1-month trial with an additive-free diet, 25.6% experienced a recurrence of urticaria when challenged with 0.1, 0.5, and 1 mg of tartrazine.[22]

In a series of 70 patients with chronic urticaria or angioedema with suspected food intolerance placed on an oligoallergic diet for 4 weeks, double-blind challenge with tartrazine 1 and 10 mg produced positive immediate reactions in six patients and delayed reactions in three patients for a total incidence of 12.8%.[23]

A study of 24 adults with chronic urticaria revealed only one positive response to oral challenge with tartrazine 25 or 50 mg. These patients were tested during an active phase of the disease, with attempts to quantify an increase in affected skin. The lower incidence in this study may be related to difficulties in assessing exacerbation of active lesions.[16]

Ingestion of tartrazine may exacerbate atopic dermatitis by induction of erythema and urticaria. In a series of 24 children with severe atopic dermatitis, double-blind placebo-controlled challenge with tartrazine 0.1 mg produced pruritus and skin redness within 10 min in 2 children; reactions resembled immediate mast cell degranulation.[24]

D. Hyperkinesis

The Feingold theory, which related hyperkinesis and learning disabilities to a variety of natural salicylates and artificial food colors and flavorings, was widely publicized in the 1970s, resulting in many children being placed on an additive-free diet.[25] Many attempts to confirm this hypothesis using double-blind studies with tartrazine challenges have failed to demonstrate a relationship of tartrazine to behavior.[26-30] In one study of 24 children with a parental history of acute behavioral reactions to tartrazine, 22 were able to return to a normal diet after tartrazine challenges failed to induce behavioral changes in a controlled inpatient setting.[31]

E. Eosinophilia

An anecdotal report described a patient who developed intense eosinophilia (38%) several weeks after the addition of a green food dye containing tartrazine and sulfiting agents to tube feedings. Discontinuation of the dye resulted in normalization of the eosinophil count in 5 d which rebounded to 19% 24 h after rechallenge. No attempt was made to investigate which component of the food dye was causative. Subsequently, another patient was identified to have eosinophilia (8%) after tube feedings containing the same dye with discontinuation again resulting in normalization; rechallenge was not performed.[32]

Eosinophilia (9%) was also described in a 50-year-old woman with tartrazine-induced urticaria provoked by ingestion of estradiol tablets.[21]

Application of tartrazine to abraded skin demonstrated local eosinophilia of 8% in a confirmed tartrazine-sensitive patient with angioedema and asthma, which could be passively transferred by the patient's serum to a nonallergic volunteer.[33]

F. Anaphylactoid Reactions

Several cases of anaphylactoid reactions have been reported associated with tartrazine.[34-36]

G. Systemic Lupus Erythematosus

One clinician has suggested a relationship between tartrazine and systemic lupus erythematosus, which was attributed to the metabolite sulfophenylhydrazine.[37] This relationship has not been proven in controlled studies.

H. Purpura

Tartrazine has been reported to cause allergic vasculitis with purpura in eight patients; generalized symptoms of fever, malaise, abdominal pain, and arthralgia were described in some patients.[38-40] Deposits of the C3 component of complement, IgA, and IgM were reported in the vessel walls in one case.[41]

I. Contact Dermatitis

Tartrazine has been rarely reported to cause contact dermatitis. One case was attributed to yellow dye in a shirt and was confirmed by positive patch test to tartrazine.[42] Another case involved a 40-year-old woman experiencing dermatitis at sites of application of an ECG paste containing tartrazine; a positive patch test to tartrazine was shown.[43]

J. Orofacial Granulomatosis

Symptoms of recurrent upper lip and gum swelling and a fissured tongue "Melkersson-Rosenthal syndrome" were found to be triggered by sodium benzoate 50 mg and tartrazine 5 mg during oral double-blind challenges in a 34-year-old man. An elimination diet excluding these two additives resulted in complete remission lasting at least 1 year.[44]

K. Mechanism

Several mechanisms have been postulated, most concluding that one or more dose-related nonimmunologic responses are most likely to be responsible for clinical intolerance. The most convincing evidence suggests that tartrazine induces histamine release from mast cells in a dose-related manner. A double-blind study in ten healthy adults with no history of food intolerance showed an increase in the mean plasma histamine levels in nine of ten subjects from 0.27 to 0.92 ng/ml after ingestion of 150 mg of tartrazine, peaking at 30 to 200 min postingestion. No increase in histamine was seen after ingestion of placebo or lower doses of tartrazine, 5 or 50 mg. *In vitro* release of histamine from leukocytes was only demonstrated in one subject.[45]

Histamine release from gastric mucosa mast cells was demonstrated following administration of 10 mg of tartrazine under endoscopic control in three patients with a history of food additive intolerance with positive open oral tartrazine challenges. Symptoms related to tartrazine in these patients were migraine, atopic eczema, and stomatitis.[46]

A dose-related activation of the C3 component of complement was demonstrated in one patient,[47] and C3 deposition in vessel walls was documented in a case of tartrazine-induced allergic vasculitis.[41]

IgE-related immune responses have generally been excluded when investigated in tartrazine-sensitive individuals.[48] A report of specific IgE antibodies to tartrazine in five patients with a history of immediate hypersensitivity reactions was not confirmed by double-blind oral challenge.[49]

VII. CLINICAL RELEVANCE

A. Dermatitis/Urticaria

Tartrazine is a well-known cause of urticaria, presumably via a nonimmunologic degranulation of mast cells, with release of histamine and other mediators.[50] It has been estimated that urticarial-producing food additives, such as tartrazine or benzoic acid, will exacerbate dermatitis in at least 5% of children with chronic atopic dermatitis.[51]

Studies evaluating patients with chronic urticaria or angioedema in a quiescent phase with suspected food intolerance have demonstrated an incidence of immediate or delayed urticarial responses following double-blind tartrazine challenge of 8 to 34%.[52-56] Some of these reactions may have been due to a rebound phenomenon or breakthrough of urticaria following discontinuation of long-term antihistamine therapy prior to the tartrazine challenges.

The prevalence of tartrazine intolerance in the overall population is more difficult to estimate. Extrapolation of data collected in series of selected high-risk populations, such as urticaria, angioedema, asthma, and rhinitis, adjusted for prevalence of those patients in the general population, have suggested an overall incidence of around 0.01 to 0.6%.[57,58] A higher incidence of intolerance was suggested in patients treated with tricyclic antidepressants, which often contain tartrazine, in a retrospective review of 170 patients. Although five reported urticaria (2.9%), not all of the offending antidepressants were confirmed to contain tartrazine, and double-blind challenges were not done.[59]

Tartrazine-induced urticaria may often be a transient phenomenon, failing to be subsequently reproduced even in patients with positive double-blind challenges.[50,55,60] A cohort of 12 patients with tartrazine-exacerbated urticaria were retested at a later time. The two patients with active chronic urticaria were still reactive to tartrazine. Three patients with recent clearing of urticaria (3 to 8 weeks) reacted to a 10 mg dose, but not to a 1 mg dose. The remaining 7 patients had been clear of all but minor episodic urticaria for 1.5 to 7 years and had no reaction to either dose of tartrazine.[56]

B. Asthma

The long-held assertion that a substantial number of patients with aspirin-exacerbated asthma are also sensitive to tartrazine has not been substantiated by well-controlled double-blind studies using objective acceptable criteria for documentation of clinical deterioration. The true incidence of cross-intolerance is probably less than 3%.

Aspirin-sensitive asthmatics may have more clinically labile disease and are therefore more likely to exhibit false-positive tartrazine challenges, especially when withdrawn from bronchodilators prior to testing. Because of the ubiquitous presence of tartrazine in foods, dietary avoidance is difficult and should be reserved for those patients with documented double-blind challenges on at least two occasions.

REFERENCES

1. **Food and Drug Administration,** *Inactive Ingredients Guide,* FDA, Washington, D.C., March 1990.
2. **Michaelsson, G. and Juhlin, L.,** Urticaria induced by preservatives and dye additives in food and drugs, *Br. J. Dermatol.,* 88, 525, 1973.
3. **Ted Tse, C. S.,** Food products containing tartrazine, *N. Engl. J. Med.,* 306, 681, 1982.
4. **Borzelleca, J. F. and Hallagan, J. B.,** Chronic toxicity/carcinogenicity studies of FD&C Yellow No. 5 (tartrazine) in rats, *Food Chem. Toxicol.,* 26, 179, 1988.
5. **Borzelleca, J. F. and Hallagan, J. B.,** Chronic toxicity/carcinogenicity studies of FD&C Yellow No. 5 (tartrazine) in mice, *Food Chem. Toxicol.,* 26, 189, 1988.
6. **Samter, M. and Beers, R. F.,** Concerning the nature of the intolerance to aspirin, *J. Allergy,* 40, 281, 1967.
7. **Stenius, B. S. M. and Lemola, M.,** Hypersensitivity to acetylsalicylic acid (ASA) and tartrazine in patients with asthma, *Clin. Allergy,* 6, 119, 1976.
8. **Spector, S. L., Wangaard, C. H., and Farr, R. S.,** Aspirin and concomitant idiosyncrasies in adult asthmatic patients, *J. Allergy Clin. Immunol.,* 64, 500, 1979.
9. **Weber, R. W., Hoffman, M., Raine, D. A., and Nelson, H. S.,** Incidence of bronchoconstriction due to aspirin, azo dyes, non-azo dyes and preservatives in a population of perennial asthmatics, *J. Allergy Clin. Immunol.,* 64, 32, 1979.
10. **Tarlo, S. M. and Broder, I.,** Tartrazine and benzoate challenge and dietary avoidance in chronic asthma, *Clin. Allergy,* 12, 303, 1982.
11. **Juhlin, L., Michaelsson, G., and Zetterstrom, O.,** Urticaria and asthma induced by food and drug additives in patients with aspirin hypersensitivity, *J. Allergy Clin. Immunol.,* 50, 92, 1972.
12. **Miller, K.,** Sensitivity to tartrazine, *Br. Med. J.,* 285, 1597, 1982.
13. **Simon, R. A.,** Adverse reactions to drug additives, *J. Allergy Clin. Immunol.,* 74, 623, 1984.

14. **Virchow, C., Szczeklik, A., Bianco, S., et al.,** Intolerance to tartrazine in aspirin-induced asthma: results of a multicenter study, *Respiration*, 53, 20, 1988.

15. **Morales, M. C., Basomba, A., Pelaez, A., et al.,** Challenge tests with tartrazine in patients with asthma associated with intolerance to analgesics (ASA-triad), *Clin. Allergy*, 15, 55, 1985.

16. **Stevenson, D. D., Simon, R. A., Lumry, W. R., and Mathison, D. A.,** Adverse reactions to tartrazine, *J. Allergy Clin. Immunol.*, 78, 182, 1986.

17. **Gerber, J. G., Payne, N. A., Oelz, O., Nies, A. S., and Oates, J. A.,** Tartrazine and the prostaglandin system, *J. Allergy Clin. Immunol.*, 63, 289, 1979.

18. **Vargaftig, B. B., Bessot, J. C., and Pauli, G.,** Is tartrazine-induced asthma related to inhibition of prostaglandin biosynthesis?, *Respiration*, 39, 276, 1980.

19. **Williams, W. R., Pawlowicz, A., and Davies, B. H.,** Aspirin-like effects of selected food additives and industrial sensitizing agents, *Clin. Exp. Allergy*, 19, 533, 1989.

20. **Lockey, S. D.,** Allergic reactions due to F.D. and C. yellow No. 5 tartrazine, an aniline dye used as a coloring agent in various steroids, *Ann. Allergy*, 17, 719, 1959.

21. **Baumgardner, D. J.,** Persistent urticaria caused by a common coloring agent, *Postgrad. Med.*, 85, 265, 1989.

22. **Supramaniam, G. and Warner, J. O.,** Artificial food additive intolerance in patients with angio-oedema and urticaria, *Lancet*, 2, 907, 1986.

23. **Ortolani, C., Pastorello, E., Luraghi, M. T., Della Torre, F., Bellani, M., and Zanussi, C.,** Diagnosis of intolerance to food additives, *Ann. Allergy*, 53, 587, 1984.

24. **Van Bever, H. P., Docx, M., and Stevens, W. J.,** Food and food additives in severe atopic dermatitis, *Allergy*, 44, 588, 1989.

25. **Feingold, B. F.,** Hyperkinesis and learning disabilities linked to artificial food flavors and colors, *Am. J. Nurs.*, 75, 801, 1975.

26. **Goyette, C. H., Connors, C. K., Petti, T. A., and Curtis, L. E.,** Effects of artificial colors on hyperkinetic children: a double-blind challenge study, *Psychopharmacol. Bull.*, 14, 39, 1978.

27. **Levy, F., Dumbrell, S., Hobbes, G., Ryan, M., Wilton, N., and Woodhill, J. M.,** Hyperkinesis and diet: a double-blind crossover trial with a tartrazine challenge, *Med. J. Aust.*, 1, 61, 1978.

28. **Weiss, B., Williams, J. H., Margen, S., et al.,** Behavioural responses to artificial food colors, *Science*, 207, 1487, 1980.

29. **Mattes, J. A. and Gittelman, R.,** Effects of artificial food colorings in children with hyperactive symptoms, *Arch. Gen. Psychiatry*, 38, 714, 1981.

30. **Thorley, G.,** Pilot study to assess behavioural and cognitive effects of artificial food colours in a group of retarded children, *Dev. Med. Child Neurol.*, 26, 56, 1984.

31. **David, T. J.,** Reactions to dietary tartrazine, *Arch. Dis. Child.*, 62, 119, 1987.

32. **Bell, R. T. and Fishman, S.,** Eosinophilia from food dye added to enteral feedings, *N. Engl. J. Med.*, 322, 1822, 1990.

33. **Chafee, F. H. and Settipane, G. A.,** Asthma caused by FD&C approved dyes, *J. Allergy*, 40, 65, 1967.

34. **Desmond, R. E. and Trautlein, J. J.,** Tartrazine (FD&C Yellow #5) anaphylaxis: a case report, *Ann. Allergy*, 46, 81, 1981.

35. **Zlotlow, M. J. and Settipane, G. A.,** Allergic potential of food additives: a report of a case of tartrazine sensitivity without aspirin intolerance, *Am. J. Clin. Nutr.*, 30, 1023, 1977.

36. **Trautlein, J. J. and Mann, W. J.,** Anaphylactic shock caused by yellow dye (FD&C #5 and FD&C #6) in an enema (case report), *Ann. Allergy*, 41, 28, 1978.

37. **Pereyo, N.,** Hydrazine derivatives and induction of systemic lupus erythematosus, *J. Am. Acad. Dermatol.*, 14, 514, 1986.

38. **Criep, L. H.,** Allergic vascular purpura, *J. Allergy Clin. Immunol.*, 48, 7, 1971.

39. **Michaelsson, G., Pettersson, L., and Juhlin, L.,** Purpura caused by food and drug additives, *Arch. Dermatol.*, 109, 49, 1974.

40. **Kubba, R. and Champion, R. H.,** Anaphylactoid purpura caused by tartrazine and benzoates, *Br. J. Dermatol.*, 93 (Suppl. 2), 61, 1975.

41. **Parodi, G., Parodi, A., and Rebora, A.,** Purpuric vasculitis due to tartrazine, *Dermatologica*, 171, 62, 1985.

42. **Roeleveld, C. G. and van Ketel, W. G.,** Positive patch test to the azo-dye tartrazine, *Contact Dermatitis*, 2, 180, 1976.

43. **Fisher, A. A.,** *Contact Dermatitis*, 3rd ed., Lea & Febiger, Philadelphia, 1986.

44. **Pachor, M. L., Urbani, G., Cortina, P., Lunardi, C., Nicolis, F., Peroli, P., Coorrocher, R., and Gotte, P.,** Is the Melkersson-Rosenthal syndrome related to the exposure to food additives?, *Oral Surg. Oral Med. Oral Pathol.,* 67, 393, 1989.

45. **Murdoch, R. D., Pollock, I., and Naeem, S.,** Tartrazine induced histamine release in vivo in normal subjects, *J. R. Coll. Physicians London,* 21, 257, 1987.

46. **Schaubschlager, W. W., Zabel, P., and Schlaak, M.,** Tartrazine-induced histamine release from gastric mucosa, *Lancet,* 2, 800, 1987.

47. **Voigtlander, V., Moll, I., and Stach, C.,** Intolerance to aspirin and tartrazine: evidence of complement activation in a patient following oral challenge with tartrazine, *Arch. Dermatol. Res.,* 274, 359, 1982.

48. **Weltman, J. K., Szaro, R. P., and Settipane, G. A.,** An analysis of the role of IgE in intolerance to aspirin and tartrazine, *Allergy,* 34, 273, 1978.

49. **Weiner, M. and Bernstein, I. L.,** *Adverse Reactions to Drug Formulation Agents,* Marcel Dekker, New York, 1989.

50. **Murdoch, R. D., Pollock, I., Young, E., and Lessof, M. H.,** Food additive-induced urticaria: studies of mediator release during provocation tests, *J. R. Coll. Physicians London,* 21, 262, 1987.

51. **David, T. J.,** Food additives, *Arch. Dis. Child.,* 63, 582, 1988.

52. **Settipane, G. A., Chafee, F. H., Postman, M., and Levine, M. I.,** Significance of tartrazine sensitivity in chronic urticaria of unknown etiology, *J. Allergy Clin. Immunol.,* 57, 541, 1976.

53. **Mikkelson, H., Larsen, J. C., and Tarding, F.,** *Arch. Toxicol.,* 1 (Suppl.), 141, 1978.

54. **Juhlin, L.,** Recurrent urticaria: clinical investigation of 330 patients, *Br. J. Dermaol.,* 104, 369, 1981.

55. **Gibson, A. and Clancy, R.,** Management of chronic idiopathic urticaria by the identification and exclusion of dietary factors, *Clin. Allergy,* 10, 699, 1980.

56. **Warin, R. P. and Smith, R. J.,** Role of tartrazine in chronic urticaria, *Br. Med. J.,* 284, 1443, 1982.

57. **Juhlin, L.,** Incidence of intolerance to food additives, *Int. J. Dermatol.,* 19, 548, 1980.

58. **Young, E., Patel, S., Stoneham, M., Rona, R., and Wilkinson, J. D.,** The prevalence of reaction to food additives in a survey population, *J. R. Coll. Physicians London,* 21, 241, 1987.

59. **Pohl, R., Balon, R., Berchou, R., and Yeragani, V. K.,** Allergy to tartrazine in antidepressants, *Am. J. Psychiatry,* 144, 237, 1987.

60. **Pollock, I. and Warner, J. O.,** A follow-up study of childhood food additive intolerance, *J. R. Coll. Physicians London,* 21, 248, 1987.

FD&C Yellow No. 6

I. REGULATORY CLASSIFICATION

FD&C Yellow No. 6 is a monoazo certified dye approved for use as a color additive since 1986 in foods, drugs, and cosmetics.

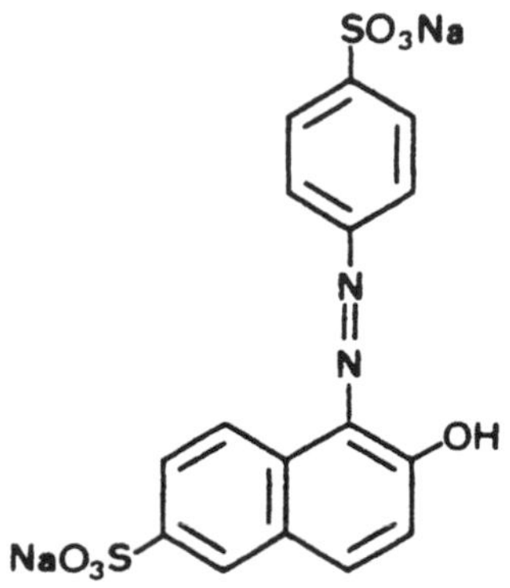

FIGURE 32. FD&C Yellow No. 6.

II. SYNONYMS

C.I. Food Yellow 3
C.I. 15985
E110
Orange Yellow S
Sunset yellow FCF

III. AVAILABLE FORMULATIONS

A. Foods
Sunset yellow is the third most commonly used artificial food coloring.[1]

B. Cosmetics

Sunset yellow is commonly used in cosmetics, including shampoos, colognes, toilet waters, perfumes, creams and lotions, aftershave lotions, bubble baths, hair conditioners and rinses, and toothpastes.[2]

C. Foods that May Contain Azo Dyes[3]

1. Candy

> Caramels
> Life Savers
> Fruit drops
> Filled chocolates

2. Beverages

> Soft drinks
> Fruit drinks
> Lemonade

3. Desserts

> Butterscotch puddings
> Chocolate puddings
> Caramel custard
> Dessert whips and sauces
> Flavored gelatin

4. Baked Goods

> Crackers
> Cheese puffs
> Chips
> Cookie and cake mixes
> Waffle and pancake mixes
> Macaroni and spaghetti
> Cereals

5. Condiments

> Mayonnaise
> Salad dressings
> Catsup
> Mustard
> Bearnaise sauce
> Hollandaise sauces
> Other sauces (curry, fish, onion, tomato, white cream)

6. Seafood

Canned anchovies
Canned herring
Canned sardines
Canned fish balls
Canned caviar

7. Soups

Packaged soups
Canned soups

IV. TABLE OF COMMON DRUG PRODUCTS (INCLUDES LAKES)

A. Oral Drug Products

Trade name	Manufacturer
Achromycin V capsule	Lederle
Accutane capsule 40 mg	Roche
Adalat capsule 10, 20 mg	Miles
Aldoclor tablet 125, 250 mg	Merck Sharp & Dohme
Alu-tab tablet	3M Riker
Alu-cap capsule	3M Riker
Alurate elixir	Roche
Amen tablet	Carnrick
Ancoban capsule	Roche
Anexsia 7.5/650	Beecham
Anspor capsules 250, 500 mg	Smith Kline & French
Anspor oral suspension 125 mg/5 ml	Smith Kline & French
Antivert tablet 25 mg	Roerig
Apresoline tablet 100 mg	Ciba
Apresoline-Esidrix tablet	Ciba
Aralen with Primaquine tablet	Winthrop
Asendin tablet 50, 150 mg	Lederle
Atromid S capsule	Wyeth-Ayerst
Bactrim tablet, suspension	Roche
Bancap HC capsule	Forest
Basaljel	Wyeth-Ayerst
Bayer Children's Cold tablet	Glenbrook
Beelith tablet	Beach
Bentyl syrup	Lakeside
Berocca tablet	Roche
Bontril slow-release capsule	Carnrick
Bromfed capsule	Muro
Bronkodyl capsule 100, 200 mg	Winthrop

Trade name (cont'd)	Manufacturer (cont'd)
Butisol sodium tablet 100 mg	Wallace
Cafergot PB tablet	Sandoz
Calcidrine syrup	Abbott
Caltrate 600 + Vitamin D	Lederle
Caltrate Jr. chewable tablet	Lederle
Capozide tablet	Squibb
Cardizem tablet 60, 120 mg	Marion
Catapres tablet 0.1, 0.2, 0.3 mg	Boehringer Ingelheim
Celontin capsule 150, 300 mg	Parke-Davis
Centrax capsule 5 mg	Parke-Davis
Cephulac syrup	Merrell Dow
Cerose DM	Wyeth-Ayerst
Chloraseptic menthol lozenge	Richardson-Vicks
Choledyl tablet 100 mg	Parke-Davis
Choloxin tablet 1 mg	Boots-Flint
Chronulac syrup	Merrell Dow
Cinobac capsule	Dista
Colace capsule 100 mg	Mead Johnson
Comhist LA capsule	Norwich Eaton
Compazine tablet, spansule	Smith Kline & French
Contac Allergy capsule	Smith Kline Beecham
Coumadin tablet 5, 7.5 mg	DuPont
Cylert tablet 37.5 mg	Abbott
Cylert chewable tablet 37.5 mg	Abbott
Dalmane capsule 15, 30 mg	Roche
Danocrine capsule 100 mg	Winthrop
Dantrium capsule 25, 50, 100 mg	Norwich Eaton
Darbid tablet	Smith Kline & French
Darvocet N-50 tablet	Lilly
Darvocet N-100 tablet	Lilly
Darvon-N suspension	Lilly
Darvon-N with A.S.A. tablet	Lilly
Darvon compound pulvule	Lilly
Darvon compound 65 pulvule	Lilly
Darvon pulvule with A.S.A.	Lilly
Decadron tablet 0.25, 0.5 mg	Merck Sharp & Dohme
Declomycin capsule	Lederle
Deconamine SR capsule	Berlex
Delsym liquid	McNeil Consumer
Deltasone tablet 20 mg	Upjohn
Demulen 1/50 pink tablet	Searle
Depakene capsule 250 mg	Abbott
Depakote tablet 250 mg	Abbott
Desoxyn tablet 10 mg gradumet	Abbott

Trade name (cont'd)	Manufacturer (cont'd)
Desyrel tablet 50, 150 mg	Mead Johnson
Dexedrine tablet, spansule, elixir	Smith Kline & French
Dibenzyline capsule	Smith Kline & French
Didrex tablet 50 mg	Upjohn
Diethylstilbestrol enseals	Lilly
Dilantin capsule 100 mg	Parke-Davis
Dilantin infatabs	Parke-Davis
Dilantin oral suspension 125 mg/5 ml	Parke-Davis
Dilaudid tablet 2 mg	Key
Dimetane DX cough syrup	Robins
Diulo tablet 10 mg	Schiapparelli Searle
Dolobid tablet	Merck Sharp & Dohme
Donnatal capsule	Robins
Donnazyme tablet	Robins
Doral tablet 7.5, 15 mg	Baker Cummins
Doryx capsule	Parke-Davis
Dramamine chewable tablet	Richardson-Vicks
Dristan tablet, caplet	Whitehall
Dristan maximum strength	Whitehall
Duricef oral suspension	Mead Johnson
Dyazide capsule	Smith Kline & French
Dyrenium capsule	Smith Kline & French
Ecotrin tablet	Smith Kline Beecham
Edecrin tablet 50 mg	Merck Sharp & Dohme
E.E.S. 400 liquid	Abbott
Elavil tablet 25, 50, 75, 150 mg	Stuart
Endep tablet 10, 25, 50, 75, 100, 150 mg	Roche
Enduron tablet 2.5 mg	Abbott
Enduronyl tablet	Abbott
Enovid tablet 5, 10 mg	Searle
Entex capsule	Norwich Eaton
Entex LA capsule	Norwich Eaton
Entex liquid	Norwich Eaton
Entozyme tablet	Robins
Ergotrate sublingual tablet	Parke-Davis
Eryc capsule	Parke-Davis
Esidrix tablet 25 mg	Ciba
Eskalith capsule	Smith Kline & French
Estinyl tablet 0.02, 0.5 mg	Schering
Estratest tablet	Reid-Rowell
Ethatab	Allen & Hanburys
Etrafon tablet 2-10, 2-25	Schering
Etrafon A tablet 4-10	Schering
Etrafon forte tablet 4-25	Schering

Trade name (cont'd)	Manufacturer (cont'd)
Eulexin capsule	Schering
Extra Strength Tylenol adult liquid	McNeil consumer
Fedahist gyrocap	Schwarz Pharma
Feosol elixir	Smith Kline Beecham
Fero-gradumet filmtab	Abbott
Fioricet capsule	Sandoz
Fiorinal with codeine #3 capsule	Sandoz
Gantanol tablet	Roche
Geocillin tablet	Roerig
Grifulvin V suspension	Ortho
Grisactin capsule 125, 250 mg	Wyeth-Ayerst
Halotestin tablet 2 mg	Upjohn
Hycomine syrup	DuPont
Hycotuss expectorant	DuPont
Hydrodiuril tablet 25, 50, 100 mg	Merck Sharp & Dohme
Hydropres tablet	Merck Sharp & Dohme
Hytrin tablet 2 mg	Abbott
Iberet-500 filmtab	Abbott
Ilosone pulvule 250 mg	Dista
Ilosone suspension 125 mg/5 ml	Dista
Ilosone drops	Dista
Imodium capsule 2 mg	Janssen
Inderal tablet 10, 40, 80 mg	Wyeth-Ayerst
Inversine tablet	Merck Sharp & Dohme
Janimine tablet 10, 50 mg	Abbott
Kaon-Cl-10	Adria
Keflet tablet 500, 1000 mg	Dista
K-lor powder	Abbott
Klonopin tablet 0.5 mg	Roche
Klor-Con 10	Upsher-Smith
Klotrix tablet	Mead Johnson
K-lyte	Bristol
K-lyte DS orange	Bristol
K-lyte/Cl	Bristol
K-lyte/Cl 50 fruit punch	Bristol
Kutrase capsule	Schwarz Pharma
Kuzyme capsule	Schwarz Pharma
Lanoxin tablet 0.125 mg	Burroughs Wellcome
Larobec tablet, capsule 250, 500 mg	Roche
Lasix oral solution	Hoechst-Roussel
Levlin 21 tablet	Berlex
Levlin 28 tablet	Berlex
Levsin elixir, drops	Schwarz Pharma
Levsinex timecaps	Schwarz Pharma

Trade name (cont'd)	Manufacturer (cont'd)
Librium capsule 5, 10 mg	Roche
Lithium carbonate capsule	Roxane
Lithobid tablet 300 mg	Ciba
Loestrin 1.5/30 green tablet	Parke-Davis
Loestrin FE 1.5/30 green tablet	Parke-Davis
Lomotil liquid	Searle
Lopressor HCT tablet 100/25	Geigy
Lortab 7.5/500	Russ
Loxitane capsule 25 mg	Lederle
Ludiomil tablet 25, 50 mg	Ciba
Lufyllin GG elixir	Wallace
Luride lozitab 0.25, 1 mg	Colgate-Hoyt
Macrodantin capsule	Norwich Eaton
Mandelamine granules	Parke-Davis
Marplan tablet	Roche
Matulane capsule	Roche
Meclomen capsule	Parke-Davis
Mellaril tablet 10, 100, 150 mg	Sandoz
Mellaril suspension 100 mg	Sandoz
Menrium tablet 5-2, 5-4	Roche
Meprospan capsule 200, 400 mg	Wallace
Metamucin powder orange	Procter & Gamble
Methadone HCL oral solution	Roxane
Mexitil capsule 150 mg	Boehringer Ingelheim
Micro-K, Micro-K 10 extencaps	Robins
Midrin capsule	Carnrick
Milontin capsule	Parke-Davis
Minocin capsule 50, 100 mg	Lederle
Moban tablet 5, 25, 100 mg	DuPont
Modane plus tablet	Adria
Moduretic tablet	Merck Sharp & Dohme
Motrin tablet 400, 600, 800 mg	Upjohn
Mulvidren F softab	Stuart
Mycostatin tablet	Squibb
Mysoline suspension	Wyeth-Ayerst
Naldecon DX adult liquid	Bristol
Naldecon DX children's syrup	Bristol
Naldecon DX pediatric drops	Bristol
Nalfon capsule 300 mg	Dista
Nalfon tablet 600 mg	Dista
Naprosyn suspension	Syntex
Nardil tablet	Parke-Davis
Naturetin tablet 10 mg	Princeton
Navane capsules	Roerig

Trade name (cont'd)	Manufacturer (cont'd)
Nembutal sodium capsule 50, 100 mg	Abbott
Nordette tablet 21, 28	Wyeth-Ayerst
Norlestrin 1/50 yellow tablet	Parke-Davis
Norlutate tablet	Parke-Davis
Normodyne tablet 100 mg	Schering
Normozide tablet 100/25	Schering
Norpace capsule 100, 150 mg	Searle
Norpace CR capsule 150 mg	Searle
Norpramin tablet 10, 25, 50, 75, 100 mg	Merrell Dow
Norzine tablet 10 mg	Purdue Frederick
Novafed capsule	Lakeside
Novafed A capsule	Lakeside
Nucofed capsule	Beecham
Nuprin tablet, caplet	Bristol
Ogen tablet 0.625, 1.25 mg	Abbott
Omnipen oral suspension 125	Wyeth-Ayerst
Ornade spansule	Smith Kline & French
Ortho-Novum 7/7/7 peach tablets	Ortho
Ortho-Novum 10/11 peach tablet	Ortho
Ortho-Novum 1/35 peach tablet	Ortho
Orudis capsules	Wyeth-Ayerst
Ovcon 50	Mead Johnson
Ovcon 35	Mead Johnson
Pamelor capsule 10, 25, 75 mg	Sandoz
Panwarfin tablet 2.5, 5 mg	Abbott
Paradione capsule 150 mg	Abbott
Paraflex tablet	McNeil
Parnate tablet	Smith Kline & French
Pavabid HP capsule	Marion
Pentids for syrup	Squibb
Pen Vee K oral solution 250	Wyeth-Ayerst
Percogesic tablet	Richardson-Vicks
Peritinic tablet	Lederle
Persantine tablet 25, 50, 75 mg	Boehringer Ingelheim
Phenaphen with codeine No. 2, 3, 4	Robins
Phenergan syrup plain	Wyeth-Ayerst
Phenergan tablet 12.5 mg	Wyeth-Ayerst
Phenergan with codeine syrup	Wyeth-Ayerst
Phenergan with dextromethorphan syrup	Wyeth-Ayerst
Phenergan D tablet	Wyeth-Ayerst
Phenergan VC syrup	Wyeth-Ayerst
Phenergan VC with codeine syrup	Wyeth-Ayerst
Phillips laxcaps	Glenbrook
Placidyl capsule 750 mg	Abbott

Trade name (cont'd)	Manufacturer (cont'd)
Plegine tablet	Wyeth-Ayerst
PMB 200 tablet	Wyeth-Ayerst
Polaramine repetab 4, 6 mg	Schering
Polaramine tablet 2 mg	Schering
Pondimin tablet	Robins
Ponstel capsule 250 mg	Parke-Davis
Preludin tablet	Boehringer Ingelheim
Prelu-2	Boehringer Ingelheim
Premarin tablet 0.3, 1.25 mg	Wyeth-Ayerst
Premarin with methyltestosterone	Wyeth-Ayerst
Primatene tablet	Whitehall
Primatene P formula	Whitehall
Primatene M formula	Whitehall
Principen oral suspension 250	Squibb
Principen with probenecid capsule	Squibb
Procan SR tablet 250, 500, 750, 1000 mg	Parke-Davis
Procardia capsule	Pfizer
Prolixin elixir	Princeton
Pronestyl tablet	Princeton
Pronestyl capsule 250, 275, 500 mg	Princeton
Proventil syrup	Schering
Provera tablet 2.5 mg	Upjohn
Pyridium tablet	Parke-Davis
Quarzan capsule 5 mg	Roche
Quinidex extentabs	Robins
Raudixin tablet	Princeton
Reglan syrup	Robins
Renese tablet 2 mg	Pfizer
Ridaura capsule	Smith Kline & French
Rimactane capsule 300 mg	Ciba
Robaxin, Robaxin-750 tablet	Robins
Rocaltrol capsule 0.25, 0.5 mcg	Roche
Rondec orals drops, syrup, tablets	Ross
Ru-Tuss with hydrocodone liquid	Boots-Flint
Sectral capsules	Wyeth-Ayerst
Septra suspension	Burroughs Wellcome
Ser-ap-es tablet	Ciba
Serax capsule 15 mg	Wyeth-Ayerst
Seromycin capsule	Lilly
Serpasil-Esidrix tablet	Ciba
Sinemet tablet 25-100	Merck Sharp & Dohme
Sinulin tablet	Carnrick
Soma compound tablet	Wallace
Stelazine tablets, concentrate	Smith Kline & French

Trade name (cont'd)	Manufacturer (cont'd)
Surmontil capsule 25, 50 mg	Wyeth-Ayerst
Synthroid tablet 25, 125 mcg	Boots-Flint
Tace capsule 25 mg	Merrell Dow
Tagamet tablets, oral liquid	Smith Kline & French
Taractan tablet 10, 25, 50 mg	Roche
Taractan concentrate	Roche
Tegretol suspension	Geigy
Teldrin capsules	Smith Kline Beecham
Temaril tablet, spansule	Herbert
Theo-24 capsule 100 mg	Searle
Thorazine tablets, spansules	Smith Kline & French
Tofranil PM capsule 75, 100, 125, 150 mg	Geigy
Tolectin DS capsule	McNeil
Tolectin tablet 600 mg	McNeil
Torecan tablet 10 mg	Roxane
Trancopal caplet 100 mg	Winthrop
Trandate tablet 100, 300 mg	Allen & Hanburys
Trandate HCT tablet 100/25, 300/25	Allen & Hanburys
Tranxene tablet 7.5 mg	Abbott
Trecator SC tablet	Wyeth-Ayerst
Triaminic expectorant DH	Sandoz
Triaminic oral infant drops	Sandoz
Triaminic TR tablet	Sandoz
Triavil tablet 2-25, 4-50, 4-25	Merck Sharp & Dohme
Tridione solution	Abbott
Tridione capsule 300 mg	Abbott
Trilisate tablet 500, 1000 mg	Purdue Frederick
Trimox capsules	Squibb
Trinalin tablet	Schering
Trinsicon	Russ
Tri-vi-flor chewable tablet 1 mg	Mead Johnson
Tuss Ornade spansule	Smith Kline & French
Tussionex suspension	Fisons
Tylenol liquid	McNeil consumer
Tylenol cold tablet, caplet	McNeil consumer
Tylenol maximum strength allergy sinus	McNeil consumer
Tylenol with codeine elixir	McNeil
Unipen capsule, oral suspension	Wyeth-Ayerst
Urecholine tablet 25, 50 mg	Merck Sharp & Dohme
Urobiotic 250	Roerig
Valium tablet 5 mg	Roche
Valpin 50	DuPont
Valrelease capsule	Roche
Vari-Flavors orange flavor	Ross

Trade name (cont'd)	Manufacturer (cont'd)
Veetids tablet 250	Squibb
Velosef oral suspension 125	Squibb
Ventolin syrup	Allen & Hanburys
Vermox chewable tablet	Janssen
Vibramycin tablet	Pfizer
Vicon forte capsule	Russ
Vi-Daylin + iron chewable tablet	Ross
Vi-Daylin F chewable tablet	Ross
V-Daylin F + iron chewable tablet	Ross
Vistaril capsule	Pfizer
Vivactil tablet 5, 10 mg	Merck Sharp & Dohme
Vi-zac capsule	Allen & Hanburys
Vontrol tablet	Smith Kline & French
Wyamycin S tablet	Wyeth-Ayerst
Wygesic tablet	Wyeth-Ayerst
Xanax tablet 0.5 mg	Upjohn
Zarontin syrup	Parke-Davis
Zyloprim tablet 300 mg	Burroughs Wellcome

V. ANIMAL TOXICITY DATA

An increased incidence of adrenal medulla adenomas was found in chronic feeding studies in rats, but these effects were not confirmed by other studies.[4] The WHO has estimated the acceptable daily dietary intake of sunset yellow to be up to 2.5 mg/kg/d.[5]

A slight increase in chromosome aberrations was seen in hamsters given extremely large doses (1.5 g/kg), which was not dose related and could not be reproduced in the mouse model. It was concluded that sunset yellow does not have significant mutagenic hazard for humans.[6]

VI. HUMAN TOXICITY DATA

A. Pseudoallergic Reactions

Ten of 36 children with recurrent urticaria challenged with oral dose of sunset yellow 0.1 mg in a double-blind fashion reacted with urticaria or angioedema. Atopy was not found to be a risk factor for additive-related urticaria in this study.[7] In a series of 45 patients with moderately severe perennial asthma, none of the patients developed bronchoconstriction after double-blind challenge with 20 mg orally of FD&C Yellow No. 6.[8]

Positive oral challenges were observed with sunset yellow 0.1 mg in 2 patients and 1 mg in 8 patients in a series of challenges performed in 504 patients with asthma or rhinitis. Symptoms were similar to those observed in the classic aspirin allergy triad (asthma, urticaria, rhinitis), but generally of a milder nature. All but one of the positive responses occurred in patients with concurrent hypersensitivity to tartrazine (one patient was not tested for tartrazine), and negative responses to sunset yellow were only seen in patients who also had negative responses to tartrazine.[9]

In contrast, another study found no definite pattern or relationship between reactors to tartrazine and sunset yellow or other azo dyes. Eleven of 27 patients with chronic urticaria or angioedema reacted to sunset yellow; 4 of these did not react to tartrazine, and 6 of the tartrazine-sensitive subjects did not react to sunset yellow.[3]

Anaphylaxis was attributed to hypersensitivity to sunset yellow, tartrazine, or both in a patient exposed to these dyes in an enema preparation.[10]

Schonlein-Henoch allergic purpura was related to sunset yellow hypersensitivity in a 32-year-old woman with recurrent purpuric lesions on the lower legs for 12 years. Oral provocation tests with sunset yellow and tartrazine were both strongly positive 8 to 10 h after the dose.[11]

A chronic granulomatous reaction of the upper lip, gum, and oral mucosa was attributed to sunset yellow sensitivity in an 8.5-year-old girl. Double-blind challenge with sunset yellow produced a rapid and severe reaction with pronounced lip and cheek swelling and oral mucosal ulceration.[12]

B. Gastrointestinal Intolerance

Gastrointestinal intolerance, manifested by vomiting, abdominal pain, and belching, was described in a 53-year-old woman within 6 h following ingestion of a sunset yellow-containing rifampin/isoniazid tablet which were reproduced on rechallenge with 1 mg of sunset yellow powder.[13]

Recurrent severe abdominal cramps, diarrhea, hives, and eosinophilia, requiring hospitalization and parenteral narcotics, in a 43-year-old physician were associated with sunset yellow, the only additive in common with an elimination diet implicating seven different foods or drugs. Complement levels (C3 and C4) levels were decreased and the ANA test was reactive at 1:80. Open challenge with sunset yellow 8 mg twice a day for 4 d resulted in cramps, hives, and jitteriness, which was confirmed on subsequent double-blind rechallenge.[1]

VII. CLINICAL RELEVANCE

Sunset yellow is a monoazo dye that has been implicated in causing hypersensitivity reactions with about the same frequency as tartrazine, but has been less thoroughly studied. Most patients intolerant to sunset yellow will also be intolerant to tartrazine; many will also be intolerant to aspirin. As with tartrazine, an immunologic mechanism has not been fully documented, and the reactions are believed to be pseudoallergic. Asthmatics with exacerbations due to aspirin should also be challenged under controlled circumstances with tartrazine and sunset yellow.

REFERENCES

1. **Gross, P. A., Lance, K., Whitlock, R. J., and Blume, R. S.,** Additive allergy: allergic gastroenteritis due to yellow dye #6, *Ann. Intern. Med.,* 111, 87, 1989.
2. **Nikitakis, J. M.,** *CTFA Cosmetic Ingredient Handbook,* 1st ed., The Cosmetic, Toiletry and Fragrance Association, Washington, D.C., 1988.
3. **Michaelsson, G. and Juhlin, L.,** Urticaria induced by preservatives and dye additives in food and drugs, *Br. J. Dermatol.,* 88, 525, 1973.
4. **Reynolds, J. E. F.,** *Martindale: The Extra Pharmacopoeia,* (CD-ROM Version), Micromedex, Inc, Denver, Colorado, 1990.
5. **World Health Organization,** Twenty-sixth Report of the Joint FAO/WHO Expert Committee on Food Additives, Tech. Rep. Ser. No. 683, WHO, Geneva, 1982.

6. **Wever, J., Munzner, R., and Renner, H. W.,** Testing of sunset yellow and orange II for genotoxicity in different laboratory animal species, *Environ. Mol. Mutagen.,* 13, 271, 1989.

7. **Supramaniam, G. and Warner, J. O.,** Artificial food additive intolerance in patients with angio-oedema and urticaria, *Lancet,* 2, 907, 1986.

8. **Weber, R. W., Hoffman, M., Raine, D. A., Jr., and Nelson, H. S.,** Incidence of bronchoconstriction due to aspirin, azo dyes, non-azo dyes, and preservatives in a population of perennial asthmatics, *J. Allergy Clin. Immunol.,* 64, 32, 1979.

9. **Rosenhall, L.,** Evaluation of intolerance to analgesics, preservatives and food colorants with challenge tests, *Eur. J. Respir. Dis.,* 63, 410, 1982.

10. **Desmond, R. E. and Trautlein, J. J.,** Tartrazine (FD&C Yellow #5) anaphylaxis: a case report, *Ann. Allergy,* 46, 81, 1981.

11. **Michaelsson, G., Pettersson, L., and Juhlin, L.,** Purpura caused by food and drug additives, *Arch. Dermatol.,* 109, 49, 1974.

12. **Sweatman, M. C., Tasker, R., Warner, J. O., Ferguson, M. M., and Mitchell, D. N.,** Oro-facial granulomatosis: response to elemental diet and provocation by food additives, *Clin. Allergy,* 16, 331, 1986.

13. **Jenkins, P., Michelson, R., and Emerson, P. A.,** Adverse drug reaction to sunset-yellow in rifampicin/isoniazid tablet, *Lancet,* 2, 385, 1982.

GERANIOL

I. REGULATORY CLASSIFICATION

Geraniol is classified as a fragrance ingredient.

II. AVAILABLE FORMULATIONS

Geraniol is a naturally occurring straight chain terpene alcohol found in many flowering plants, including *Pelargonium* species (Geraniums), *Jasminum* species, *Rosa* species, *Michelia* species, and *Canangium* species. Geraniol is also found in the peels of some citrus fruits, including lemon and grapefruit.[1]

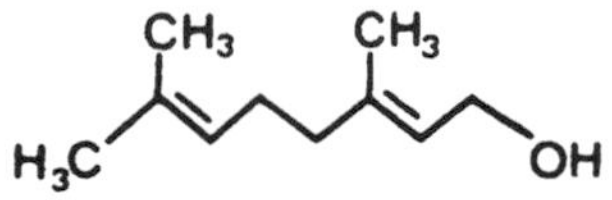

FIGURE 33. Geraniol.

III. HUMAN TOXICITY DATA

A. Allergic Contact Dermatitis

In 17 patients with positive reactions to the European Fragrance Mix, 2 were sensitive to geraniol when tested individually.[2] Occupational contact dermatitis was reported in a bartender who handled various citrus peels. Patch testing showed positive reactions to citral and geraniol, but not to D-limonene.[1]

Geraniol was implicated in 8 of 713 (1.1%) patients with cosmetic-related contact dermatitis in a 5-year study done in the U.S.[3] A similar study of 179 Dutch patients showed a positive reaction to geraniol 10% in petrolatum in 6.1% of patients with cosmetic dermatitis.[4]

Medication-related contact dermatitis was attributed to the geraniol component of perfumed topical creams and ointments in a 53-year-old woman with chronic leg ulcers,[5] in a 66-year-old man with chronic leg ulcers, and a 34-year-old woman with a grafted skin burn.[6]

Pigmented contact dermatitis (Riehl's melanosis) was reported in a 27-year-old woman following the use of a compact face powder for 2 months. Patch testing was positive to many of the components of fragrance mix, including geraniol, hydroxycitronellal, and lemon oil. The reaction to hydroxycitronellal was considered to be a cross-reaction with geraniol. Complete clearing of the lesions was evident 6 months after discontinuation of facial cosmetics.[7]

B. Contact Urticaria

Open patch testing with geraniol 5% in petrolatum produced nonimmunologic immediate contact urticaria in 35 of 50 patients consisting of Stage 1 reactions (localized macular erythema or wheal-and-flare).[8]

C. Primary Irritation

Geraniol concentrations of 5% or greater may cause primary irritant reactions and were capable of producing sensitization in 20 of 25 patients undergoing human maximization testing.[9]

IV. CLINICAL RELEVANCE

Geraniol may cause nonimmunologic contact urticaria and allergic eczematous contact dermatitis, occasionally with hyperpigmentation. The patch test concentration of geraniol in the International Contact Dermatitis Fragrance Mix is 1%, while the North American Fragrance Mix contains 2%. Higher concentrations, 5 to 10% in petrolatum, have been recommended for testing of individual components.[4]

REFERENCES

1. **Cardullo, A. C., Ruszkowski, A. M., and DeLeo, V. A.**, Allergic contact dermatitis resulting from sensitivity to citrus peel, geraniol, and citral, *J. Am. Acad. Dermatol.*, 21, 395, 1989.
2. **Roesyanto-Mahadi, I. D., Geursen-Reitsma, A. M., Van Joost, T., and Van Den Akker, T. W.**, Sensitization to fragrance materials in Indonesian cosmetics, *Contact Dermatitis*, 22, 212, 1990.
3. **Adams, R. M. and Maibach, H. I.**, A five-year study of cosmetic reactions, *J. Am. Acad. Dermatol.*, 13, 1062, 1985.
4. **De Groot, A. C., Liem, D. H., Nater, J. P., and Van Ketel, W. G.**, Patch tests with fragrance materials and preservatives, *Contact Dermatitis*, 12, 87, 1985.
5. **Guerra, P., Aguilar, A., Urbina, F., Cristobal, M. C., and Garcia-Perez, A.**, Contact dermatitis to geraniol in a leg ulcer, *Contact Dermatitis*, 16, 298, 1987.
6. **Romaguera, C., Grimalt, F., and Vilaplana, J.**, Geraniol dermatitis, *Contact Dermatitis*, 14, 185, 1986.
7. **Serrano, G., Pujol, C., Cuadra, J., Gallo, S., and Aliaga, A.**, Riehl's melanosis: pigmented contact dermatitis caused by fragrances, *J. Am. Acad. Dermatol.*, 21, 1057, 1989.
8. **Emmons, W. W. and Marks, F. G., Jr.**, Immediate and delayed reactions to cosmetic ingredients, *Contact Dermatitis*, 13, 258, 1985.
9. **Malten, K. E., Van Ketel, W. G., Nater, J. P., and Liem, D. H.**, Reactions in selected patients to 22 fragrance materials, *Contact Dermatitis*, 11, 1, 1984.

GLUTEN

I. REGULATORY CLASSIFICATION

Wheat flour and wheat starch are listed as GRAS food ingredients.

II. SYNONYMS

Wheat gluten

III. AVAILABLE FORMULATIONS

A. Foods

Gluten is the protein-containing portion of wheat flour milled from the grain endosperm. Wheat flour contains 6 to 17% gluten. Wheat starch contains not more than 0.5% protein and therefore has insignificant amounts of gluten.[1] Gliadin (wheat prolamin) is found in the alcohol-soluble fraction of gluten and is present as a macromolecular complex with glutenin. Prolamins with related N-terminal amino acid sequences are found in rye and barley and are termed secalins and hordeins, respectively. Prolamins from other grains include avenins from oats and zeins from corn.[2]

Wheat flour is commonly found in breads, pastries, and pasta products. A typical diet provides 15 to 20 g/d of gluten.[3] An evaluation of 14 different Holy Communion wafers found 2.2 to 4.9 mg of gliadin in ten brands. Brands labeled as "gluten-free" contained 0.28 to 0.66 mg.[4]

B. Cosmetics

A review of cosmetic formulations published in 1980 found only one type of product containing gluten, a mascara containing less than 0.1%.[5]

IV. TABLE OF COMMON PRODUCTS

Information on gluten-free diets and pharmaceuticals is available from the Celiac Sprue Society, Natick, MA (508-651-3230). Because gluten is generally not specifically mentioned as an inactive ingredient and may be included in the general category of "starch", it is difficult

195

to assess gluten content of pharmaceuticals without consultation with the manufacturer. Articles have been published with lists of gluten-free and gluten-containing pharmaceuticals for Australian,[6] Canadian,[7] and American products.[8,9]

V. HUMAN TOXICITY DATA

A. Immediate Hypersensitivity Reactions

Anaphylactic shock was described in a 3-month-old infant following feeding of a wheat rusk after introduction of this food at the age of 2 months. Enteropathy was excluded by small intestinal mucosal biopsy. Rechallenge with a small amount of wheat rusk resulted in vomiting, diarrhea, bronchospasm, and shock in 2 h. A wheat RAST test was negative at 14 months, and the infant was successfully rechallenged with a wheat-containing diet.[10]

There were six cases of exercise-induced immediate hypersensitivity reactions to wheat. The reactions consisted of generalized urticaria and angioedema and anaphylaxis within 30 min after ingestion of wheat products and during vigorous exercise. Skin tests were positive to wheat, bread, gluten, gliadin, and glutenin. A RAST test for wheat flour was negative in all five patients tested; a positive RAST for gluten and gliadin were found in one of two patients tested.[11]

B. Gluten-Sensitive Enteropathy

Gluten-sensitive enteropathy, also called celiac disease and idiopathic sprue, is thought to be due to an immunologic defect, resulting in an enhanced response to gluten protein. The resulting immune cytotoxic effect causes two types of syndromes: one involving primarily the gastrointestinal tract and presenting with diarrhea, steatorrhea, weight loss, and abdominal bloating; and the other predominantly involving the skin, also known as Dermatitis herpetiformis.

The gastrointestinal variant affects the mature villous epithelial cells. The immature crypt cells predominate in these patients, resulting in impaired absorptive function. The villous atrophy may predispose to development of gastrointestinal cancer. Dermatitis herpetiformis presents with vesicular, pruritic lesions on extensor surfaces or sun-exposed areas, appearing histologically as a chronic lymphocytic cell infiltration with IgA and complement deposition in the dermal papillae. Gastrointestinal villous atrophy occurs, but is milder, in patients with primarily skin manifestations.[12]

The precipitating factor for the initiation of gluten sensitivity has been proposed to involve generation of a cross-reacting antibody to shared amino acid sequences in gluten and an adenovirus surface protein.[13] A genetically determined cell-mediated response has also been convincingly argued.[3]

VI. CLINICAL RELEVANCE

Gluten-sensitive enteropathy is a permanent condition requiring strict avoidance of ingested sources of gluten. The prevalence of gluten intolerance is between 1:300 and 1:10,000 in European populations.[14] Cereal grains containing gluten or related prolamines include wheat, rye, triticale, barley, and oats. Malt and hydrolyzed vegetable protein should also be avoided. Up to 10 mg of gliadin per day may be tolerated. Very few pharmaceutical products contain gluten, but these should be avoided.

Wheat protein is a rare cause of immediate Type I hypersensitivity reactions. Anaphylaxis, exercise-induced anaphylaxis, and exercise-induced urticaria with angioedema have been reported. In contrast to gluten-sensitive enteropathy, immediate hypersensitivity may be a transient problem limited to early childhood.

REFERENCES

1. **Cosmetic Ingredient Review Expert Panel,** Final report of the safety assessment for wheat flour and wheat starch, *J. Environ. Pathol. Toxicol.,* 4, 19, 1980a.
2. **Ciclitira, P. J. and Ellis, H. J.,** Investigation of cereal toxicity in coeliac disease, *Postgrad. Med. J.,* 63, 767, 1987.
3. **van der Meer, J. B.,** Gluten-free diet and elemental diet in dermatitis herpetiformis, *Int. J. Dermatol.,* 29, 679, 1990.
4. **Moriarty, K. J., Loft, D., Marsh, M. N., Brookds, S. T., Gordon, D., and Garner, G. V.,** Holy communion wafers and celiac disease, *N. Engl. J. Med.,* 321, 332, 1989.
5. **Cosmetic Ingredient Review Expert Panel,** Final report of the safety assessment for wheat germ glycerides and wheat gluten, *J. Environ. Pathol. Toxicol.,* 4, 5, 1980.
6. **Challen, R. G. and O'Shannassy, R. M.,** Gluten content of Australian pharmaceutical products, *Med. J. Aust.,* 146, 91, 1987.
7. **Patel, D. G., Krogh, C. M. E., and Thompson, W. G.,** Gluten in pills: a hazard for patients with celiac disease, *Can. Med. Assoc. J.,* 133, 114, 1985.
8. **Olson, G. B. and Gallo, G. R.,** Gluten in pharmaceutical and nutritional products, *Am. J. Hosp. Pharm.,* 40, 121, 1983.
9. **Pence, R. A. and Garrison, T. J.,** Gluten-containing pharmaceutical products, *Am. J. Hosp. Pharm.,* 44, 2254, 1987.
10. **Rudd, P., Manuel, P., and Walker-Smith, J.,** Anaphylactic shock in an infant after feeding with a wheat rusk. A transient phenomenon, *Postgrad. Med. J.,* 57, 794, 1981.
11. **Kushimoto, H. and Aoki, T.,** Masked Type I wheat allergy, *Arch. Dermatol.,* 121, 355, 1985.
12. **Strober, W.,** Gluten-sensitive enteropathy: a nonallergic immune hypersensitivity of the gastrointestinal tract, *J. Allergy Clin. Immunol.,* 78, 202, 1986.
13. **Kagnoff, M. F., Austin, R. K., Hubert, J. J., et al.,** Possible role for a human adenovirus in the pathogenesis of celiac disease, *J. Exp. Med.,* 160, 1544, 1984.
14. **Skerritt, J. H. and Hill, A. S.,** Self-management of dietary compliance in coeliac disease by means of ELISA "home test" to detect gluten, *Lancet,* 1337, 379, 1991.

GLYCERIN

I. REGULATORY CLASSIFICATION

Glycerin is classified as a humectant, plasticizer, solvent, and tonicity agent in pharmaceutical products.

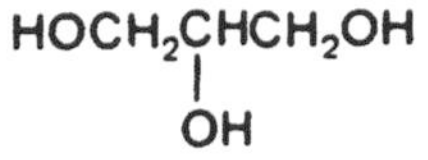

$$HOCH_2CHCH_2OH$$
$$|$$
$$OH$$

FIGURE 34. Glycerin.

II. SYNONYMS

E422
Glicerol
Glycerol
Propane-1,2,3-triol
1,2,3-propanetriol
Trihydroxypropane glycerol

III. AVAILABLE FORMULATIONS

A. Foods
Glycerin occurs naturally in foods, constituting about 10% by weight of the triglyceride content in animal or vegetable fats and oils. Glycerin is also used as a food additive with humectant, plasticizer, rehydration, and crystallization modification properties. It is an ingredient in casing solutions sprayed on tobacco during cigarette manufacturing and is on the GRAS list.

B. Drugs
Glycerin is used as an emollient and humectant in 81 FDA-registered topical pharmaceuticals in concentrations of 0.2 to 65.7%. Liquid oral pharmaceuticals may contain 1 to 50% as a sweetener or preservative; 149 products were registered in 1990. Parenteral formulations may

contain up to 50% as a solvent, but usually do not exceed 15%; 21 products were registered in 1990. Oral film-coated tablets contain variable amounts as a plasiticizer. Dentifrices and dental solutions contain 7 to 10%. There are 11 ophthalmic products registered containing 0.5 to 3%.[1]

C. Cosmetics

Glycerin is used in cosmetics as a denaturant and humectant in a wide variety of products, including hair dyes and conditioners, makeup products, mouthwashes, breath fresheners, aftershave lotions, shaving creams, skin fresheners, creams, lotions, and mud packs.[2]

IV. TABLE OF COMMON PRODUCTS

A. Oral Liquid Drug Products

Trade name	Manufacturer
Actifed with codeine syrup	Burroughs Wellcome
Aldomet suspension	Merck Sharp & Dohme
Aludrox suspension	Wyeth-Ayerst
Ambenyl cough syrup	Forest
Amphojel	Wyeth-Ayerst
Anacin-3 Children's liquid, drops	Whitehall
Calcidrine syrup	Abbott
Cerose DM	Wyeth-Ayerst
Children's Advil suspension	Whitehall
Chloromycetin oral suspension	Parke-Davis
Choledyl elixir	Parke-Davis
Decadron elixir	Merck Sharp & Dohme
Deconamine syrup	Berlex
Depakene syrup	Abbott
Dilantin suspension	Parke-Davis
Dilor-G liquid	Savage
Dimetane DC cyrup	A. H. Robins
Dimetane DX syrup	A. H. Robins
Diuril oral suspension	Merck Sharp & Dohme
Dramamine liquid	Richardson-Vicks
Elixophyllin GG	Forest
Elixophyllin KI	Forest
Elixophyllin elixir	Forest
Emetrol solution	Adria
Entex liquid	Norwich-Eaton
Hycomine syrup	DuPont
Hycotuss syrup	DuPont
Imodium AD liquid	Janssen
Lasix oral solution	Hoechst-Roussel
Levsin elixir, drops	Schwartz Pharma
Lomotil liquid	Searle

Trade name (cont'd)	Manufacturer (cont'd)
Mellaril concentrate	Sandoz
Mycostatin oral suspension	Squibb
Naldecon syrup	Bristol
Naldecon DX children's syrup	Bristol
Naldecon EX children's syrup	Bristol
Novahistine DH	Lakeside
Novahistine expectorant	Lakeside
Nucofed expectorant, syrup	Beecham
Nucofed pediatric expectorant	Beecham
Organidin expectorant	Wallace
Pediaprofen liquid	McNeil
Pediacare cough-cold liquid	McNeil
Pediacare Nightrest cough-cold liquid	McNeil
Pediacare infant's oral decongestant drops	McNeil
Periactin syrup	Merck Sharp & Dohme
Phenobarbital elixir	Lilly
Phenergan DM syrup	Wyeth-Ayerst
Phenergan VC syrup	Wyeth-Ayerst
Phenergan syrup	Wyeth-Ayerst
Phenergan VC with codeine syrup	Wyeth-Ayerst
Prolixin oral concentrate	Princeton
Quelidrine syrup	Abbott
Reglan syrup	A. H. Robins
Robitussin DAC, AC	A. H. Robins
Rondec DM	Ross
Rynatan	Wallace
Rynatuss	Wallace
Septra suspension	Burroughs Wellcome
Taractan concentrate	Roche
Temaril syrup	Herbert

B. Topical Drug Products

Trade name	Manufacturer
Aristocort A cream	Lederle
Benzac AC	Owen/Galderma
Cleocin T lotion	Upjohn
Cyclocort cream, lotion	Lederle
Dermaide aloe cream	Dermaide Research
Dermoplast lotion	Whitehall
Eucerin lotion	Beiersdorf
Furacin cream	Norwich Eaton
Kwell cream	Reed & Carnrick

Trade name (cont'd)	Manufacturer (cont'd)
Lac-Hydrin Five	Westwood-Squibb
Lac-Hydrin 12%	Westwood-Squibb
Moisturel cream, lotion	Westwood-Squibb
Sulfamylon cream	Winthrop
Tridesilon cream	Miles

V. ANIMAL TOXICITY DATA

Short-term genotoxicity tests, including the Ames mutagenicity assay, Chinese hamster ovary chromosome aberration assay, sister chromatid assay, and mammalian mutagenesis assay, have been negative. Glycerin has been shown to protect against DNA damage induced by tumor promoters, ultraviolet light, and radiation, presumably via free radical scavenger activity.[3]

VI. HUMAN TOXICITY DATA

A. Hyperosmolality

Glycerin is an osmotically active compound that causes predictable increases in plasma osmolality resulting in expansion of the intravascular fluid volume. Glycerin is filtered and almost completely reabsorbed by the renal tubules. The reabsorptive capacity is exceeded at plasma concentrations of about 0.15 mg/ml. With larger concentrations, glycerol is excreted unchanged in the urine, creating an osmotic diuresis.[4]

Increased plasma and urine osmolality was observed in a 3900-g neonate given progressively increasing doses of an oral phenobarbital elixir containing glycerin. At a dose of 15 ml every 6 h of the elixir, containing 40% v/v of glycerin and 40% v/v of 90% ethanol, marked hyperosmolality occurred associated with a plasma glycerin level of 2.5 mg/ml. The total amount of glycerin received was 7.5 mg/kg/d. Three other neonates receiving the same medication were subsequently evaluated and found to be hyperosmolar. The only clinical effect attributed to the glycerin was diuresis.[5]

Topical application of a burn cream containing 5% povidone iodine and 95% v/v glycerin to a 40-year-old man with a 60% total body surface area partial-thickness burn resulted in hyperosmolality within 36 h, peaking at 560 mOsm/l. The serum triglyceride level rose from 171 to 9700 mg/dl. At this time, the patient developed status epilepticus and remained comatose until death from sepsis 4 weeks after admission. The elevated triglyceride level was considered spurious, secondary to contamination of the triglycerides with the excess glycerin. The glycerin level was calcutated by dividing the reported triglyceride level by 9.62, corresponding to a serum glycerin level of 10 mg/ml.[6]

B. Hemolysis

Extremely large parenteral doses of glycerin, used therapeutically in doses of 70 to 80 g over 30 to 60 min in adults, have resulted in intravascular hemolysis, presumably due to osmolality changes in the red blood cell environment.[7] Extravascular hemolysis was described in a premature neonate undergoing continuous arteriovenous hemofiltration using a glycerin-primed hemofilter that had not been rinsed with normal saline prior to use. It was estimated that 1 g of glycerin was infused into the infant who weighed 950 g. Serum glycerin levels and osmolality were not measured.[8]

C. Dermatologic Reactions

Glycerin is nonirritating and an extremely rare sensitizer.[9] One reported case described an allergy laboratory staff member who developed bilateral hand eczema following the use of a mixture of one part glycerin and nine parts 70% ethanol after handwashing. Delayed-type hypersensitivity to glycerin was confirmed by 3+ reactions at 48 and 72 h to 1, 5, and 10% glycerin.[10]

Another case documented by the same investigator developed a generalized dermatitis following the use of a cream containing 10% glycerin. A peroral challenge with 5 ml of glycerin produced an exanthem within 2 h.[11]

VII. CLINICAL RELEVANCE

Along with other osmotically active substances such as propylene glycol, glycerin has only been associated with clinical effects from excipient usage in the setting of low-birth-weight premature infants and burn patients treated topically over large body surface areas. Other known side effects of large oral or parenteral therapeutic doses of glycerin include hyperosmolar nonketotic coma, diabetic acidosis, pulmonary edema, and minor symptoms of headache, nausea, vomiting, and dizziness.[12-14] These effects have not been reported following ingestion or administration of glycerin as an excipient or food additive.

Patients with rare genetic defects in utilization of glycerin may develop severe hypoglycemia after ingestion. The amount present naturally in a glass of milk (about 1 g) did not produce symptoms.[15]

REFERENCES

1. **Food and Drug Administration,** *Inactive Ingredients Guide,* FDA, Washington, D.C., March 1990.
2. **Nikitakis, J. M.,** *CTFA Cosmetic Ingredient Handbook,* 1st ed., The Cosmetic, Toiletry and Fragrance Association, Washington, D.C., 1988.
3. **Doolittle, D. J., Lee, D. A., and Lee, C. K.,** The genotoxic activity of glycerol in an *in vitro* test battery, *Food Chem. Toxicol.,* 26, 631, 1988.
4. **Frank, M. S. B., Nahata, M. C., and Hilty, M. D.,** Glycerol: a review of its pharmacology, pharmacokinetics, adverse reactions, and clinical use, *Pharmacotherapy,* 1, 147, 1981.
5. **Van Der Westhuyzen, J. H., Berger, G. M. B., Beyers, N., and Moosa, A.,** Iatrogenic hyperosmolality in a neonate, *S. Afr. Med. J.,* 60, 996, 1981.
6. **Hershey, S. D. and Gursel, E.,** Hyperosmolality caused by percutaneously absorbed glycerin in a burned patient, *J. Trauma,* 22, 250, 1982.
7. **Hagnevik, K., Gordon, E., Lins, L. E., Whlhelmsson, S., and Forster, D.,** Glycerol-induced haemolysis with haemoglobinuria and acute renal failure: report of three cases, *Lancet,* 1, 75, 1974.
8. **Feng, C. S., Lam, T. K., Lee, N., Fok, T. F., and Lai, F. M. M.,** Glycerin-induced haemolysis associated with the use of haemofilter, *J. Paediatr. Child. Health,* 26, 166, 1990.
9. **Fisher, A. A.,** Reactions to popular cosmetic humectants. Part III. Glycerin, propylene glycol, and butylene glycol, *Cutis,* 26, 243, 1980.
10. **Hannuksela, M. and Forstrom, L.,** Contact hypersensitivity to glycerol, *Contact Dermatitis,* 2, 291, 1976.
11. **Hannuksela, M.,** Allergic and toxic reactions caused by cream bases in dermatologic patients, *Int. J. Cosmet. Sci.,* 1, 257, 1979.

12. **Almog, Y., Geyer, O., and Lazar, M.,** Pulmonary edema as a complication of oral glycerol administration, *Ann. Ophthalmol.,* 18, 38, 1986.

13. **Oakley, D. E. and Ellis, P. P.,** Glycerol and hyperosmolar nonketotic coma, *Am. J. Ophthalmol.,* 81, 469, 1976.

14. **D'Alena, P. and Ferguson, W.,** Adverse effects after glycerol orally and mannitol parenterally, *Arch. Ophthalmol.,* 75, 201, 1966.

15. **Wapnir, R. A., Lifshitz, F., Sekaran, C., Teichberg, S., and Moak, S. A.,** Glycerol-induced hypoglycemia: a syndrome associated with multiple liver enzyme deficiencies. Clinical and in vitro studies, *Metabolism,* 31, 1057, 1982.

Imidazolidinyl Urea

I. REGULATORY CLASSIFICATION

Imidazolidinyl urea is a heterocyclic-substituted urea used as a preservative in topical pharmaceutical and cosmetic products.

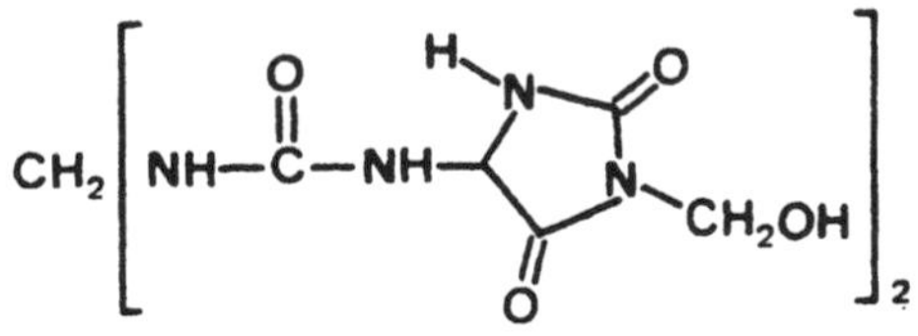

$$CH_2 \left[NH-\overset{\overset{\displaystyle O}{\|}}{C}-NH-\underset{\underset{\displaystyle O}{}}{\overset{\overset{\displaystyle H}{N}}{\underset{N-CH_2OH}{\diamondsuit}}} \right]_2$$

FIGURE 35. Imidazolidinyl urea.

II. SYNONYMS

Germall 115®
Imidurea

III. AVAILABLE FORMULATIONS

A. Cosmetics

Imidazolidinyl urea is used as a preservative in cream, lotions, makeup foundations and powders, eye cosmetics, shampoos, and dusting powders.[1] It is used in concentrations of 0.02 to 0.5%.[2] Next to methyl and propyl parabens, it is the most frequently used cosmetic preservative in the U.S., present in 1254 of 18,500 cosmetics listed.[3]

The antimicrobial spectrum includes activity primarily against gram-positive and gram-negative bacteria; some efficacy has been demonstrated for selective molds and yeasts. It has synergistic activity when combined with other antimicrobials and is frequently found with parabens to enhance anti-Pseudomonal activity.[4]

IV. TABLE OF COMMON PRODUCTS

A. Topical Drug Products

Trade name	Manufacturer
Cutivate cream	Glaxo
Dermaide aloe cream	Dermaide Research
Eucerin cleansing lotion	Beiersdorf
Extra Strength Arthritis Balm	Blair
Flexall 454 gel	Chattem
Fototar cream	ICN
Gelocast unna boot	Beiersdorf
Lacticare lotion	Stiefel
Neutrogena moisture	Neutrogena
Nix cream rinse	Burroughs Wellcome
Osti-derm lotion	Pedinol
P & S shampoo	Baker-Cummins
T/Gel shampoo	Neutrogena
Vanseb shampoo	Herbert
Vanseb T shampoo	Herbert
X-Seb shampoo	Baker-Cummins
Zeasorb AF	Stiefel

V. ANIMAL TOXICITY DATA

Imidazolidinyl urea releases formaldehyde, which is responsible for development of sensitization. The guinea pig maximization test using Freund's complete adjuvant to enhance immunogenicity has classified this preservative as a strong sensitizer.[5] A study using an open epicutaneous test in guinea pigs may more closely represent actual human exposure to imidazolidinyl urea. In this study, only weak sensitization potential was demonstrated. No animals were sensitized to 0.3 and 1% concentrations, and only 3 of 24 animals were sensitized to higher concentrations, up to 50% in aqueous solution, after the 4-week induction phase. Most of the sensitized animals were also sensitized to formaldehyde by this procedure.[2]

VI. HUMAN TOXICITY DATA

While the germicidal activity of imidazolidinyl urea is attributed to the parent molecule, sensitization and contact dermatitis may occur to both the parent compound and the formaldehyde released on decomposition. The incidence of positive delayed hypersensitivity responses after patch testing with 2% in petrolatum was 0.7% in 2298 unselected British subjects; 37.5% of those also reacted to formaldehyde.[6] A prospective study of cosmetic dermatitis by the North America Contact Dermatitis Group attributed 3% to imidazolidinyl urea.[7] Patch tests done in over 2000 unselected dermatitis patients by this group showed an incidence of 0.9%.[8]

Eyelid dermatitis secondary to imidazolidinyl urea may be difficult to diagnose with traditional patch test methods. A 41-year-old woman with a positive repeated open application test (ROAT) on the arm to a cream containing this preservative had negative patch tests to concentrations as high as 10% in water. Subsequent ROATs to 2% aqueous imidazolidinyl urea and to formaldehyde 1% aqueous were positive.[9]

VII. CLINICAL RELEVANCE

Imidazolidinyl urea releases about 90 ppm of formaldehyde at a concentration of 2%. The highest amount of formaldehyde present in a commercially available product would therefore be 22.5 ppm, which is less than the threshold of 30 ppm for elicitation of contact dermatitis in formaldehyde-sensitive patients.[10] This undoubtedly accounts for the low prevalence of reported reactions to this preservative and the apparent safety when used in patients with formaldehyde allergy.

The methods used to detect formaldehyde release have been criticized, since most require heat and extraordinary conditions. When tested at normal temperatures using the USP method, formaldehyde was not released by 1% imidazolidinyl urea.[8]

At the 2% concentration used for patch testing, interpretation is difficult in the patient allergic to formaldehyde. The commonly recommended petrolatum vehicle has also been questioned. Imidazolidinyl urea is more soluble in water than in oil. Positive patch tests to 0.3 to 2% aqueous solutions have been found in patients with clinically relevant dermatitis and negative patch tests to both 2% in petrolatum and to formaldehyde.[11,12]

REFERENCES

1. **Nikitakis, J. M.**, *CTFA Cosmetic Ingredient Handbook*, 1st ed., The Cosmetic, Toiletry and Fragrance Association, Washington, D.C., 1988.
2. **Ziegler, V., Ziegler, B., and Kipping, D.**, Dose-response sensitization experiments with imidazolidinyl urea, *Contact Dermatitis*, 19, 236, 1988.
3. **Decker, R. L., Jr.**, Frequency of preservative use in cosmetic formulas as disclosed to FDA-1984, *Cosmet. Toilet.*, 100, 65, 1985.
4. **Fisher, A. A.**, Cosmetic dermatitis. Part II. Reactions to some commonly used preservatives, *Cutis*, 26, 136, 1980.
5. **Andersen, K., E., Boman, A., Haman, K., and Wahlberg, J. E.**, Guinea pig maximization tests with formaldehyde releasers, *Contact Dermatitis*, 10, 257, 1984.
6. **Ford, G. P. and Beck, M. H.**, Reactions to Quaternium 15, Bronopol and Germall 115 in a standard series, *Contact Dermatitis*, 14, 271, 1986.
7. **Adams, R. M. and Maibach, H. I.**, A five-year-study of cosmetic reactions, *J. Am. Acad. Dermatol.*, 13, 1062, 1985.
8. **Fisher, A. A.**, *Contact Dermatitis*, 3rd ed., Lea & Febiger, Philadelphia, 1986.
9. **De Groot, A. C. and Weyland, J. W.**, Hidden contact allergy to formaldehyde in imidazolidinyl urea, *Contact Dermatitis*, 17, 124, 1987.
10. **Jordan, W. P., Sherman, W. T., and King, S. E.**, Threshold responses in formaldehyde-sensitive subjects, *J. Am. Acad. Dermatol.*, 1, 44, 1979.
11. **De Groot, A. C., Bruynzeel, D. P., Jagtman, B. A., and Weyland, J. W.**, Contact allergy to diazolidinyl urea (Germall 115), *Contact Dermatitis*, 18, 202, 1988.
12. **Foussereau, J. and Cavelier, C.**, Water versus petrolatum for testing imidazolidinyl urea, *Contact Dermatitis*, 21, 54, 1989.

ISOPROPYL MYRISTATE

I. REGULATORY CLASSIFICATION

Isopropyl myristate is classified as an oleaginous vehicle.

II. SYNONYMS

Ceraphyl IPL®
IPM
Promyr®
Propal®

III. AVAILABLE FORMULATIONS

A. Constituents
Isopropyl myristate is the ester of isopropyl alcohol myristic acid, a saturated high molecular weight fatty acid.

$$CH_3(CH_2)_{12}\overset{\overset{\displaystyle O}{\|}}{C}-OCH(CH_3)_2$$

FIGURE 36. Isopropyl myristate.

B. Drugs
Isopropyl myristate is used as a solvent for topical pharmaceuticals.

C. Cosmetics
Isopropyl myristate is absorbed readily by the skin and is commonly used in nongreasy emollient creams, lotions, and ointments. Types of products include bath oils, bath tablets,

bath salts, eye makeup, dusting powders, face makeup, lipsticks, personal cleanliness products, cleansing products, moisturizers, and night creams or lotions.[1]

IV. TABLE OF COMMON PRODUCTS

A. Topical Drug Products

Trade name	Manufacturer
Axsain cream	GalenPharma
Barseb HC scalp lotion	Barnes-Hind
Barseb scalp lotion	Barnes-Hind
Barseb theraspray	Barnes-Hind
Caldecort light cream, spray	Fisons
Carmol 20 cream	Syntex
Carmol HC cream	Syntex
Clearasil adult cream	Richardson-Vicks
Clearasil maximum strength cream	Richardson-Vicks
Cruex antifungal powder, spray	Fisons
Cutivate cream	Glaxo
Decaderm	Merck Sharp & Dohme
Decaspray	Merck Sharp & Dohme
Desenex powder	Fisons
Eucerin moiturizing lotion	Beiersdorf
Exelderm cream	Westwood-Squibb
Fluoroplex cream	Herbert
Foille Plus cream	Blistex
Hytone cream, lotion	Dermik
Massengill medicated towelettes	Beecham
Naftin cream	Herbert
Nivea creme lotion	Beiersdorf
Nizoral cream	Janssen
Retin A cream	Ortho
SSD cream	Boots
Topicort cream	Hoechst-Roussel
Topicort LP cream, gel	Hoechst-Roussel
Vytone cream	Dermik
Zostrix cream	GenDerm

V. ANIMAL TOXICITY DATA

A. Irritation

Isopropyl myristate exhibits species-variable primary irritation. Application of saturated skin patches to rabbits produced moderate-severe irritant effects. Identical exposures to human skin resulted in only slight irritation.[2]

B. Comedogenicity

Isopropyl myristate and its analogs, such as isopropyl palmitate, isopropyl isostearate, butyl stearate, isostearyl neopentanoate, myristyl myristate, decyl oleate, octyl stearate, octyl palmitate, and isocetyl stearate, are highly comedogenic in the rabbit ear model.[3]

VI. HUMAN TOXICITY DATA

A. Hypersensitivity

Allergic contact dermatitis was reported in six patients using an antibiotic-corticosteroid spray containing isopropyl myristate.[4] Fisher found only one patient with positive patch test to 10% in alcohol or as is.[5]

VII. CLINICAL RELEVANCE

Isopropyl myristate is highly comedogenic. Cosmetics containing this ingredient and related substances should be avoided in the patient with acne vulgaris. Allergic contact dermatitis is very rarely reported.

REFERENCES

1. **Nikitakis, J. M.,** *CTFA Cosmetic Ingredient Handbook,* 1st ed., The Cosmetic, Toiletry and Fragrance Association, Washington, D.C., 1988.
2. **Campbell, R. L. and Bruce, R. D.,** Comparative dermatotoxicology. 1. Direct comparison of rabbit and human primary skin irritation responses to isopropyl myristate, *Toxicol. Appl. Pharmacol.,* 59, 555, 1981.
3. **Fulton, J. E., Jr., Pay, S. R., and Fulton, J. E., III,** Comedogenicity of current therapeutic products, cosmetics, and ingredients in the rabbit ear, *J. Am. Acad. Dermatol.,* 10, 96, 1984.
4. **Calnan, C. D.,** Isopropyl myristate sensitivity, *Contact Dermatitis Newsletter,* 2, 38, 1968.
5. **Fisher, A. A.,** *Contact Dermatitis,* 3rd ed., Lea & Febiger, Philadelphia, 1986.

Kathon CG®

I. REGULATORY CLASSIFICATION

The ingedients contained in Kathon CG® are classified as antimicrobial preservatives.

II. SYNONYMS

A. 5-Chloro-2-Methyl-4-Isothiazolin-3-One

Chloromethylisothiazolinone

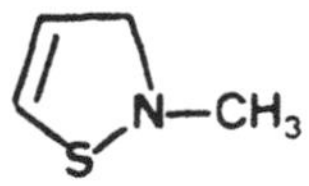

FIGURE 37. 5-Chloro-2-methyl-4-isothiazolin-3-one.

B. 2-Methyl-4-Isothiazolin-3-One

Methylisothiazolinone

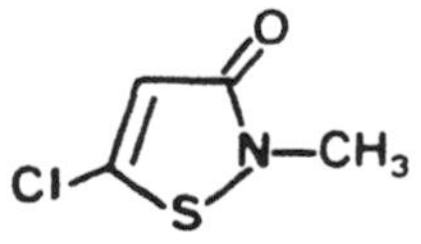

FIGURE 38. 2-Methyl-4-isothiazolin-3-one.

III. AVAILABLE FORMULATIONS

A. Constituents

Kathon CG® is a relatively recently introduced preservative that was first marketed in the U.S. in 1980. It is manufactured by Rohm & Haas and contains an aqueous solution of 5-chloro-2-methyl-4-isothiazolin-3-one and 2-methyl-4-isothiazolin-3-one in a 3:1 ratio with two magnesium salts as stabilizers, $MgCl_2$ and $Mg(NO_3)_2$.

B. Cosmetics

Kathon CG® is found used as a cosmetic preservative in products such as shampoos, creams, lotions, hair conditioners, bubble baths, and cleansers.[1,3] Available products contain 0.02 to 0.1% of Kathon CG® which supplies 3 to 15 ppm of the active ingredients. It was listed as a preservative in 135 of 20,183 cosmetics registered with the FDA in 1982.[4]

C. Other

Other household and industrial uses include household cleaners, swimming pool water, paper mills, cooling-tower water, metal-working fluid, and latex emulsions.[2,3]

IV. TABLE OF COMMON PRODUCTS

A. Topical Drug Products

Trade name	Manufacturer
Eucerin cream, lotion	Beiersdorf
Eurax cream	Westwood-Squibb
Head & Shoulders lotion	Procter & Gamble
Lac-Hydrin Five	Westwood-Squibb
Moisturel cream	Westwood-Squibb
Moisturel skin cleanser	Westwood-Squibb

V. ANIMAL TOXICITY DATA

A. Sensitization

Delayed contact dermatitis was produced in 50% of guinea pigs following induction applications of 88 ppm three times a week for 3 weeks and a challenge concentration of 2000 ppm, a level which was found to be nonirritating.[2]

Guinea pig maximization tests have shown the chlorinated derivative to be the active sensitizer. Cross-reactions may occur to other isothiazolinone compounds, such as 4,5-dichloro-2-methyl-4-isothiazolin-3-one.[5]

VI. HUMAN TOXICITY DATA

A. Allergic Contact Dermatitis

1. Experimental Sensitization

A series of repeat insult patch tests were performed in 1450 subjects with varying concentrations of Kathon CG®. Induction was similar to that described for the guinea pig model with application 3 times per week for 3 weeks. No evidence of skin sensitization occurred with concentrations of 5, 6, or 10 ppm in 1121 subjects or with 15 ppm in 200 subjects. One of 84 subjects tested with 12.5 ppm and 2 of 45 tested with 20 ppm reacted with a delayed hypersensitivity response.[6]

Daily application of a leave-on body lotion containing 15 ppm of Kathon CG® for 13 weeks in a double-blind study of 209 healthy subjects was not associated with development of irritation or sensitization.[7]

2. *Product Use Experience*

Case reports of allergic contact dermatitis to Kathon CG® are numerous and include exposure via industrial sources, such as cutting oils and water-cooling biocides.[8-10] The incidence of positive delayed patch test reactions to Kathon CG® has varied, depending on the nature of the product exposure, concentration in the patch test, year studied, geographic location, and patient population.

Prevalence (%)	N	Population	Concentration (ppm)	Country	Year	Ref.
13 (0.8%)	1511	Contact dermatitis	100	Denmark	1983	11
18 (1.3%)	1396	Contact dermatitis	100	Denmark	1985–1988	12
0	300	Contact dermatitis	15	Hungary	1983	13
16 (2%)	749	Eczema	100	Finland	1985	14
14 (4.9%)	285	Eczema	100	Finland	1986	14
7 (1.4%)	501	Contact dermatitis	100	Netherlands	1985	15
8 (11%)	75	Cosmetic dermatitis	100	Netherlands	1981–1986	16
33 (27.7%)	119	Cosmetic dermatitis	100	Netherlands	1986–1987	17
43 (4.4%)	976	Occupational dermatitis	300	Sweden	1986	18
50 (1.3%)	3744	Dermatitis	100	Italy	1983–1986	19
52 (8.3%)	620	Contact dermatitis	100	Italy	1985–1987	20
13 (3.6%)	365	Contact dermatitis	100	U.S.	1983–1986	21
20 (3.1%)	655	Contact dermatitis	100	U.S.	1986–1987	22
7 (2%)	358	Contact dermatitis	100	U.S.	1987–1988	22
2 (1.1%)	174	Contact dermatitis	100	U.S.	1986	23
20 (1.7%)	1182	Dermatitis	100, 250	U.S.	1985–1987	24
7 (1.1%)	653	Contact dermatitis	100	England	1986–1987	25
23 (5.5%)	420	Dermatitis	100	Switzerland	1986–1987	26
6 (0.8%)	718	Dermatitis	100	Czechoslovakia	1988–1989	27
6 (1.1%)	540	Dermatitis	100	France	1986–1988	28
22 (3.5%)	626	Contact dermatitis	200	Spain	1988–1989	29

The most active allergen in this combination product was the chlorinated derivative, chloromethylisothiazolinone, which produced positive patch test reactions in all of 22 patients with documented Kathon CG® sensitivity; only 2 reacted to methylisothiazolinone.[30]

VII. CLINICAL RELEVANCE

When consecutive series of patients presenting to dermatology clinics have been patch tested over many years, a trend toward an increasing prevalence of sensitization to Kathon CG® has been found in Finland, the Netherlands, Sweden, and Italy, while a more stable low level of sensitivity of around 1% was demonstrated in England.[31] The higher incidence in some countries appears to be related to the availability of cosmetics containing concentrations exceeding the recommended 15 ppm,[20] but sensitization has occurred to products containing as low as 7 ppm.[32] The maximum permitted concentration in European cosmetics was reduced in 1989 from 30 to 15 ppm.[22] The Cosmetic Ingredient Review Expert Panel has suggested a tentative recommendation to exclude Kathon CG® from leave-on cosmetics.[32]

In most series of patients tested in the U.S., the prevalence of hypersensitivity has remained fairly constant between 1 and 2%; however, in series of elderly patients with a history of chronic, generalized, often refractory dermatitis, 2 to 3.5% had relevant sensitivity to Kathon CG®. The sensitization was most frequently traced to two leave-on moisturizing products. The high incidence in this series may reflect a highly selected population in a tertiary care facility.[21]

Interpretation of patch test results is confounded by the similarity of these reactions to irritant responses. The typical lesion is erythematous, edematous, and vesicular, with sharply demarcated borders and limited spreading outside of the patch test site. An aqueous vehicle containing 100 ppm of active ingredients is currently recommended.

REFERENCES

1. **Nikitakis, J. M.,** *CTFA Cosmetic Ingredient Handbook,* 1st ed., The Cosmetic, Toiletry and Fragrance Association, Washington, D.C., 1988.
2. **Chan, P. K., Baldwin, R. C., Parsons, R. D., Moss, J. N., Stiratelli, R., Smith, J. M., and Hayes, A. W.,** Kathon biocide: manifestation of delayed contact dermatitis in guinea pigs is dependent on the concentration for induction and challenge, *J. Invest. Dermatol.,* 81, 409, 1983.
3. **de Groot, A. C. and Weyland, J. W.,** Kathon CG: a review, *J. Am. Acad. Dermatol.,* 18, 350, 1988.
4. **Decker, R. L. and Wenninger, J. A.,** Frequency of preservative use in cosmetic formulas as disclosed to FDA-1982 update, *Cosmet. Toilet.,* 97, 57, 1982.
5. **Bruze, M., Fregert, S., Gruvberger, B., and Persson, K.,** Contact allergy to the active ingredients of Kathon CG in the guinea pig, *Acta Derm. Venereol.,* 67, 315, 1987.
6. **Cardin, C. W., Weaver, J. E., and Bailey, P. T.,** Dose-response assessments of Kathon biocide (II). Threshold prophetic patch testing, *Contact Dermatitis,* 15, 10, 1986.
7. **Schwartz, S. R., Weiss, S., Stern, E., Morici, I. J., Moss, J. N., Goodman, J. J., and Scarborough, N. L.,** Human safety study of body lotion containing Kathon CG, *Contact Dermatitis,* 16, 203, 1987.
8. **Pilger, C., Nethercott, J. R., and Weksberg,** Allergic contact dermatitis due to a biocide containing 5-chloro-2-methyl-4-isothiazolin-3-one, *Contact Dermatitis,* 14, 210, 1986.
9. **O'Driscoll, J. B. and Beck, M. H.,** Occupational allergic contact dermatitis from Kathon WT, *Contact Dermatitis,* 19, 63, 1988.
10. **de Boer, E. M., van Ketel, W. G., and Bruynzeel, D. P.,** Dermatoses in metal workers (II). Allergic contact dermatitis, *Contact Dermatitis,* 20, 280, 1989.
11. **Hjorth, N. and Roed-Petersen, J.,** Patch test sensitivity to Kathon CG, *Contact Dermatitis.* 14, 155, 1986.
12. **Menne, T. and Hjorth, N.,** Kathon CG reactivity in 1396 consecutively patch tested patients in the Copenhagen area, *Contact Dermatitis,* 19, 260, 1988.
13. **Husz, S. and Simon, N.,** Epicutaneous patch test with the preservative Kathon CG, *Contact Dermatitis,* 15, 245, 1986.
14. **Hannuksela, M.,** Rapid increase in contact allergy to Kathon CG in Finland, *Contact Dermatitis,* 15, 211, 1986.
15. **de Groot, A. C., Bos, J. D., Jagtman, B. A., Bruynzeel, D. P., van Joost, T., and Weyland, J. W.,** Contact allergy to preservatives-II, *Contact Dermatitis,* 15, 218, 1986.
16. **de Groot, A. C.,** Contact allergy to cosmetics: causative ingredients, *Contact Dermatitis,* 17, 26, 1987.
17. **de Groot, A. C., Bruynzeel, D. P., Bos, J. D., van der Meeren, H. L. M., van Joost, T., Jagtman, B. A., and Weyland, J. W.,** The allergens in cosmetics, *Arch. Dermatol.,* 124, 1525, 1988.
18. **Bjorkner, B., Bruze, M., Dahlquist, I., Fregert, S., Gruvberger, B., and Persson, K.,** Contact allergy to the preservative Kathon CG, *Contact Dermatitis,* 14, 85, 1986.
19. **Meneghini, C. L., Angelini, G., and Vena, G. A.,** Contact allergy to Kathon CG, *Contact Dermatitis,* 17, 247, 1987.
20. **Tosti, A.,** Prevalence and sources of Kathon CG sensitization in Italy, *Contact Dermatitis,* 18, 173, 1988.
21. **Fransway, A. F.,** Sensitivity to Kathon CG: findings in 365 consecutive patients, *Contact Dermatitis,* 19, 342, 1988.

22. **Fransway, A. F.,** Isothiazolinone sensitivity, *Lancet,* 1, 910, 1989.
23. **Cronin, E., Hannuksela, M., Lachapelle, J. M., Maibach, H. I., Malten, K., and Meneghini, C. L.,** Frequency of sensitisation to the preservative Kathon CG, *Contact Dermatitis,* 18, 274, 1988.
24. **Rietschel, R. L., Nethercott, J. R., Emmett, E. A., Maibach, H. I., Storrs, F. J., Larsen, W. G., Adams, R. M., Taylor, J. S., Marks, J. G., Jr., Mitchell, J. C., Fisher, A. A., Kanof, N. B., Clendenning, W. E., and Schorr, W. F.,** Methylchloroisothiazolinone-methylisothiazolinone reactions in patients screened for vehicle and preservative hypersensitivity, *J. Am. Acad. Dermatol.,* 22, 734.
25. **Cox, N. H. and Shuster, S.,** Risk of sensitisation to Kathon, *Contact Dermatitis,* 18, 54, 1988.
26. **Pasche, F. and Hunziker, N.,** Sensitization to Kathon CG in Geneva and Switzerland, *Contact Dermatitis,* 20, 115, 1989.
27. **Machackova, J., Kalensky, J., and Vocilkova, A.,** Patch test sensitivity to Kathon CG in Prague, *Contact Dermatitis,* 22, 189, 1990.
28. **Foussereau, J.,** An epidemiological study of contact allergy to 5-chloro-3-methyl isothiazolone/3-methyl isothiazolone in Strasbourg, *Contact Dermatitis,* 22, 68, 1990.
29. **Hasson, A., Guimaraens, D., and Conde-Salazar, L.,** Patch test sensitivity to the preservative Kathon CG in Spain, *Contact Dermatitis,* 22, 257, 1990.
30. **Bruze, M., Dahlquist, I., Fregert, S., Gruvberger, B., and Persson, K.,** Contact allergy to the active ingredients of Kathon CG, *Contact Dermatitis,* 16, 183, 1987.
31. **Shuster, S. and Spiro, J.,** Measurement of risk of sensitisation and its application to Kathon, *Contact Dermatitis,* 17, 299, 1987.
32. **de Groot, A. C. and Herxheimer, A.,** Isothiazolinone preservative: cause of a continuing epidemic of cosmetic dermatitis, *Lancet,* 1, 314, 1989.

LACTOSE

I. REGULATORY CLASSIFICATION

Lactose is classified as a tablet and/or capsule diluent.

II. SYNONYMS

Lactosum
Milk sugar
Saccharum lactis

III. AVAILABLE FORMULATIONS

A. Foods
Lactose is a disaccharide obtained from milk whey and is less sweet than sucrose. Milk contains about 50 g of lactose per liter.

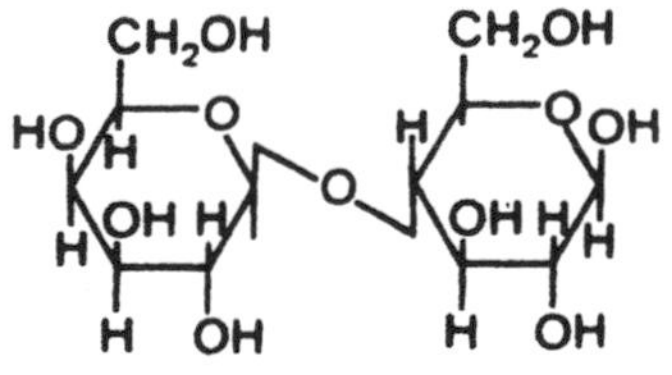

FIGURE 39. Lactose.

B. Drugs
Lactose is an extremely common excipient in solid oral dosage forms, present in over 2300 products approved by the FDA. It is a better diluent than starch for release of soluble drugs in capsule formulations. Other types of pharmaceutical products that may contain lactose include intramuscular, intravenous, and subcutaneous injectables; inhalation capsules; rectal syrups and tablets; topical creams, ointments, and powders; vaginal creams, suppositories, and tablets; and transdermal ointments and patches.[1]

C. Cosmetics

Lactose is used as a skin conditioning agent and humectant. Products include hair dyes and douches.[2]

IV. TABLE OF COMMON PRODUCTS

A. Lactose-Free Oral Drug Products

Trade name	Manufacturer
Achromycin V oral suspension	Lederle
Actifed syrup	Burroughs Wellcome
Allerest chewable, tablet, headache	Fisons
Ambenyl cough syrup, Ambenyl-D	Forest
Atarax syrup	Roerig
Azo-Gantanol tablets	Roche
Azo-Gantrisin tablets	Roche
Azulfidine tablets	Pharmacia
Bayer Aspirin products	Glenbrook
Benadryl products	Parke-Davis
Betapen VK oral solution	Squibb
Bronkolixir	Winthrop
Bronkotabs	Winthrop
Bufferin products	Bristol-Myers
Ceclor oral suspension	Lilly
Chlor-Trimeton syrup	Schering-Plough
Cleocin pediatric suspension	Upjohn
Colace products	Mead Johnson
Colymycin S oral suspension	Parke-Davis
Compazine spansules	Smith Kline & French
Congesprin products	Bristol-Myers
Corgard tablets	Bristol
CoTylenol products	McNeil
Cytoxan	Bristol-Myers
Declomycin products	Lederle
Dilantin suspensions	Parke-Davis
Dimetapp products	A.H. Robins
Dimetane elixir, expectorant, extentab	A.H. Robins
Diuril suspension	Merck Sharp & Dohme
Donnagel products	A.H. Robins
Donnatal elixir, tablet, extentab	A.H. Robins
Dorcol Childrens cough syrup	Dorsey
Duricef products	Mead Johnson
Dynapen oral suspension	Squibb
Empirin compound products	Burroughs Wellcome
Erythrocin stearate products	Abbott
Excedrin products	Bristol-Myers

Trade name (cont'd)	Manufacturer (cont'd)
Feosol products	Smith Kline Beecham
Fero-Folic 500 filmtab	Abbott
Gantanol products	Roche
Gantrisin pediatric suspension, syrup	Roche
Gelusil products	Parke-Davis
Iberet Folic 500 filmtab	Abbott
Inderal LA capsules	Wyeth-Ayerst
Klotrix	Bristol
K-lyte, K-lyte Cl products	Bristol
Lanoxin pediatric elixir	Burroughs Wellcome
Ledercillin VK products	Lederle
Lomotil products	Searle
Maalox products	Rorer
Maalox plus products	Rorer
Marax tablets, DF syrup	Roerig
Minipress capsules	Pfizer
Mintezol suspension	Merck Sharp & Dohme
Mucomyst	Mead Johnson
Mycostatin oral suspension	Squibb
Mylanta and Mylanta II products	Stuart
Naldecon syrup, pediatric drops	Bristol
Natalins, Natalins Rx	Mead Johnson
Nilstat oral suspension	Lederle
Norpramin tablets	Marion Merrell Dow
Omnipen oral suspensions	Wyeth-Ayerst
Ornade capsules	Smith Kline & French
Pentids syrup	Squibb
Pen-Vee-K oral solution	Wyeth-Ayerst
Periactin syrup	Merck Sharp & Dohme
Peri-Colace products	Mead Johnson
Phenergan syrup, expectorants	Wyeth-Ayerst
Phenergan VC expectorants	Wyeth-Ayerst
Polycillin suspension, drops	Squibb
Poly-Vi-Sol products	Mead Johnson
Principen oral suspension	Squibb
Prostaphlin oral solution	Squibb
Questran	Mead Johnson
Quibron products	Bristol
Robitussin products	A.H. Robins
Sinarest tablets	Fisons
Sine-Aid tablets	McNeil
Sudafed 30 mg tablet, syrup	Burroughs Wellcome
Sumycin syrup, tablets	Squibb
Suprax products	Lederle

Trade name (cont'd)	Manufacturer (cont'd)
Tegopen oral suspension	Squibb
Teldrin capsules	Smith Kline Beecham
Temaril products	Herbert
Tempra products	Mead Johnson
Terramycin products	Pfizer
Thorazine products	Smith Kline & French
Triaminic syrup, cough, expectorant	Sandoz
Triaminicol syrup	Sandoz
Trimox capsules	Squibb
Tylenol products	McNeil
Unipen oral solution	Wyeth-Ayerst
Veetids oral solution	Squibb
Velosef oral suspension	Squibb
Vibramycin products	Pfizer
Vi-Daylin products	Ross
Vistaril products	Pfizer
Zarontin products	Parke-Davis

V. HUMAN TOXICITY DATA

A. Lactose Intolerance
1. Foods

Lactose intolerance is a clinical condition resulting from malabsorption and inability to digest lactose. The usual route for lactose absorption involves generation of glucose and galactose after enzymatic digestion by lactase on the brush border of the small intestine. Undigested lactose is fermented by bacteria in the large intestine producing various gases that cause clinical symptoms of abdominal discomfort, flatulence, and borborygmi. Diarrhea results when the digestive capacity of the bacteria is exceeded due to an osmotic effect of the undigested solute.[3]

The ability to digest lactose is a dominantly inherited trait that occurs most frequently in adult habitants of Northern Europe, Finland, India, Hungary, and Mongolia. Unless intermixed with these races, other populations have been found to have an incidence of lactose malabsorption approaching 100%. Children are less likely to manifest lactose intolerance even in populations with a high incidence of lactase deficiency and usually do not experience adverse reactions before the age of 5 to 7 years and occasionally not until adolescence.[3]

An adaptive response occurs in adults who continue to consume lactose, allowing consumption of up to 240 ml of whole milk, presumably due to a protective effect of the enzyme. Tolerance can be increased by consumption of fermented dairy products such as yogurt, cheeses, and acidophilus milk. Pasteurization of these products decreases the inherent lactase content, but does not significantly affect tolerance.[4] In addition, there are wide interpatient variations in the extent of lactase deficiency and the amount of lactose required to produce symptoms of intolerance.

Transient lactose intolerance may occur as a result of intestinal mucosal injury in the setting of acute viral (particularly rotavirus), parasitic, or bacterial infections. Patients with altered brush border tissue due to chronic gluten enteropathy, chronic inflammatory gastrointestinal diseases, protein-energy malnutrition, or recent gastrointestinal surgery may develop secondary lactose intolerance.

2. Drugs

Intolerance of pharmaceuticals containing lactose is unlikely in most lactose-intolerant individuals due to the small amount of lactose present in most formulations. In one case report, a 24-year-old woman developed bloating, abdominal cramps, and flatulence 2 h after inhalation of an Intal® capsule containing 20 mg of lactose. The same reaction occurred on four subsequent occasions after using the Intal® turbo-inhaler. The reaction was confirmed by a double-blind challenge with an Intal® capsule dissolved in saline, a nebulized nonlactose containing Intal® solution, and isotonic saline. Mild nausea and abdominal cramps only occurred with the lactose challenge.[5]

Two lactose-intolerant individuals developed abdominal cramps and diarrhea after starting therapy with lactose-containing lithium carbonate and flurazepam hydrochloride. The amount of lactose was not determined.[6]

B. Asthma

Oral administration of a "placebo" capsule containing 300 mg of lactose was reported to cause severe dyspnea and wheezing 1 h after ingestion, with an unobtainable FEV_1 requiring epinephrine treatment, in a 42-year-old asthmatic woman being evaluated for possible aspirin intolerance. The reaction was repeated after a double-blind challenge.[7]

A 52-year-old asthmatic woman with lactose intolerance developed dyspnea and asthma following administration of a theophylline tablet containing lactose. Double-blind oral challenge confirmed bronchospasm following vanillin and a placebo capsule containing 500 mg of lactose. The peak expiratory flow decreased by 32 and 46% 4 and 4 h, respectively, after two lactose challenges.[8]

C. Bioavailability

Changes in pharmaceutical formulations to add or delete lactose as a diluent may result in significant changes in the bioavailability of the active drug. An outbreak of phenytoin toxicity in 51 Australian patients was attributed to a change in the diluent of one brand of phenytoin capsules from calcium sulfate dihydrate to lactose. Measurement of serial phenytoin blood concentrations in three of these patients switched to the lactose-containing products showed a marked increase in the phenytoin level at the same dose of phenytoin taken previously.[9]

VI. CLINICAL RELEVANCE

A. Lactose Intolerance

1. Foods

The American Academy of Pediatrics recommended in 1990 that children with an inherited predisposition to lactase deficiency continue to ingest milk products unless a definite intolerance or severe diarrhea develop. Tolerance can be encouraged by ingestion of fermented dairy products. The addition of commercially available lactase supplements to milk products will predigest the lactose and allow consumption by intolerant patients.[3] Although the suggested procedure involves incubating the milk product for 12 to 24 h to allow complete digestion of the lactose, some studies have demonstrated benefits from lactase when added immediately prior to ingestion of the milk or ingested separately in tablet form prior to lactose ingestion.[10,11] Lactrase® capsules are designed to be taken with food or sprinkled directly onto the food. Dairy-Ease® is a chewable tablet to be taken with or immediately after ingestion of lactose-containing products.

2. Drugs

The amount of lactose used to demonstrate clinical lactose intolerance in the majority of lactase-deficient individuals, 1 g/kg up to 50 g, is not found in pharmaceutical products and is unlikely to result in significant symptoms. A dose of 10 g in an adult produces symptoms in about 50%.[12] Lactose is present in amounts varying from 1 to 500 mg per tablet or capsule.[1]

REFERENCES

1. **Food and Drug Administration,** *Inactive Ingredients Guide,* FDA, Washington, D.C., March 1990.
2. **Nikitakis, J. M.,** *CTFA Cosmetic Ingredient Handbook,* 1st ed., The Cosmetic, Toiletry and Fragrance Association, Washington, D.C., 1988.
3. **American Academy of Pediatrics Committee on Nutrition,** Practical significance of lactose intolerance in children: supplement, *Pediatrics,* 86, 643, 1990.
4. **Savaiano, D. A., AbouElAnouar, A., Smith, D. E., and Levitt, M. D.,** Lactose malabsorption from yogurt, pasteurized yogurt, sweet acidophilus milk, and cultured milk in lactase-deficient individuals, *Am. J. Clin. Nutr.,* 40, 1219, 1984.
5. **Brandstetter, R. D., Conetta, R., and Glazer, B.,** Lactose intolerance associated with Intal capsules, *N. Engl. J. Med.,* 315, 1613, 1986.
6. **Lieb, J. and Kazienko, D. J.,** Lactose filler as a cause of "drug-induced" diarrhea, *N. Engl. J. Med.,* 299, 314, 1978.
7. **Zeiss, C. R. and Lockey, R. F.,** Refractory period to aspirin in a patient with aspirin-induced asthma, *J. Allergy Clin. Immunol.,* 57, 440, 1976.
8. **Van Assendelft, A. H. W.,** Bronchospasm induced by vanillin and lactose, *Eur. J. Respir. Dis.,* 65, 468, 1984.
9. **Tyrer, J. H., Eadie, M. J., Sutherland, J. M., and Hooper, W. D.,** Outbreak of anticonvulsant intoxication in an Australian city, *Br. Med. J.,* 2, 271, 1970.
10. **Biller, J. A., King, S., Rosenthal, A., and Grand, R. J.,** Efficacy of lactase-treated milk for lactose-intolerant pediatric patients, *J. Pediatr.,* 111, 91, 1987.
11. **Medow, M. S., Thek, K. D., Newman, L. J., Berezin, S., Glassman, M. S., and Schwarz, S. M.,** B-galactosidase tablets in the treatment of lactose intolerance in pediatrics, *Am. J. Dis. Child.,* 144, 1261, 1990.
12. **Davidson, G. P.,** Lactase deficiency: diagnosis and management, *Med. J. Aust.,* 141, 442, 1984.

LANOLIN

I. REGULATORY CLASSIFICATION

Lanolin and lanolin anhydrous are classified as ointment bases. Lanolin alcohol is used as an emulsifying and/or solubilizing agent.

II. SYNONYMS

Wool fat

III. AVAILABLE FORMULATIONS

A. Cosmetics

Lanolin is obtained by refining the unctuous sebaceous secretion (wool fat) of sheep. It is a complex mixture of esters of fatty acids and high molecular weight alcohols, including aliphatic, steroid, or triterenoid alcohols. Up to 12% consists of free fatty alcohols.[1,2] Many alterations of plain lanolin have been developed for specific cosmetic uses. Lanolin alcohol is extracted by hydrolysis of lanolin. It has strong water-in-oil emulsifying properties, absorbing up to 2000 times its own weight, and is used as a binder, emulsion stabilizer, skin conditioner, hair conditioner, and viscosity-increasing agent in products such as eye products, lipsticks, makeup, and creams and lotions.[3] Lanolin alcohol is a wax-like substance used as a stabilizer in stick formulations.[2]

Liquid lanolin is obtained by solvent fractionation of lanolin. It is odorless, tasteless, and soluble in mineral oil, making it useful in bath oil products. The other component of fractionation is lanolin wax. This wax-like substance is also odorless and tasteless and is a better water-in-oil emulsifier than plain lanolin. It is used primarily in lipsticks and glosses.[2]

Hydrogenated lanolin is a more saturated derivative with a large percentage of free alcohols. It has increased water absorption and mineral oil solubility, is odorless and tasteless, and is used in lip products, scented moisturizing and night creams and lotions, cleansing products, and emulsion formulas.[2,3]

Acetylated lanolin is a more hydrophobic product with a high ester and low alcohol content. It is useful in water-resistant products such as sunscreens, baby products, and hand and body creams.[2]

Ethoxylated lanolins, also known as polyoxyethylene lanolins, are reacted with ethylene oxide, resulting in a more hydrophilic product containing 6 to 75 mol of ethylene oxide. They are used as solubilizers in sunscreens, perfumes, pharmaceuticals, hair products, shampoos, soaps, dishwashing detergents, aftershave lotions, preshave lotions, astringent lotions, and colognes.[2]

Transesterified lanolin is partially saponified and is more soluble and spreadable than lanolin. It is used to create a lighter velvety-feeling emulsion in nongreasy formulations.

IV. TABLE OF COMMON PRODUCTS

A. Topical Drug Products

Trade name	Manufacturer	Lanolin type
A & D ointment	Schering	Anhydrous lanolin
Aquaphor	Beiersdorf	Lanolin alcohol
Balnetar	Westwood-Squibb	Lanolin oil
CaldeCort cream	Fisons	Lanolin alcohol
Cruex cream	Fisons	Anhydrous lanolin
Decaderm gel 0.1%	Merck Sharp & Dohme	Lanolin alcohol
Elimite cream	Herbert	Lanolin alcohol
Eucerin creme, lotion	Beiersdorf	Lanolin alcohol
Florone ointment	Dermik	Lanolin alcohol
Kwell cream	Reed & Carnrick	Lanolin
Lubriderm cream, lotion	Warner-Lambert	Alcohol, oil, anhydrous
Maxifor ointment	Herbert	Lanolin alcohol
Medicone Derma-HC ointment	Medicone	Lanolin
Nitro-Bid ointment	Marion	Lanolin
Pramosone cream, ointment	Ferndale	Lanolin alcohol
Proctocream HC	Reed & Carnrick	Lanolin alcohol
Sultrin cream	Ortho	Lanolin
Sween cream	Sween	Lanolin oil
Topicort cream 0.25%	Hoechst-Roussel	Lanolin alcohol
Topicort LP cream 0.05%	Hoechst-Roussel	Lanolin alcohol
Vanoxide HC lotion	Dermik	Lanolin alcohol

V. HUMAN TOXICITY DATA

A. Hypersensitivity

Lanolin is well known to cause delayed hypersensitivity contact dermatitis. The incidence is greatest in patients with leg ulcers and stasis eczema. The prevalence of positive patch tests to lanolin in this group of patients has ranged from 10 to 51% in a large series.[4-8]

In patients with contact allergy without leg dermatitis, the occurrence of positive patch tests to lanolin is relatively low. Various series have reported an incidence of 2 to 4% in this population.[8] The prevalence of positive patch tests in various populations (leg dermatitis patients excluded) is summarized in the following:

Prevalence (%)	N	Population	Country	Ref.
1.7	298	Contact dermatitis	U.S.	9
2–3	Unknown	Contact dermatitis	Sweden	8
2.6	270	Eczema	Norway	10
2	502	Contact dermatitis	Japan	11
2.7	1230	Eczema	Denmark	12
4.1	487	Cosmetic dermatitis	U.S.	13
4.4	899	Eczema	Denmark	12
9.5	330	Drug-induced dermatitis	Belgium	5

Contact hypersensitivity to lanolin may diminish with time and careful avoidance of exposure. When 57 patients with a history of lanolin hypersensitivity were retested 1 to 4 years later, only 20 (35%) tested positive. The incidence of positive reactions decreased as the time since the initial test increased. Testing with a purified lanolin, with free fatty alcohols removed, produced a positive reaction in one patient.[8]

Identification of the allergenic component of lanolin has been the subject of considerable debate. The lanolin alcohols have been the most frequently implicated culprit.[9] While wool alcohols are definitely allergic, patch testing with this substance exclusively will miss up to 50% of patients with lanolin-related hypersensitivity.[12] Caution should be used in evaluating reactions to testing with the lanolin alcohols.

False-positive reactions due to primary irritation have been described after patch testing with a 30% wool alcohol mixture.[14] Preliminary studies using a 10% wool alcohol product have demonstrated a lower false-positive rate.[9] There is also evidence that the detergent fraction of lanolin, used to separate the grease from the wool, is at least partially responsible for allergenicity. Removing the detergent from lanolin reduced the incidence of positive patch tests from 33 to 18% in patients with lanolin hypersensitivity. Removing both detergent and lanolin alcohols further decreased the incidence to 1%.[1]

Altered forms of lanolin, such as hydrogenated lanolin, may contain additional potential allergens, including saturated organic low molecular weight substances and trace amounts of contaminants of the hydrogenation process, nickel, copper, and chromium. The incidence of positive patch tests in a series of 502 contact dermatitis patients to hydrogenated lanolin was 5.2% compared to an incidence of 2% to plain lanolin.[11] Vollum[15] identified two patients with strongly positive patch tests to hydrogenated lanolin that tested negative to lanolin and wool alcohols.

Hydrogenated lanolin contains a higher percentage of free alcohols than plain lanolin, which may account for its increased allergenicity.[2] In a series of patients in England, the incidence of positive patch tests to hydrogenated lanolin was similar to that of lanolin alcohol (61 vs. 67%).[1] All products labeled hydrogenated lanolin may not be created equal; a Belgian author reported a lower incidence of hypersensitivity to hydrogenated lanolin than to other lanolin derivatives in a series of 330 patients.[5]

Alterations in lanolin that result in decreased free alcohol content, such as acetylated, transesterified, and ethoxylated lanolin, are associated with decreased allergenicity.[2,16] However, these alterations do not totally eliminate the risk of sensitization. Positive patch test results decreased from 33 to 12% in lanolin-sensitive patients tested with acetylated lanolin.[1]

Hypersensitivity to acetylated lanolin alcohol has been demonstrated in a patient with a negative reaction to acetylated and ethoxylated lanolin.[16]

The incidence of lanolin hypersensitivity in a general population of healthy individuals has been demonstrated to be less than 1 in 1000.[14] Estimates as low as 1 in 5.5 million have been made.[17]

B. Pesticide Contamination

Residues of pesticides used as insect repellant sheep dips have been found in lanolin preparations in Australia and the U.S. These dips are formulated to withstand rain and thus do not wash off readily when the wool is washed.[18] An investigation of four lots of anhydrous lanolin found trace amounts of diazinon, DDE, lindane, alpha-BHC, beta-BHC, chlorpyrifos, and dieldrin. The two compounds found in the highest concentrations were beta-benzene hexachloride 35 ppm and diazinon 21 ppm. Although these amounts of pesticides probably do not pose a threat to adult users of topical products, there is some concern over the popular use of lanolin to treat sore nipples in breast-feeding mothers where the infant may chronically ingest the pesticide.[19]

VI. CLINICAL RELEVANCE

A. Hypersensitivity

The incidence of delayed contact hypersensitivity to lanolin is highest in patients with leg stasis dermatitis, followed by drug-related dermatitis and chronic eczema. The incidence in the normal healthy population is extremely low. Controversy exists over the best method to test for documentation of suspected lanolin allergy. Patch testing with 30% wool alcohol is associated with a high degree of false positive reactions and requires confirmation with the product reported to be used by the patient. Some sources recommend testing initially with plain lanolin and/or with the suspected product.[20]

REFERENCES

1. **Clark, E. W., Cronin, E., and Wilkinson, D. S.,** Lanolin with reduced sensitizing potential, *Contact Dermatitis,* 3, 69, 1977.
2. **Schlossman, M. L. and McCarthy, J. P.,** Lanolin and derivatives chemistry: relationship to allergic contact dermatitis, *Contact Dermatitis,* 5, 65, 1979.
3. **Nikitakis, J. M.,** *CTFA Cosmetic Ingredient Handbook,* 1st ed., The Cosmetic, Toiletry and Fragrance Association, Washington, D.C., 1988.
4. **Breit, R.,** Allergen change in stasis dermatitis, *Contact Dermatitis,* 3, 309, 1977.
5. **Oleffe, J. A., Blondeel, A., and Boschmans, S.,** Patch testing with lanolin, *Contact Dermatitis,* 4, 233, 1978.
6. **Le Roy, R., Grosshans, E., and Foussereau, J.,** Recherche d'allergie de contact dans 100 cas d'ulcere de jambe, *Dermatosen,* 29, 168, 1981.
7. **Wilkinson, S., Wilkinson, J. D., and Wilkinson, D. S.,** Medicament contact dermatitis: risk sites, *Boll. Dermat. Allergol. Prof.,* 2, 21, 1987.
8. **Edman, B. and Moller, H.,** Testing a purified lanolin preparation by a randomized procedure, *Contact Dermatitis,* 20, 287, 1989.
9. **Epstein, E.,** The detection of lanolin allergy, *Arch. Dermatol.,* 106, 678, 1972.
10. **Wereide, K.,** Contact allergy to wool-fat ("lanolin"), *Acta Derm. Venereol.,* 45, 15, 1965.
11. **Sugai, T. and Higashi, J.,** Hypersensitivity to hydrogenated lanolin, *Contact Dermatitis,* 1, 146, 1975.
12. **Mortensen, T.,** Allergy to lanolin, *Contact Dermatitis,* 5, 137, 1979.
13. **Adams, R. M. and Maibach, H. I.,** A five-year-study of cosmetic reactions, *J. Am. Acad. Dermatol.,* 13, 1062, 1985.

14. **Kligman, A.M.,** Lanolin allergy: crisis or comedy, *Contact Dermatitis*, 9, 99, 1983.

15. **Vollum, D. I.,** Sensitivity to hydrogenated lanolin, *Arch. Dermatol.*, 100, 774, 1969.

16. **Giorgini, S., Melli, M. C., and Sertoli, A.,** Comments on the allergenic activity of lanolin, *Contact Dermatitis*, 9, 425, 1983.

17. **Clark, E. W., Blondeel, A., Cronin, E., et al.,** Lanolin of reduced sensitizing potential, *Contact Dermatitis*, 7, 80, 1981.

18. **Rosanove, R.,** Dangers of the application of lanolin, *Med. J. Aust.*, 46, 232, 1987.

19. **Copeland, C. A., Raebel, M., and Wagner, S. L.,** Pesticide residue in lanolin, *JAMA*, 261, 242, 1989.

20. **Fisher, A. A.,** *Contact Dermatitis*, 3rd ed., Lea & Febiger, Philadelphia, 1986.

MINERAL OIL

I. REGULATORY CLASSIFICATION

Mineral oil and light mineral oil are oleaginous vehicles. Light mineral oil is also used as a tablet and/or capsule lubricant.

II. SYNONYMS

Liquid Paraffin
Liquid Petrolatum
White Mineral Oil

III. AVAILABLE FORMULATIONS

A. Constituents
Heavy mineral oil is a mixture of liquid saturated hydrocarbons obtained from petroleum with a viscosity of not less than 34.5 cSt at 40°C. Light mineral oil is a similar mixture with a lower viscosity. These formulations are refined to remove all aliphatic (unsaturated straight chain), naphthenic, and aromatic hydrocarbons.

B. Foods
Mineral oil is allowed in small amounts in food products.

C. Drugs
Mineral oil or light mineral oil are listed as ingredients in at least 49 ophthalmic ointments or suspensions; 50 oral capsules or tablets; and 154 topical creams, lotions, or ointments.[1]

D. Cosmetics
Mineral oil is widely used in cosmetics such as eye shadows, lipsticks, lip glosses, facial makeup, cleansing products, creams, and lotions.[2]

IV. TABLE OF COMMON PRODUCTS

A. Ophthalmic Lubricants

Trade name	Manufacturer
AKWA tears ointment	Akorn
Duolube ointment	Muro
Duratears ointment	Alcon
Hypotears ointment	Coopervision
Lacri-Lube ointment	Allergan
Lacri-Lube NP ointment	Allergan
Refresh P.M. ointment	Allergan
Tears Renewed ointment	Akorn

V. ANIMAL TOXICITY DATA

Cosmetic grade mineral oil was found to lack comedogenicity in the rabbit ear model, producing no visible follicular keratosis.[3,4]

VI. HUMAN TOXICITY DATA

A. Lipoid Pneumonia

1. Cosmetics

Cosmetic-related lipoid pneumonia is unusual. In one such case, an 18-year-old girl with osteogenic sarcoma had a history of using large quantities of flavored lip gloss that contained mineral oil as the primary ingredient. A routine chest X-ray revealed a soft-tissue density which was resected. Microscopic examination revealed an organizing lipoid pneumonia with lipid-containing macrophages.[5]

2. Drugs

Most cases of excipient-related lipoid pneumonia have occurred following the chronic use of oily nose drops containing a mineral oil vehicle.[6-8] Risk factors in these cases included advanced age, very young age, debilitating illness, central nervous system disorders (Parkinson's disease, multiple sclerosis, cerebral palsy, hemiplegia), and functional or structural disorders of the pharynx or esophagus (achalasia, hiatal hernia). Only half of these cases have been symptomatic at the time of discovery with chest pain, dyspnea, cough, and fever; the remainder have been discovered after a routine chest radiography.

In a review of 136 published cases, the majority were reported in infants aged 2 years or younger (63 cases). The other age group predominantly affected were adults 50 years and older.[9] Mineral oil was administered as a component of nose drops in about 15% of these cases.

In a survey of 389 chronically ill patients, 57 (14.6%) were found to have evidence of lipoid pneumonia. Of these 57 patients, 5 had a history of the use of oily nose drops as the only mineral oil source, while 45 had a history of chronic ingestion of mineral oil.[6]

An unusual case of lipoid pneumonia was reported in a 67-year-old woman who had been using a topical ophthalmic ointment for the treatment of xerophthalmia. Constituents of the ointment included mineral oil 42.5%, petrolatum 55%, lanolin 2%, and chlorobutanol 0.5%. The ointment was applied to both eyes at bedtime and had been used for approximately 2 weeks before the onset of nocturnal coughing associated with production of a "slimy" mucus. A pulmonary infiltrate was found in the right middle lobe of the lung 4 months later after seeking consultation. Bronchial washings confirmed lipid-laden macrophages. Within a few weeks of discontinuing the ophthalmic product, the cough had disappeared. A repeat chest radiograph 19 months later was normal.[10]

B. Lipogranulomas

Lipogranulomas of the liver have been attributed to the routine ingestion of mineral oil in foods. This finding was noted in 9.2% of 824 liver biopsy specimens analyzed during the period 1978 to 1980. Gas liquid chromatography comparison of the biopsied sections and various food sources of mineral oil, such as polished apple or cucumber skins and candies, revealed mineral oil in both tissue and food sources. The incidence of this finding prior to the approval of use of mineral oil in foods was 1.7% in nonfatty livers.[11]

C. Allergic Contact Dermatitis

One case of allergic contact dermatitis was attributed to mineral oil in a series of 713 cases of cosmetic dermatitis.[12]

D. Irritation

Mineral oils are hygroscopic substances that may cause a drying effect on the horny layer surface of the skin, resulting in chronic skin damage.[13]

E. Photosensitivity

Photoallergic contact dermatitis was described in a 39-year-old man with a 2-year history of an eczematous eruption on the hands, forearms, and neck, which improved during the winter. He had been using a cutting oil in the workplace for 20 years. Patch tests were negative to ingredients of the cutting oil, but photopatch tests were positive for the cutting oil and mineral oil.[14]

F. Acneiform Eruptions

Although mineral oil is noncomedogenic in the rabbit ear model, mild grade I pomade acne was observed in 46% of male black subjects who regularly used mineral oil daily for at least 1 year. Experimental continuous occlusive exposure to the back in six subjects did not produce clinical acne, but slight follicular hyperkeratosis and acanthosis was found in three subjects.[15]

VII. CLINICAL RELEVANCE

Chronic topical application of high concentrations of mineral oil may result in hygroscopic damage. Mineral oil is generally nonacneigenic and has a very low incidence of allergic contact dermatitis.

The major hazard associated with mineral oil is the production of lipoid pneumonia from aspiration into the lungs, generally after chronic excessive use of products capable of gaining access to this site. Patients with debilitating illness, infants, and elderly patients are predisposed to develop lipid pneumonia. With the removal of mineral oil vehicles from nose drops, the most common source of this problem in the past, the incidence of reported cases has dropped. Novel sources of exposure, such as lip glosses and ophthalmic ointments, have been depicted in recent cases. In many cases, these were subclinical and only discovered during routine chest radiographs.

REFERENCES

1. **Food and Drug Administration**, *Inactive Ingredients Guide*, FDA, Washington, D.C., March 1990.
2. **Nikitakis, J. M.**, *CTFA Cosmetic Ingredient Handbook*, 1st ed., The Cosmetic, Toiletry and Fragrance Association, Washington, D.C., 1988.
3. **Kligman, A. M. and Mills, O.H.**, Acne Cosmetica, *Arch. Dermatol.*, 106, 843, 1972.
4. **Fulton, J. E., Jr., Pay, S. R., and Fulton, J. E., III**, Comedogenicity of current therapeutic products, cosmetics, and ingredients in the rabbit ear, *J. Am. Acad. Dermatol.*, 10, 96, 1984.
5. **Becton, D. L., Lowe, J. E., and Falletta, J. M.**, Lipoid pneumonia in an adolescent girl secondary to use of lipgloss, *J. Pediatr.*, 105, 421, 1984.
6. **Volk, B. W., Nathanson, L., Losner, S., Slade, W. R., and Jacobi, M.**, Incidence of lipoid pneumonia in a survey of 389 chronically ill patients, *Am. J. Med.*, 10, 316, 1951.
7. **Olsen, A. M.**, The spectrum of aspiration pneumonitis, *Ann. Otol. Rhinol. Laryngol.*, 79, 875, 1970.
8. **Spatafora, K. M., Bellia, V., Ferrara, G., and Genova, G.**, Diagnosis of a case of lipoid pneumonia by bronchoalveolar lavage, *Respiration*, 52, 154, 1987.
9. **Bishop, P. G. C.**, Oil aspiration pneumonia and pneumolipoidosis, *Ann. Intern. Med.*, 13, 1327, 1940.
10. **Prakash, U. B. S. and Rosenow, E. C.**, Pulmonary complications from ophthalmic preparations, *Mayo Clin. Proc.*, 65, 521, 1990.
11. **Dincsoy, H. P., Weesner, R. E., and MacGee, J.**, Lipogranulomas in non-fatty human livers: a mineral oil induced environmental disease, *Am. J. Clin. Pathol.*, 78, 35, 1982.
12. **Adams, R. M. and Maibach, H. I.**, A five-year-study of cosmetic reactions, *J. Am. Acad. Dermatol.*, 13, 1062, 1985.
13. **Zesch, A.**, Adverse reactions of externally applied drugs and inert substances, *Dermatosen*, 36, 128, 1988.
14. **Sakakibara, S., Kawabe, Y., and Mizuno, N.**, Photoallergic contact dermatitis due to mineral oil, *Contact Dermatitis*, 20, 291, 1989.
15. **Plewig, G., Fulton, J. E., and Kligman, A. M.**, Pomade acne, *Arch. Dermatol.*, 101, 580, 1970.

MONOSODIUM GLUTAMATE

I. REGULATORY CLASSIFICATION

Monosodium glutamate (MSG) is classified as a food additive and is on the GRAS list. It is infrequently used in pharmaceuticals and is not used in cosmetics. The FDA lists MSG as an inactive ingredient in one approved oral syrup pharmaceutical product.[1]

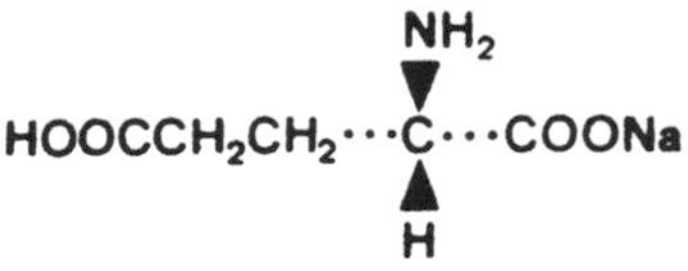

$$HOOCCH_2CH_2 \cdots \overset{\overset{\displaystyle NH_2}{\blacktriangledown}}{\underset{\underset{\displaystyle H}{\blacktriangle}}{C}} \cdots COONa$$

FIGURE 40. Monosodium glutamate.

II. SYNONYMS

Chinese seasoning
MSG
Sodium glutamate

III. AVAILABLE FORMULATIONS

Monosodium glutamate is a white, almost odorless, crystalline powder with a slightly sweet or salty taste. Each gram contains 5.5 meq of sodium. MSG is used as a flavor enhancer, imparting a meaty flavor, commonly in oriental foods. The FDA allows labeling of MSG or MSG-containing ingredients under several terms, which create difficulties in identifying MSG-containing products. Terms that usually indicate MSG include hydrolyzed vegetable protein, hydrolyzed plant protein, natural flavor, flavoring, and Kombu extract. Hydrolyzed vegetable protein obtained from wheat or soy typically contains 10 to 30% MSG.[2]

IV. TABLE OF COMMON PRODUCTS

A. Foods
1. Condiments

> Accent®
> Ajinomoto®
> Beef bouillon cubes and granules
> Chicken bouillon cubes and granules
> Seaweeds
> Seatangles
> Soy sauces
> Tenderquick®

2. Prepared Foods[2]

> Canned soups and gravies
> Beef stews, hot pies, beef burgers
> Lunchmeats
> English sausage
> Liver sausage and pates
> Meat curing brines
> Poultry products

3. Natural Foods[3]

> Cheeses
> Mushrooms
> Tomatoes

V. ANIMAL TOXICITY DATA

Both central and peripheral mechanisms for MSG toxicity have been investigated in animal studies. One group of investigators found hypothalamic neuronal lesions with doses of 500 mg/kg and larger in immature mice and primates.[4,5] All other studies in mature animals have failed to demonstrate neurotoxicity with daily intakes as high as 42 g/kg.[6]

An *in vitro* study using rabbit aorta demonstrated a concentration-related increased contractility following exposure to MSG, which was antagonized by the glutamate-receptor antagonist, ketamine. Tyramine was found to enhance MSG-induced vasoconstriction. In contrast, vessels that were previously constricted by adrenergic agonists, serotonin, prostaglandin F2a, or histamine were transiently relaxed by MSG, followed by a slight (histamine) or significant (serotonin) rebound vasoconstriction. These observations may have significance in the mechanism of MSG-induced vascular headache.[7]

The fact that many Oriental foods contain both tyramine and histamine may indicate a multifactoral mechanism for the "Chinese Restaurant Syndrome" that may not be mimicked by challenge with MSG alone. The food product most frequently associated with this syndrome, wonton soup, contained a high amount of histamine, 950 mcg per portion. Soy sauce contained 220 mcg/ml.[8]

VI. HUMAN TOXICITY DATA

A. Chinese Restaurant Syndrome

The first clinical syndrome to be associated with MSG was dubbed "Chinese Restaurant Syndrome" and was an anecdotal self-reported case of a physician who described numbness radiating from the neck to the arms, weakness, and palpitations whenever the first course of Chinese food was consumed in a restaurant setting. These effects were noticed 15 to 20 min after beginning the meal and lasted about 2 h.[9]

After a deluge of similar anecdotal reports of adverse reactions to MSG, a triad of dose-related symptoms was suggested. This triad included (1) paresthesias or a skin burning sensation, (2) facial pressure or tightness sensation, and (3) substernal chest pressure. Headache was a fourth symptom that occurred in a minority of individuals (two of eight with a prior history of vascular headache challenged in a double-blind manner). The oral dose threshold for development of one or more of these symptoms in 56 individuals tested with a single-blind procedure ranged widely from 1.5 to 12 g. The intensity of the response corresponded with the increase in dosage. Only one individual did not respond to an oral dose of 21 g. Many subjects experienced only a partial response to one or two components of the proposed syndrome.[10]

In a single-blind, placebo-controlled study of 98 healthy male volunteers, administration of MSG 5 g in aqueous solution or broth was associated with facial tightness in 12% and lightheadedness in 18%. The full triad could not be produced, even after subsequent double-blind challenges in 11 subjects with doses of up to 12 g.[11]

A double-blind crossover study of 24 healthy volunteers given MSG 3 g in beef broth demonstrated no difference in subjective symptoms between placebo and MSG.[12]

Kenney and Tidball[13] reported a placebo-controlled study on 77 normal volunteers given MSG 5 g in tomato juice. Of 25 subjects reporting symptoms from MSG, 22 were further studied with graded doses. A dose-response relationship was demonstrated for symptoms of facial stiffness and tightness and sensations of pressure, warmth, and tingling, with a threshold between 2 and 3 g. Nonspecific symptoms of heartburn, gastric discomfort, weakness, and lightheadedness were not dose related and were more frequent in the tomato juice placebo group. Plasma glutamate levels were measured and were not significantly different between reactors and the group as a whole.

Reports of Chinese Restaurant Syndrome in children are rare. In a case report of apparent symptoms of abdominal distress 10 min after ingestion of wonton soup in an 18-month-old boy, examination at 1 h revealed irritability, facial flushing, and injected conjunctivae. He was asymptomatic within 2 h and was not challenged with MSG.[14] Anecdotal reports in children have described "shudder attacks", with symptoms of shivering, shuddering, irritability, crying, tinnitus, confusion, ataxia, and delirium, which resolved after starting an MSG-free diet.[15-18] A study of eight normal 1-year-old infants given 54 mg/kg of MSG in a beef consomme indicated no difference in plasma glutamate levels compared to adult subjects. No adverse symptoms were reported in the infants.[19]

As the legend grew with time, so did the myriad of symptoms attributed to this food additive. Symptoms that have been associated with MSG include flushing,[20-22] vascular headache,[10,23] dyspepsia,[6,24] asthma,[25,26] psychosis or behavioral abnormalities,[18,27] ventricular tachycardia,[28] oro-facial granulomatosis,[29] and angioedema.[30]

Very few of these additional symptoms have been confirmed by adequate double-blind, placebo-controlled challenges. The flushing reaction was examined in six normal healthy subjects given up to 285.7 mg/kg of L-glutamate and up to 71.4 mg/kg of L-pyroglutamate. No patient developed flushing or changes in facial cutaneous blood flow, including three with a prior history of flushing reactions to Chinese food. The maximum dose induced a sensation of chest and hand tightness in one subject. A retrospective review of oral challenges with 3 g of MSG in aqueous solution in 15 subjects who reported flushing reactions again failed to induce flushing.[31] The facial burning or tightness sensation may be perceived as "flushing" subjectively.

Angioedema was a delayed effect, confirmed by placebo-controlled challenge with 250 mg or more of MSG in a 50-year-old man. A prodromal sensation of imminent swelling was reported within a few hours, with clinically evident angioedema at 16 to 24 h.[30] Angioedema accompanied by urticaria was reported in two children with positive oral capsule provocation tests with MSG 50 and 100 mg. MSG induced urticaria alone in two additional children.[32] Urticaria was described 1 to 6 h after oral challenge with 5 to 200 mg of sodium glutamate in 4 of 17 adults with chronic idiopathic urticaria and/or angioneurotic edema suspected to be related to food additives.[33]

Chronic granulomatous lesions of the upper lip, gum, and face were reported in an atopic 8.5-year-old girl, which worsened with double-blind challenge to sunset yellow, carmoisine, and less markedly to MSG.[29]

Many mechanisms have been proposed to explain the relationship of one or more of these symptoms to Chinese food. These include the high sodium content,[20] concomitant ingestion of ethanol,[23] additional ingredients in soy sauce (histamine, tyramine, unknown factors),[8,20] and cyclization of L-glutamate to form L-pyroglutamate during repeated boiling of soups and broths.[31] Substances shown to provoke the "classic triad" include monosodium L-glutamate, monopotassium L-glutamate, L-glutamic acid, and monosodium DL-glutamate. Similar reactions have not been produced with monosodium D-glutamate, sodium chloride alone, monosodium L-aspartate, or glycine.[10]

B. Asthma

The term "Chinese Restaurant Asthma" was coined in 1981 following the observation of a delayed onset of bronchospasm 11 to 12 h after ingestion of MSG in two patients with a history of food additive sensitivity. The challenge procedure that confirmed this reaction was single-blind but not placebo-controlled.[25] A subsequent report of 12 additional patients from the same investigators was performed in a placebo-controlled protocol, but objective quantified changes in pulmonary function tests were not described.[34]

In a later study of 32 subjects with a history of MSG-related asthma (14 subjects), unexplained severe sudden asthma episodes (18 subjects), or history of sensitivity to other ingested chemicals (15 subjects), oral single-blind, placebo-controlled challenges with 500 mg capsules of MSG were performed. Doses were determined individually for each patient, based on the asthma severity, and ranged from 500 mg to 5 g. Patients were followed by hourly peak expiratory flow rate measurements for 14 h after the challenge. Positive challenges were observed in 13 patients, 9 in the group with a previous history of MSG-related asthma. The onset of asthma was 1 to 2 h in seven patients and delayed for 6 to 14 h in the remaining patients.[3]

All of these reports concerned patients that were specifically selected because of a previous history of food sensitivity and did not provide data regarding the incidence of MSG sensitivity in an unselected population of asthmatics. A double-blind, randomized, crossover study of 12 unselected patients with chronic stable nonsteroid-dependent asthma showed no evidence of bronchoconstriction following oral capsule administration of 25 mg/kg of MSG, as measured by FEV_1. Pulmonary function tests were measured for only 4 h after the challenge in all but one patient, thus delayed subclinical effects may not have been detected. No patient reported subjective effects when questioned 24 and 48 h after the challenge. This small group of negative responders included one patient with a history of wheezing after eating Chinese food.[35]

A case report of a patient who had a respiratory arrest following ingestion of wonton soup illustrated a dose-dependent change in FEV_1 following a double-blind, placebo-controlled oral challenge with MSG. Challenge with MSG in a liquid form produced a more rapid bronchoconstrictive response (less than 1 h) and required a lower provocative dose (1 g) than did challenge with encapsulated MSG (onset 3 h, dose 3 g).[26] The earlier observation that Chinese Restaurant Syndrome occurs only with the first course of the meal, which is usually a soup,[10] is consistent with the increased response to an oral liquid formulation in this patient.

The mechanism of asthmatic reactions to MSG is open to speculation. A peripheral stimulation of irritant lung receptors has been suggested. Glutamate is a central nervous system neurotransmittor, and central augmentation of the irritant reflex activity may be responsible for delayed reactions.[3]

VII. CLINICAL RELEVANCE

A. Chinese Restaurant Syndrome

While most well-designed, placebo-controlled studies have concluded that the full triad of the originally described "Chinese Restaurant Syndrome" cannot be reproduced and may be related to simple dyspepsia, it is clear that some dose-related symptoms can be attributed to MSG. Well-designed studies using MSG dissolved in a liquid vehicle, such as broth, tomato juice, or soft drink, indicate that sensations specific to MSG (warmth, burning, tingling, numbness, tightness, pressure) could be provoked by about 30% of individuals tested with concentrations of greater than 3% or doses of 5 g.[6]

The typical onset of symptoms is within 10 to 20 min. The reactions are uniformly transient and self limited, dissipating spontaneously within 2 to 3 h. Although various treatments such as hydroxyzine and ketamine have been discussed in relation to various presumed mechanisms, due to the mild and transient nature of the reactions, treatment is unlikely to be necessary.

Symptoms attributed to MSG are most likely related to a nonspecific irritant effect on upper esophageal nerve endings in susceptible individuals and may be elicited more frequently by similar irritants, such as orange or tomato juice. Patients with underlying reflux esophagitis may be at increased risk.[6] The overall prevalence of Chinese Restaurant Syndrome in the general adult population has been estimated to be 1 to 2%.[21]

B. Asthma

Severe asthmatics or those with a history of food additive sensitivity may be at increased risk for dose-dependent exacerbation of asthma, particularly following ingestion of liquids, such as soups, containing MSG. Both immediate and delayed reactions have been observed. Careful double-blind challenges with MSG, preferably in both capsule and liquid media, should be performed to confirm suspected reactions. Pulmonary function tests should be measured for at least 12 h to detect delayed reactions. Patients with stable, less severe, asthma, without a history of food additive sensitivity, are at low risk for induction of bronchospasm after ingestion of MSG in usual quantities found in foods.

C. Angioedema/Urticaria

Angioedema and urticaria have been clearly documented to be related to MSG or other food additives in 21 to 67% of cases. In a series of 17 cases suspected to be related to food additives, 23% were exacerbated by MSG challenge. Doses of MSG required to induce urticaria or angioedema are much less than those associated with the Chinese Restaurant Syndrome, in the range of 5 to 200 mg.[33]

REFERENCES

1. **Food and Drug Administration,** *Inactive Ingredients Guide,* FDA, Washington, D.C., March 1990.
2. **Scopp, A. L.,** MSG and hydrolyzed vegetable protein induced headache: review and case studies, *Headache,* 31, 107, 1990.
3. **Allen, D. H., Delohery, D., and Baker, G.,** Monosodium L-glutamate-induced asthma, *J. Allergy Clin. Immunol.,* 80, 530, 1987.
4. **Olney, J. W.,** Brain lesions, obesity and other disturbances in mice treated with monosodium glutamate, *Science,* 164, 719, 1969.
5. **Olney, J. W., Sharpe, L. G., and Geigin, R. D.,** Glutamate-induced brain damage in infant primates, *J. Neuropathol. Exp. Neurol.,* 31, 464, 1972.
6. **Kenney, R. A.,** The Chinese restaurant syndrome: an anecdote revisited, *Food Chem. Toxicol.,* 24, 351, 1986.
7. **Merritt, J. E. and Williams, P. B.,** Vasospasm contributes to monosodium glutamate-induced headache, *Headache,* 30, 575, 1990.
8. **Chin, K. W., Garriga, M. M., and Metcalfe, D. D.,** The histamine content of oriental foods, *Food Chem. Toxicol.,* 27, 283, 1989.
9. **Kwok, H. M.,** Chinese restaurant syndrome, *N. Engl. J. Med.,* 278, 796, 1968.
10. **Schaumburg, H. H., Byck, R., Gerstl, R., and Mashman, J. H.,** Monosodium glutamate: its pharmacology and role in the Chinese restaurant syndrome, *Science,* 163, 826, 1969.
11. **Rosenblum, I., Bradley, J. D., and Coulston, F.,** Single and double blind studies with oral monosodium glutamate in man, *Toxicol. Appl. Pharmacol.,* 18, 367, 1971.
12. **Morselli, P. L. and Garattini, S.,** Monosodium glutamate and the Chinese restaurant syndrome, *Nature (London),* 227, 611, 1970.
13. **Kenney, R. A. and Tidball, C. S.,** Human susceptibility to oral monosodium L-glutamate, *Am. J. Clin. Nutr.,* 25, 140, 1972.
14. **Asnes, R. S.,** Chinese restaurant syndrome in an infant, *Clin. Pediatr.,* 10, 705, 1980.
15. **Reif-Lehrer, L. and Stemmermann, M. G.,** Monosodium glutamate intolerance in children, *N. Engl. J. Med.,* 293, 1204, 1975.
16. **Reif-Lehrer, L.,** A search for children with possible MSG intolerance, *Pediatrics,* 58, 771, 1976.
17. **Andermann, F., Vanasse, M., and Wolfe, L.,** Shuddering attacks in children: essential tremor and monosodium glutamate, *N. Engl. J. Med.,* 295, 174, 1976.
18. **Cochran, J. W. and Cochran, A. H.,** Monosodium glutamania: the Chinese restaurant syndrome revisited, *JAMA,* 252, 899, 1984.
19. **Stegink, L. D., Filer, L. J., Baker, G. L., and Bell, E. F.,** Plasma glutamate concentrations in 1-year-old infants and adults ingesting monosodium L-glutamate in consomme, *Pediatr. Res.,* 20, 53, 1986.
20. **Reif-Lehrer, L.,** Possible significance of adverse reactions to glutamate in humans, *Fed. Proc.,* 35, 2205, 1976.
21. **Kerr, G. R., Wu-Lee, M., El-Lozy, M., et al.,** Prevalence of the "Chinese restaurant syndrome", *J. Am. Diet. Assoc.,* 75, 29, 1979.

22. **Ghadimi, H., Kumar, S., and Abaci, F.,** Studies on monosodium glutamate ingestion, *Biochem. Med.,* 5, 447, 1971.

23. **Sauber, W. J.,** What is Chinese restaurant syndrome?, *Lancet,* 1, 721, 1980.

24. **Smith, S. J., Markadu, N. D., Roteller, C., Elder, D. M., and MacGregor, G. A.,** A new or old Chinese restaurant syndrome?, *Br. Med. J.,* 285, 1205, 1982.

25. **Allen, D. H. and Baker, G. H.,** Chinese restaurant asthma, *N. Engl. J. Med.,* 305, 1154, 1981.

26. **Koepke, J. W. and Selner, J. C.,** Combined monosodium glutamate (MSG)/metabisulfite (MBS) induced asthma, *J. Allergy Clin. Immunol.,* 77, 158, 1986.

27. **Colman, A. D.,** Possible psychiatric reactions to monosodium glutamate, *N. Engl. J. Med.,* 299, 902, 1978.

28. **Gann, P.,** Ventricular tachycardia in a patient with "Chinese Restaurant Syndrome", *South. Med. J.,* 70, 879, 1977.

29. **Sweatman, M. C., Tasker, R., Warner, J. O., Ferguson, M. M., and Mitchell, D. N.,** Oro-facial granulomatosis: response to elemental diet and provocation by food additives, *Clin. Allergy,* 16, 331, 1986.

30. **Squire, E. N.,** Angio-oedema and monosodium glutamate, *Lancet,* 1, 988, 1987.

31. **Wilkin, J. K.,** Does monosodium glutamate cause flushing (or merely "glutamania")?, *J. Am. Acad. Dermatol.,* 15, 225, 1986.

32. **Botey, J., Cozzo, M., Marin, A., and Eseverri, J. L.,** Monosodium glutamate and skin pathology in pediatric allergology, *Allergol. Immunopathol.,* 16, 425, 1988.

33. **Genton, C., Frei, P. C., and Pecoud, A.,** Value of oral provocation tests to aspirin and food additives in the routine investigation of asthma and chronic urticaria, *J. Allergy Clin. Immunol.,* 76, 40, 1985.

34. **Allen, D. H., Van Nunen, S., Loblay, R., and Clarke, L.,** Adverse reactions to food, *Med. J. Aust.,* 141 (Suppl. 5), S37, 1984.

35. **Schwartzstein, R. M., Kelleher, M., Weinberger, S. E., Weiss, J. W., and Drazen, J. M.,** *J. Asthma,* 24, 167, 1987.

MUSK AMBRETTE

I. REGULATORY CLASSIFICATION

Musk ambrette is a synthetic nitrated tertiary butyl toluene used as an artificial musk in perfumes and fragrances.

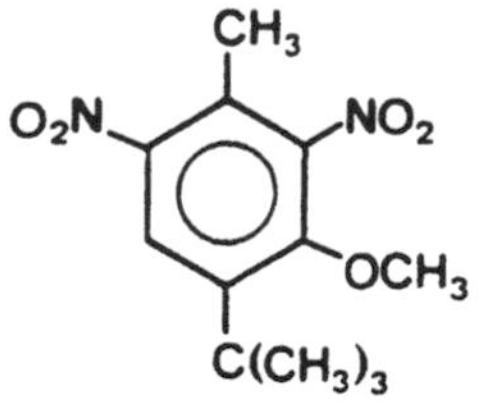

FIGURE 41. Musk ambrette.

II. SYNONYMS

2-methoxy 3,5-dinitro-4-methyl-tertiary butyl benzene

III. AVAILABLE FORMULATIONS

A. Foods

Musk ambrette is on the FDA GRAS list for use as a food flavoring in concentrations of 0.01 to 10 ppm.[1] It is used to simulate a blackberry flavor.

B. Cosmetics

Musk ambrette is an extremely commonly used fixative component of fragrances, found in concentrations of 1 to 15%. Like other fragrance components, it is infrequently, if at all, listed as an ingredient on product labels. Types of cosmetic products include colognes, perfumes, aftershave lotions, creams, lotions, dentifrices, and deodorants.[1]

IV. HUMAN TOXICITY DATA

A. Photoallergic Contact Dermatitis

Musk ambrette can cause a photoallergic dermatitis that can be detected by patch testing with concentrations as low as 0.25% followed by ultraviolet A or B irradiation. Eruptions may be transient and resolve within 1 week after discontinuation of exposure[2] or may evolve into chronic persistent light reactions with a reduced erythemal threshold to UV-B radiation.[3-6] Patients with transient reactions generally show normal responses to UV-B radiation. Because musk ambrette is present in such a wide range of products and usually not labeled, some persistent reactions may be due to unrecognized reexposure.

Persistent light reactors may develop actinic reticuloid reactions with granulomatous infiltrates.[7] The chronic eruption is characterized by scaling, erythema, thickened and lichenified skin, and postinflammatory hyperpigmentation.

Unlike most cosmetic-related contact dermatitis, these reactions are commonly seen in men due to the widespread use in men's colognes and aftershave lotions. The frequent application to abraded skin after shaving may also explain why most cases occur in men. Photodegradation of musk ambrette with generation of reactive nitrogen-containing radicals has also been implicated.[8] A typical clinical presentation is a pruritic, erythematous dermatitis of the neck and beard area following exposure to sunlight.[9]

A pigmented, lichenoid reaction, without prior dermatitis or pruritus, developed 1 month after photopatch testing with musk ambrette in a 45-year-old man who had used an aftershave lotion containing this fragrance. Persistent light reactivity was demonstrated, and a lesser reaction occurred on nonirradiated sites.[10]

The incidence of positive photopatch test reactions to musk ambrette was 12.8% in 70 patients tested; all were related to aftershave exposure.[11]

B. Cross-Sensitivity

Several related nitromusk compounds were tested and found not to produce cross-reactions in patients with musk ambrette photoallergic contact dermatitis; these included musk ketone, musk xylene, and musk tibetine.[2] However, moskene produced cross-reactions in 8 of 21 patients, and musk xylene and musk ketone produced cross-reactions in 1 case each.[12]

C. Allergic Contact Dermatitis

Nonphotoallergic contact dermatitis has also been attributed to musk ambrette. Acute contact dermatitis after exposure to aerosols, with marked eyelid swelling, has been reported.[12] Positive patch tests were found in 11 of 713 (1.5%) patients with cosmetic dermatitis.[13]

D. Inhalation Exposure

Allergic pigmented and depigmented contact dermatitis has been described from exposure to airborne musk ambrette found in incense.[14,15]

V. CLINICAL RELEVANCE

Musk ambrette has not yet been reviewed by the Cosmetic Ingredient Review Expert Panel.[16] Due to the high frequency of photoallergy and persistent, chronic light reactions to this compound, particularly when applied to abraded skin, it has been recommended by the

International Fragrance Research Institute that musk ambrette should not be used as a fragrance ingredient at a level over 4% in fragrances and at a level over 0.1% in aftershave fragrances.[17]

REFERENCES

1. **Fisher, A. A.**, Perfume dermatitis. Part II. Photodermatitis to musk ambrette and 6-methylcoumarin. *Cutis.* 26, 549, 1980.

2. **Raugi, G. J., Storrs, F. J., and Larsen, W. G.**, Photoallergic contact dermatitis to men's perfumes, *Contact Dermatitis.* 5, 251, 1979.

3. **Giovinazzo, V. J., Harber, L. C., Bickers, D. R., Armstrong, R. B., and Silvers, D. N.**, Photoallergic contact dermatitis to musk ambrette, *Arch. Dermatol.*. 117, 344, 1981.

4. **Burry, J. N.**, Persistent light reaction associated with sensitivity to musk ambrette, *Contact Dermatitis.* 7, 46, 1981.

5. **Ramsay, C. A.**, Transient and persistent photosensitivity due to musk ambrette. Clinical and photobiological studies, *Br. J. Dermatol.*. 111, 423, 1984.

6. **Ducombs, G., Abbadie, D., and Maleville, J.**, Persistent light reaction from musk ambrette, *Contact Dermatitis,* 14, 129, 1986.

7. **Cirne De Castro, J. L., Pereira, M. A., Nunes, F. P., and Pereira Dos Santos, A.**, Musk ambrette and chronic actinic dermatitis, *Contact Dermatitis,* 13, 302, 1985.

8. **Shibamoto, T.**, Photochemistry of fragrance materials. II. Aromatic compounds and phototoxicity, *J. Toxicol. Cut. Ocular Toxicol.*. 2, 267, 1983.

9. **Brandao, F., Cirne De Castro, J., and Pecegueiro, M.**, Photoallergy to musk ambrette, *Contact Dermatitis,* 9, 332, 1983.

10. **Parodi, G., Guarrera, M., and Rebora, A.**, Lichenoid photocontact dermatitis to musk ambrette, *Contact Dermatitis,* 16, 136, 1987.

11. **Menz, J., Muller, S. A., and Connolly, S. M.**, Photopatch testing: a six-year experience, *J. Am. Acad. Dermatol.*. 18, 1044, 1988.

12. **Wojnarowska, F. and Calnan, C. D.**, Contact and photocontact allergy to musk ambrette, *Br. J. Dermatol.*, 114, 667, 1986.

13. **Adams, R. M. and Maibach, H. I.**, A five-year-study of cosmetic reactions. *J. Am. Acad. Dermatol.*, 13, 1062, 1985.

14. **Hayakawa, R., Matsunaga, K., and Arima, Y.**, Airborne pigmented contact dermatitis due to musk ambrette in incense, *Contact Dermatitis.* 16, 96, 1987.

15. **Hayakawa, R., Matsunaga, K., and Arima, Y.**, Depigmented contact dermatitis due to incense. *Contact Dermatitis.* 16, 272, 1987.

16. **Bergfeld, W. F., Elder, R. L., and Schroeter, A. L.**, The Cosmetic Ingredient Review self-regulatory safety program, *Dermatol. Clin.*. 9, 105, 1991.

17. **Larsen, W. G.**, Perfume dermatitis, *Am. J. Acad. Dermatol.*. 12, 1, 1985.

OLEIC ACID

I. REGULATORY CLASSIFICATION

Oleic acid is used as an emulsifying and/or solubilizing agent in pharmaceuticals.

$$CH_3(CH_2)_7CH=CH(CH_2)_7COOH$$

FIGURE 42. Oleic acid.

II. AVAILABLE FORMULATIONS

A. Drugs
Oleic acid is used in inhalation aerosol products for nasal and pulmonary use.

B. Cosmetics
Oleic acid may be present in eye and facial makeup products and in hair dyes.

III. TABLE OF COMMON PRODUCTS

A. Inhalation Drug Products

Trade name	Manufacturer
Beclovent Inhalation Aerosol	Allen & Hanburys
Beconase inhalation aerosol	Allen & Hanburys
Proventil Inhaler	Schering
Vanceril Inhaler	Schering
Ventolin Inhalation Aerosol	Allen & Hanburys

IV. HUMAN TOXICITY DATA

A. Inhalation Drug Products
A metered-dose metaproterenol product containing soya lecithin was used in 900 well-controlled asthmatics and compared to a placebo metered-dose product containing only inert

ingredients (oleic acid) and propellants in 175 asthmatics. Bronchoconstriction was observed in 52 subjects, with an incidence of 4.4% in the drug group and 6.9% in the placebo group. Patients with bronchoconstriction from the placebo system did not improve after receiving the active metered-dose inhaler, but did improve after nebulized metaproterenol, implicating one or more of the excipients as causing the adverse response. The investigators did not determine which of the excipients was responsible.[1]

In another study of 12 asthmatic subjects with a history of severe coughing and wheezing after inhalation of beclomethasone aerosol, patients were enrolled in a double-blind crossover trial comparing a single dose of either beclomethasone aerosol or a placebo metered-dose inhaler containing oleic acid and fluorocarbon propellants. The incidence of coughing averaged 31 times after the active inhaler and 19 times after the placebo inhaler. The forced expiratory volume in 1 s (FEV_1) decreased to a similar degree with both treatments, a mean of 22.6% after the active inhaler and 22% after the placebo. Pretreatment with an inhaled bronchodilator enabled 7 of the 12 subjects to tolerate beclomethasone. The overall incidence of this adverse reaction was 20% of 70 patients prescribed the drug.[2]

V. CLINICAL RELEVANCE

The incidence of paradoxical bronchoconstriction related to the oleic acid excipient in some bronchodilator or inhaled corticosteroid products is confounded by the inherent irritability of the corticosteroid and partial reversal of the adverse effect when combined with a bronchodilator. Further studies are needed to determine the role of oleic acid, as well as other excipients, in asthmatic patients.

REFERENCES

1. **Yarborough, L., Mansfield, L., and Ting, S.,** Metered dose inhaler induced bronchospasm in asthmatic patients, *Ann. Allergy,* 55, 25, 1985.
2. **Shim, C. and Williams, M. H.,** Cough and wheezing from beclomethasone aerosol, *Chest,* 91, 207, 1987.

OLIVE OIL

I. REGULATORY CLASSIFICATION

Olive oil is considered to be an oleaginous vehicle.

II. SYNONYMS

Azeite
Gomenoleo oil
Oleum Olivae

III. AVAILABLE FORMULATIONS

A. Constituents
Olive oil is the fixed oil expressed from the fruits of *Olea eurpaea*. The major constituents are triolein, tripalmitin, trilinolein, tristearate, monostearate, triarachidin, squalene, *B*-sitosterol, and tocopherol.[1]

B. Drugs
Olive oil is infrequently used in drug products in the U.S. The FDA has only registered one oral solution and one topical solution product containing this excipient.[2]

C. Cosmetics
Olive oil is used as a solvent and skin and hair conditioner in cosmetics. Types of products include shampoos and hair conditioners, cleansing products, topical creams and lotions, and suntan products.[3]

IV. HUMAN TOXICITY DATA

A. Injectable Drug Products
Subcutaneous injection of olive oil into the forearm resulted in documentation of a calcified lipogranuloma 60 years later in a 67-year-old woman.[4]

Intramuscular injection of an estrogen-progesterone mixture in an olive oil vehicle produced an erythematous annulare lesion 9 to 10 d after injection. The reaction recurred with each subsequent injection.[5]

B. Contact Dermatitis

Positive epicutaneous tests documented delayed-type hypersensitivity to olive oil in two patients. Patch tests to the major individual constituents of the oil was negative.[6]

Contact dermatitis to topical ointments containing olive oil have been described.[7-9] Most of these patients were using the ointments to treat stasis leg ulcers, a condition known to predispose to allergic contact dermatitis.[1,8] Of 13 cases of contact allergy to olive oil seen over a 4-year period in two Swedish hospitals, 9 had a diagnosis of venous eczema. Two patients had occupational exposure to olive oil while performing pedicures and developed hand eczema. Testing to the known individual components of olive oil was negative. The prevalence of positive patch tests to olive oil in 285 patients tested during 1 year was 0.7%.[1] No cases of contact dermatitis secondary to olive oil have been reported in the U.S. since 1943.[9]

V. CLINICAL RELEVANCE

Olive oil is an infrequent sensitizer and does not appear to be a significant allergen in the U.S., possibly due to development of oral tolerance. As with other allergens, caution should be used when using as a vehicle in patients with stasis dermatitis. Occupational exposure in food handlers and pedicurists is a potential concern. The specific allergen has not been identified.

REFERENCES

1. **Padoan, S. M., Pettersson, A., and Svensson, A.,** Olive oil as a cause of contact allergy in patients with venous eczema, and occupationally, *Contact Dermatitis,* 23, 73, 1990.
2. **Food and Drug Administration,** *Inactive Ingredients Guide.* FDA, Washington, D.C., March 1990.
3. **Nikitakis, J. M.,** *CTFA Cosmetic Ingredient Handbook,* 1st ed., The Cosmetic, Toiletry and Fragrance Association, Washington, D.C., 1988.
4. **Galland, M. C., Cohen, M., Aquaron, R., Maurin, R., Duick, J. P., Bouteiller, J. C., Sauget, Y., Manez, R., Pizzi-Anselme, M., and Pelissier, J. L.,** Calcified lipogranuloma after gomenoleo oil injection:"paraffinoma" 60 years later, *Therapie,* 45, 27, 1990.
5. **Puissant, A., Saurat, J. H., Dalanoe, J., Noury-Duperrat, G., and Barrade, A.,** Anular centrifugal erythema on the site of injection of hormone preparation in oil suspension, *Med. Cutan. Ibero. Lat. Am.,* 4, 19, 1976.
6. **van Joost, T., Smitt, J. H., and van Ketel, W. G.,** Sensitization to olive oil (olea europeae), *Contact Dermatitis,* 7, 309, 1981.
7. **de Boer, E. M. and van Ketel, W. G.,** Contact allergy to an olive oil containing ointment, *Contact Dermatitis,* 11, 128, 1984.
8. **Jung, H. D. and Holzegel, K.,** Contact allergy to olive oil, *Derm. Beruf. Umwelt.,* 35, 131, 1987.
9. **Sutton, R. L.,** Contact dermatitis from olive oil, *JAMA,* 122, 34, 1943.

PARABENS

I. REGULATORY CLASSIFICATION

A. Foods

Methylparaben and propylparaben have been reconfirmed by the FDA as GRAS for direct addition to food, as well as indirect addition via packaging materials, in concentrations of up to 0.1%. Both compounds are used to inhibit molds and yeast in various foods. They are active against Gram-positive and a few Gram-negative organisms.[1]

B. Drugs

Methylparaben, ethylparaben, propylparaben, and butylparaben are classified as antimicrobial preservatives.

C. Cosmetics

Parabens are widely used cosmetic preservatives present in a wide variety of products, including face, body, and hand creams, lotions, and moisturizers; eye makeup products; foundations and other makeup products; night creams and lotions; cleansing products; hair conditioners; bubble baths; shampoos; mud packs; underarm deodorants; skin lighteners; and sachets.[2]

Parabens may be present in some cosmetics labeled as "hypoallergenic". Concentrations generally used are less than 0.3%, with the most common system containing 0.2% methylparaben and 0.1% propylparaben[3] but may range up to 1%.[4] In a survey of 20,183 cosmetics, parabens were present in 14,335.[5] In 1981, there were over 13,200 cosmetic formulations containing parabens listed with the FDA.[4]

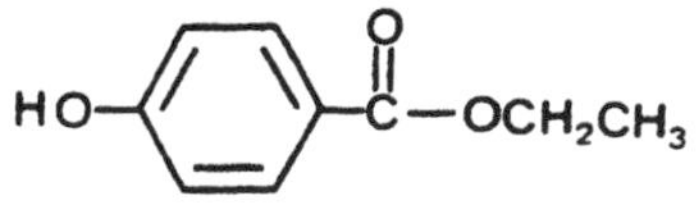

FIGURE 43. Ethylparaben.

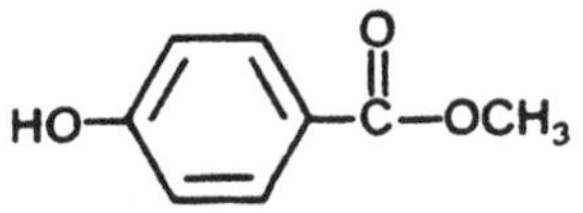

FIGURE 44. Methylparaben

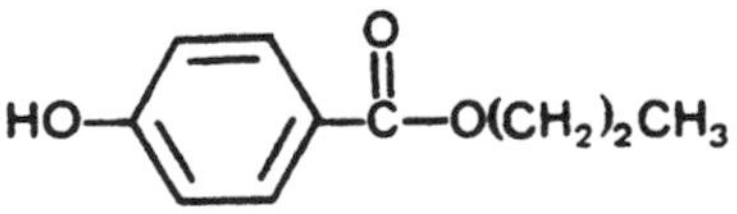

FIGURE 45. Propylparaben.

251

II. AVAILABLE FORMULATIONS

A. Foods

Types of foods that may contain parabens include alcoholic beverages, baked goods, cheese, fats and oils, frozen dairy products, gelatins, grain products, jams, jellies, marmalades, mincemeat, nonalcoholic beverages, olives, pickles, relishes, preserves, processed fruits and vegetables, tomato pulp, tomato puree, catsup, fruit juices, soft drinks, puddings, seasonings, soft candy, sugar substitutes, syrups, and sweet sauces.[1,3]

III. TABLE OF COMMON PRODUCTS AND ALTERNATIVES

A. Parenteral Drug Products

Trade name	Manufacturer	Alternative product
Adriamycin RDF	Adria	Adriamycin PFS
Aldomet	Merck Sharp & Dohme	None
Anectine	Burroughs Wellcome	None
Apresoline	Ciba	None
Aramine	Merck Sharp & Dohme	None
Bicillin CR	Wyeth-Ayerst	None
Bicillin CR 900/300	Wyeth-Ayerst	None
Bicillin LA	Wyeth-Ayerst	None
Decadron phosphate	Merck Sharp & Dohme	None
Depo-Provera 100 mg/ml	Upjohn	Depo-Provera 400 mg/ml
Elavil	Merck Sharp & Dohme	None
Garamycin	Schering	Garamycin IV
Garamycin pediatric	Schering	Garamycin IV
Haldol	McNeil	None
Hydrocortone phosphate	Merck Sharp & Dohme	None
Inapsine multidose vial	Janssen	Single-dose ampule
Levo-Dromoran 2 mg/ml ampule	Roche	10 ml vial
Lidocaine multidose vial	Abbott	Single-dose vial
Mestinon	ICN	None
Narcan	DuPont	Single-dose ampule
Nesacaine 1%, 2%	Astra	Nesacaine MPF 2%, 3%
Normodyne	Schering	None
Numorphan	DuPont	None
Oncovin	Lilly	None
Prolixin	Princeton	None
Pronestyl 500/ml	Princeton	Pronestyl 100 mg/ml
Prostigmin	ICN	Multidose vial
Sensorcaine multidose vial	Astra	Single-dose vial
Sensorcaine with epinephrine	Astra	None
Talwin multidose vial	Winthrop	Single-dose ampule
Taractan	Roche	None
Trandate	Allen & Hanburys	None

Trade name (cont'd)	Manufacturer (cont'd)	Alternative product (cont'd)
Xylocaine 0.5% multidose vial	Astra	Single-dose vial
Xylocaine 0.5%/epi 1:200,000	Astra	Single-dose vial
Xylocaine 1% multidose vial	Astra	Single-dose vial
Xylocaine 1%/epi 1:100,000	Astra	Single-dose vial
Xylocaine 2% multidose vial	Astra	Single-dose vial
Xylocaine 2%/epi 1:100,000	Astra	Single-dose vial
Wycillin	Wyeth-Ayerst	None
Wycillin and Probenecid	Wyeth-Ayerst	None

B. Topical Drug Products

Trade name	Manufacturer	Alternative products
Acid Mantle cream	Sandoz	None
Anusol HC cream	Parke-Davis	None
Aquacare lotion	Smith Kline	Carmol lotion
Aquatar gel	Herbert	Balnetar
Aspercreme cream, lotion	Thompson Medical	None
Bactine Hydrocortisone skin care cream	Miles	Hytone, Dermolate
Benadryl cream	Parke-Davis	None
Butesin picrate ointment	Abbott	None
Caladryl cream	Parke-Davis	None
Caldecort Hydrocortisone cream	Fisons	Hytone, Dermolate
Caldecort light creme	Fisons	Hytone, Dermolate
Caldesene ointment	Fisons	Desitin
Clearasil cream, lotion	Richardson-Vicks	Fostex gel
Cordran N cream	Lilly	Cordran N ointment
Cortaid cream, lotion, ointment, spray	Upjohn	Hytone, Dermolate
Cort-Dome creme, lotion	Miles	Hytone, Dermolate
Cortef Feminine itch cream	Upjohn	None
Cortef Rectal itch ointment	Upjohn	Proctocort
Corticaine	Glaxo	Hytone, Dermolate
Cortifoam	Reed & Carnrick	None
Cortisporin cream	Burroughs Wellcome	Cortisporin ointment
Cortizone-5 cream	Thompson Medical	Hytone, Dermolate
Cortril ointment	Pfizer	Hytone, Dermolate
Cruex antifungal cream	Fisons	None
Decadron cream	Merck Sharp & Dohme	Decaderm
Deep-Down pain relief rub	Beecham	Icy Hot Balm
Desenex antifungal cream, ointment	Fisons	Desenex liquid
Delacort	Medicon	Hytone, Dermolate
Dermoplast lotion	Whitehall	Solarcaine lotion
Diaperene medicated cream	Glenbrook	Diaperene ointment
Eldoquin forte cream	Elder	Eldopaque forte

Trade name (cont'd)	Manufacturer (cont'd)	Alternative products (cont'd)
Epifoam	Reed & Camrick	Analpram HC
Esoterica fade cream	Norcliff Thayer	None
Fluonid cream	Herbert	Synemol cream
Fototar cream	Elder	Balnetar
Fungizone cream, lotion	Squibb	Fungizone ointment
Garamycin cream, ointment	Schering	None
Gynecort antipruritic cream	Combe	Hytone, Dermolate
Lanacort antipruritic cream	Combe	Hytone, Dermolate
Lasan cream	Stiefel	Lasan ointment
Locoid cream	Owen	None
Mantadil cream	Burroughs Wellcome	None
Mobisyl analgesic creme	Ascher	None
Mycitracin ointment	Upjohn	Neosporin ointment
Neodecadron cream	Merck Sharp & Dohme	None
Neosporin cream	Burroughs Wellcome	Neosporin ointment
Neosynalar	Syntex	None
Noxzema clear-ups gel, lotion	Norcliff Thayer	Fostex gel
Nutracort lotion	Owen	Nutracort cream
Nystex cream	Savage	Mycostatin cream
Oxy-5 and Oxy-10 lotion	Norcliff Thayer	Fostex gel
Panoxyl AQ	Stiefel	Fostex gel
Preparation H cream	Whitehall	Preparation H ointment
Proctofoam	Reed & Camrick	Analpram HC cream
Silvadene cream	Marion	None
Sulfacet R	Dermik	None
Sulfamylon cream	Winthrop	None
Synalar cream	Syntex	Synemol cream
Tridesilon cream	Miles	Desowen cream
Tronolane cream	Ross	Anusol ointment
Tronothane cream	Abbott	Anusol ointment
Vanoxide HC	Dermik	None
Vioform hydrocortisone lotion	Ciba	Vioform HC cream

IV. ANIMAL TOXICITY DATA

In vitro studies of the vehicle of Narcan®, containing methylparaben 1.8 mg/ml and propylparaben 0.2 mg/ml, demonstrated smooth muscle relaxant properties that were not mediated via cholinergic or adrenergic mechanisms. The mechanism was shown to be related to calcium uptake or release.[6] It is not clear whether sufficient concentrations of the parabens can be achieved *in vivo* to produce a pharmacologic effect, but these studies have demonstrated caution in interpreting "placebo-controlled" studies on the benefits of naloxone in septic shock and other conditions when the vehicle is used as the placebo.

Parabens have been demonstrated to possess local anesthetic properties *in vitro*. The effect of 0.1% methylparaben was equivalent in potency to 0.05% procaine.[7]

V. HUMAN TOXICITY DATA

A. Cosmetics

Fisher[3] considers the parabens to be close to the ideal cosmetic preservative, possessing a broad spectrum of activity, being relatively nonirritating or sensitizing, and being sufficiently water soluble, with a high benefit/risk ratio. They do not appear to be photosensitizing. Primary irritant effects were not observed in human subjects tested with methylparaben 5%, ethylparaben 7%, propylparaben 12%, and butylparaben 5%.[4]

Although allergic contact dermatitis has been described, the incidence is rare. Studies involving 450 subjects without a history of dermatitis did not demonstrate positive patch testing in any of the subjects, while studies involving 27,230 patients with a history of dermatitis showed a 2.2% incidence of paraben sensitivity.[4] Patch testing can be misleading if the actual product or a "preservative" concentration of parabens is applied to normal skin in the typical test. A concentration of at least 3 to 15% of a mixed paraben preparation is recommended.[3,8]

Paraben-induced contact dermatitis displays several characteristics that are not unique to this preservative. Reapplication of the paraben to a sensitized individual is most likely to produce a reaction if applied to the site of previous dermatitis. This property is illustrated in a case report of a 49-year-old man who developed dermatitis of the left axilla following application of a paraben-containing hydrocortisone cream to an excoriated area. Subsequent application of a paraben-containing deodorant spray to both axillae produced dermatitis only in the left axilla.[8] Application of cosmetics to normal skin at a distant site may be well tolerated.[9] The incidence of cosmetic-related paraben sensitivity was 19 of 713 (2.7%) patients with cosmetic dermatitis.[10]

Immediate-type contact urticarial reactions have also been reported. Passive transfer has been demonstrated in some cases, indicating an IgE-mediated response.[11] More serious systemic reactions, such as bronchospasm, have been reported in patients receiving parabens by the intravenous route.[12]

B. Topical Drugs

Following reports of individual cases of paraben allergy to topical medications,[13-16] parabens were included in a patch test series of large numbers of patients. The North American Contact Dermatitis Group found a 3% incidence of paraben sensitivity in 1200 dermatitis patients in 1972.[17]

Allergic contact dermatitis is much more likely to occur in sensitized individuals exposed to therapeutic dermatologic products than to cosmetic products. The reason for this paradox is that therapeutic agents are usually applied to damaged or inflamed skin, while cosmetics are usually applied to normal skin. Patients with contact sensitization may develop generalized delayed eczematous reactions following systemic administration of parabens used as preservatives in injectable drugs.[18]

C. Injectable Drugs

A role of parabens in some patients with adverse reactions to local anesthetics has been suggested. Four cases of a typical Arthus reaction were attributed to methylparaben in a lidocaine solution used for mandibular block injections; however, intradermal testing could not confirm an immunologic reaction to either lidocaine or methylparaben.[19]

A stronger case was made in a 40-year-old woman who experienced an immediate hypersensitivity reaction to methylparaben in various local anesthetics. An IgE-mediated reaction was documented by intracutaneous testing and by passive transfer; subsequent procedures with paraben-free local anesthetics were uneventful. Cross-reactivity to procaine, tetracaine, and chloroprocaine was also documented.[20]

Intracutaneous testing with methylparaben in water was significantly positive (++ or greater) in 3 of 25 patients with a history of adverse reactions to local anesthetics; however, none had significant reactions to subsequent intramuscular injection of the preserved anesthetic. An immediate hypersensitivity reaction, with pruritus and maculopapular rash, was attributed to methylparaben and documented by positive intradermal tests to parabens, with cross-reactivity to the ester group of local anesthetics.[21]

Because the parabens are parahydroxybenzene acid esters, chemically similar in structure to the para-aminobenzoic acid ester group in local anesthetics, cross-reactivity has been evaluated. A few cases appear to document cross-reactivity.[20,21] In a study of 63 patients allergic to benzocaine, 20 cross-reacted with parabens.[22] Some clinicians have found that parabens do not cross-react significantly with other chemicals with a "para" group such as paraphenylenediamine and PABA,[3] although a cross-reactivity incidence of 14% was demonstrated with paraphenylenediamine.[22]

Paraben-containing drugs other than local anesthetics have also been associated with hypersensitivity reactions. An immediate systemic reaction, manifested as severe pruritus and bronchospasm, was reported in a 10-year-old boy treated with an intravenous injection of a hydrocortisone preparation reserved with methylparaben and propylparaben. The reaction was reproducible, and passive transfer was documented to the two paraben compounds.[12] An anaphylactoid reaction in a 45-year-old woman following exposure to lidocaine on two occasions and fentanyl on one occasion was suspected to be related to the methylparaben preservative present in all three drugs.[23]

D. Oral Drugs

Few cases are documented of adverse reactions to orally administered parabens. One case of an immediate anaphylactic or anaphylactoid reaction was reported in a 12-year-old boy following the second dose of an expectorant containing pseudoephedrine, amaranth, and methylparaben. No attempt was made to determine an immunologic mechanism or which of the ingredients was responsible.[24] A delayed, cell-mediated reaction, consisting of an urticarial maculopapular rash, was documented by positive macrophage migration inhibition factor test to methylparaben in a 17-year-old boy treated with oral haloperidol syrup.[25]

E. Rectal Drugs

Although the cause is poorly documented, several cases of immediate hypersensitivity reactions have been reported following the use of methylparaben-preserved barium enemas, including urticaria, laryngeal edema, supraorbital edema, hypotension, pruritus, and a sensation of heat.[26-28] In one of these cases, methylparaben was determined to be the allergen by positive prick test.[28] The incidence of these reactions is unknown, but they are not uncommon. One mail survey of radiologists, the FDA, and distributors of barium products produced reports of 106 reactions to barium enema products.[29] In one report, an incidence of 1 case in 1851 procedures was documented over a 2-year period.[30]

F. Neonates

Following a report that very high concentrations of methylparaben can displace bilirubin from albumin binding sites *in vitro*,[31] methylparaben-preserved gentamicin was suspected of causing kernicterus in hyperbilirubinemic premature neonates. Attempts to confirm this theory with single or multiple intramuscular dosing failed to demonstrate a significant effect *in vivo*.[32,33] There may be some accumulation of methylparaben in the premature neonate; the amount excreted in the urine ranged from 13.2 to 88.1% in a study of six preterm infants and was proportional to the gestational age.[33]

VI. CLINICAL RELEVANCE

A. Contact Dermatitis

The incidence of contact dermatitis in normal subjects is extremely low, estimated at 0.3%.[4] As with other contact allergens, the incidence of positive patch test reactions to parabens is greater in patients with chronic dermatitis. In a study of 653 children with chronic dermatitis, there were 3 positive tests to parabens for an incidence of 0.45%.[34] The incidence in 14,400 Italian dermatitis patients was 1.3%.[35] Consecutive series of patch tests on North American dermatitis patients have shown sensitivity rates of 0.8 to 3%.[4,17,36,37]

The documentation of hypersensitivity by a positive patch test does not preclude the use of paraben preservatives in topical substances intended for use on normal skin, such as eye, body, and facial cosmetics. These individuals should avoid the use of parabens on damaged skin, sites of previous dermatitis, or mucous membranes. Flare-ups at previous sites of dermatitis may also occur following ingestion of paraben-preserved foods.[38]

B. Local Anesthetics

Because of the possibility of cross-reactivity of parabens to ester-type local anesthetics, patients experiencing adverse reactions to this group of anesthetics should also be tested for paraben hypersensitivity.

REFERENCES

1. **Lecos, C.,** Food preservatives: a fresh report, *FDA Consumer,* April 1984.
2. **Nikitakis, J. M.,** *CTFA Cosmetic Ingredient Handbook,* 1st ed., The Cosmetic, Toiletry and Fragrance Association, Washington, D.C., 1988.
3. **Fisher, A. A.,** *Contact Dermatitis,* 3rd ed., Lea & Febiger, Philadelphia, 1986.
4. **Moore, J.,** Final report on the safety assessment of methylparaben, ethylparaben, propylparaben, and butylparabe, *J. Am. Coll. Toxicol.,* 3, 147, 1984.
5. **Decker, R. L. and Wenninger, J. A.,** Frequency of preservative use in cosmetic formulas as disclosed to FDA-1982 update, *Cosmet. Toilet.,* 97, 57, 1982.
6. **Soulioti, A. M. A., Woods, N. M., Cobbold, P. H., and Rodger, I. W.,** Preservatives in the vehicle of naloxone: pharmacological effects, *Biomed. Pharmacother.,* 43, 771, 1989.
7. **Nathan, P. W. and Sears, T. A.,** Action of methyl hydroxybenzoate on nervous conduction, *Nature (London),* 192, 668, 1961.
8. **Fisher, A. A.,** Cortaid cream dermatitis and the "paraben paradox", *J. Am. Acad. Dermatol.,* 6, 116, 1982.
9. **Fisher, A. A.,** Paraben dermatitis due to a new medicated bandage: the "paraben paradox", *Contact Dermatitis,* 5, 273, 1979.
10. **Adams, R. M. and Maibach, H. I.,** A five-year-study of cosmetic reactions, *J. Am. Acad. Dermatol.,* 13, 1062, 1985.

11. **Henry, J. C., Tschen, E. H., and Becker, L. E.,** Contact urticaria to parabens, *Arch. Dermatol.,* 115, 1231, 1979.

12. **Nagel, J. E., Fuscaldo, J. T., and Fireman, P.,** Paraben allergy, *JAMA,* 237, 1594, 1977.

13. **Schamberg, I. L.,** Allergic contact dermatitis to methyl and propyl paraben, *Arch. Dermatol.,* 95, 626, 1967.

14. **Schorr, W. P. and Mohajerin, A. H.,** Paraben sensitivity, *Arch. Dermatol.,* 93, 721, 1966.

15. **Wuepper, K. D.,** Paraben contact dermatitis, *JAMA,* 202, 127, 1967.

16. **Reed, W. B.,** Paraben allergy, a case of intractable dermatitis, *Arch. Dermatol.,* 100, 503, 1969.

17. **North American Contact Dermatitis Group,** Epidemiology of contact dermatitis in North America: 1972, *Arch. Dermatol.,* 108, 537, 1973.

18. **Carradori, S., Peluso, A. M., and Faccioli, M.,** Systemic contact dermatitis due to parabens, *Contact Dermatitis,* 22, 238, 1990.

19. **Lederman, D. A., Freedman, P. D., Kerpel, S. M., and Lumerman, H.,** An unusual skin reaction following local anesthetic injection, *Oral Surg.,* 49, 28, 1980.

20. **Aldrete, J. A. and Johnson, D. A.,** Allergy to local anesthetics, *JAMA,* 207, 356, 1969.

21. **Johnson, W. T. and DeStigter, T.,** Hypersensitivity to procaine, tetracaine, mepivacaine, and methylparaben: report of a case, *JADA,* 106, 53, 1983.

22. **Rudzki, E. and Kleniewska, D.,** Cross reactions between parabens and the para group, *Br. J. Dermatol.,* 83, 543, 1970.

23. **Fukuda, T. and Dohi, S.,** Anaphylactic reaction to fentanyl or preservative, *Can. Anaesth. Soc. J.,* 33, 826, 1986.

24. **Batty, K. T.,** Hypersensitivity to methylhydroxybenzoate: a case for additive labelling of pharmaceuticals, *Med. J. Aust.,* 144, 107, 1986.

25. **Kaminer, Y., Apter, A., Tyano, S., Livni, E., and Wijsenbeek, H.,** Delayed hypersensitivity reaction of orally administered methylparaben, *Clin. Pharm.,* 1, 469, 1982.

26. **Larsen, C. F.,** Allergic side-effects after radioscopic examination of the colon with a barium sulfate suspension, *Ugeskr. Laeg.,* 143, 937, 1981.

27. **Javors, B. R., Applbaum, Y., and Gerard, P.,** Severe allergic reaction: an unusual complication of barium enema, *Gastrointest. Radiol.,* 9, 357, 1984.

28. **Schwartz, E. E., Glick, S. N., Foggs, M. B., and Silverstein, G. S.,** Hypersensitivity reactions after barium enema examination, *Am. J. Radiol.,* 143, 103, 1984.

29. **Janower, M. L.,** Hypersensitivity reactions after barium studies of the upper and lower gastrointestinal tract, *Radiology,* 161, 139, 1986.

30. **Schuh, M., Petrelli, N. J., and Herrera, L.,** Systemic hypersensitivity reaction following a barium enema examination, *N. Y. State J. Med.,* 88, 86, 1988.

31. **Loria, C. J., Escheverria, P., and Smith, A. L.,** Effect of antibiotic formulations in serum: bilirubin interaction of newborn infants, *J. Pediatr.,* 89, 479, 1976.

32. **Woods, J. T., Bryan, L. E., Chan, G., and Schiff, D.,** *J. Pediatr.,* 89, 483, 1976.

33. **Hindmarsh, K. W., John, E., Asali, L. A., French, J. N., Williams, G. L., and McBride, W. G.,** Urinary excretion of methylparaben and its metabolites in preterm infants, *J. Pharm. Sci.,* 72, 1039, 1983.

34. **Leyden, J. J. and Kligman, A. M.,** Contact dermatitis to neomycin sulfate, *JAMA,* 242, 1276, 1979.

35. **Meneghini, C. L., Rantuccio, F., and Lomuto, M.,** Additives, vehicles and active drugs of topical medicaments as causes of delayed-type allergic dermatitis, *Dermatologica,* 143, 137, 1971.

36. **Schorr, W. F.,** Paraben allergy: a cause of intractable dermatitis, *JAMA,* 204, 107, 1968.

37. **Fisher, A. A.,** Allergic contact dermatitis due to ingredients of vehicles, *Arch. Dermatol.,* 104, 286, 1971.

38. **Fisher, A. A.,** Allergic reactions to the preservatives in over-the-counter hydrocortisone topical creams and lotions, *Cutis,* 32, 222, 1983.

PARACHLOROMETAXYLENOL

I. REGULATORY CLASSIFICATION

Parachlorometaxylenol (PCMX) is classified as an antimicrobial preservative.

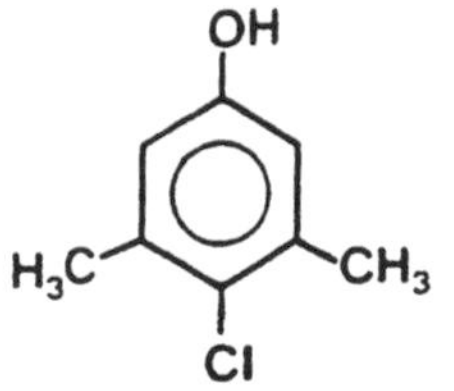

FIGURE 46. Parachlorometaxylenol.

II. SYNONYMS

PCMX
Benzytol
4-Chloro-3,5-xylenol
Chloroxylenol
p-Chloro-m-xylenol

III. AVAILABLE FORMULATIONS

A. Drugs

Parachlorometaxylenol is found as a preservative in otic and topical products in concentrations of 0.1 to 0.15%.[1] It is also present as an active ingredient in topical antifungal products in concentrations of up to 2.5%.

B. Cosmetics

Chloroxylenol is used as a biocide, deodorant, and preservative in cosmetics such as shampoos, hair conditioners, personal cleansers, creams and lotions, and hair straighteners.[2] It is a component of carbolated Vaseline®.

IV. TABLE OF COMMON PRODUCTS

A. Topical Drug Products

Trade name	Manufacturer
Absorbine Jr.	W.F. Young
Blemerase lotion	Young
Denorex shampoo	Whitehall
Foille plus cream, spray	Blistex
Fungoid creme, solution, tincture	Pedinol
Fungoid HC creme	Pedinol
Ivarest cream, lotion	Blistex
Micro-Guard cream, powder	Sween
Pedi-Pro foot powder	Pedinol
Sween Prep liquid	Sween

V. HUMAN TOXICITY DATA

A. Contact Dermatitis

Allergic contact sensitization to chloroxylenol was demonstrated in four of eight black patients with contact dermatitis following use of carbolated vaseline, which contains 0.2% phenol and 0.5% chloroxylenol.[3] Exposure to this same product resulted in contact dermatitis in another five patients; all had positive patch tests to 0.5% chloroxylenol in lanolin. An ECG paste containing chloroxylenol was responsible for dermatitis in two other patients.[4]

The incidence of hypersensitivity to chloroxylenol was second only to mercury in a review of antibacterial-related dermatitis, accounting for 24% of 220 cases.[5] Cross-reactivity to the related compound, chlorocresol, has been described.[6,7]

REFERENCES

1. **Food and Drug Administration,** *Inactive Ingredients Guide,* FDA, Washington D.C., March 1990.
2. **Nikitakis, J. M.,** *CTFA Cosmetic Ingredient Handbook,* 1st ed., The Cosmetic, Toiletry and Fragrance Association, Washington, D.C., 1988.
3. **Rubin, M. B. and Pirozzi, D. J.,** Contact dermatitis from carbolated vaseline, *Cutis,* 12, 52, 1973.
4. **Storrs, F. J.,** Para-chloro-meta-xylenol allergic contact dermatitis in seven individuals, *Contact Dermatitis,* 1, 211, 1975.
5. **Calnan, C. D.,** Contact dermatitis from drugs, *Proc. R. Soc. Med.,* 55, 39, 1962.
6. **Hjorth, N. and Trolle-Lassen, C.,** Skin reaction to ointment bases, *Trans. St. John's Hosp. Soc.,* 49, 127, 1963.
7. **Burry, J. N., Kirk, J., Reid, J. G., and Turner, T.,** Environmental dermatitis: patch tests in 1000 cases of allergic contact dermatitis, *Med. J. Aust.,* 2, 681, 1973.

PERU BALSAM

I. REGULATORY CLASSIFICATION

Peru balsam is classified as a flavor and perfume ingredient. Tolu balsam syrup NF contains 5% tolu balsam tincture in a sucrose base and is used as a flavored vehicle. Tolu balsam tincture NF contains 20% tolu balsam.

II. SYNONYMS

Balsam of Peru
Peruvian Balsam
Tolu Balsam

III. AVAILABLE FORMULATIONS

A. Constituents

Peru balsam is obtained from the exudate of the trunk of *Myroxylon balsamum* and contains benzyl salicylate, benzyl cinnamate, methyl cinnamate, benzyl benzoate, cinnamic acid, cinnamic alcohol, cinnamic aldehyde, eugenol, nerolidiol, and vanillin. Because of the presence of many constituents frequently used as individual components of fragrances, peru balsam is considered a "marker" for allergic perfume sensitivity.[1]

B. Foods

Foods that may contain peru balsam include[2]

Baked goods
Candies
Chewing gum
Citrus fruit peel
Cola-flavored soft drinks
Danish pastries
Ice cream
Juice
Marmalade
Spices (cinnamon, cloves, vanilla, curry)

C. Cosmetics

Peru balsam is used as a film former, hair conditioner, and fragrance in cosmetics, including colognes, toilet waters, hair conditioners, creams, and lotions.[3] Cosmetics may contain perfumes that have peru balsam as a component. Perfumes are used in concentrations of 0.1% or less as "masking" fragrances in products that may be labeled "unscented". Perfumes, often a mixture of 10 to 300 components, are used in concentrations of 0.5% in scented cosmetics, 4 to 5% in colognes and toilet waters, and 20% in perfumes.[1]

IV. TABLE OF COMMON PRODUCTS

A. Topical Drug Products

Trade name	Manufacturer
Balmex lotion	Macsil
Balmex ointment	Macsil
Balmex powder	Macsil
Granulex	Hickam

V. HUMAN TOXICITY DATA

A. Nonimmunologic Contact Urticaria

Stage 1 contact urticaria (localized intense macular erythema or wheal-and-flare) was produced in 74% of controls and 58% of patients with eczema or cosmetic sensitivity tested with Peru balsam 25% in petrolatum.[4] Photocontact urticaria was reported in three of ten patients with cosmetic allergy.[5]

B. Allergic Contact Dermatitis

Peru balsam is a common cause of medication and cosmetic-related delayed eczematous contact dermatitis. It is a "marker" of perfume sensitivity and is positive in about 40 to 50% of patients with perfume allergy.[6,7] Oral or parenteral administration of Peru balsam can elicit eczematous reactions in sensitized patients.[8] Summaries of the incidence of positive patch test reactions in large patient series are presented in the table below.

No. patients	No. positive (%)	Diagnosis	Country	Ref.
192	(21.4)	Stasis dermatitis	Finland	9
560	(4.6–7.6)	Medication dermatitis	Europe	10
713	3 (0.4)	Cosmetic dermatitis	U.S.	11
179	32 (17.9)	Cosmetic dermatitis	Netherlands	12
182	(21.3)	Cosmetic dermatitis	Netherlands	13
1116	47 (4.2)	Contact dermatitis	Denmark	6
2671	176 (6.6)	Contact dermatitis	Netherlands	14

VI. CLINICAL RELEVANCE

Peru balsam is a complex mixture derived from damaged tree bark and can cause nonimmunologic contact urticaria and photocontact urticaria in 30% or more individuals tested with open patch methods. Allergic contact dermatitis can be demonstrated in about 40 to 50% of patients with fragrance allergies. Like other fragrance components, it is a difficult task for an allergic individual to avoid this excipient which is likely to be labeled under the generic category of "fragrance" on food, cosmetic, and drug products.

REFERENCES

1. **Fisher, A. A.,** Perfume dermatitis. Part I. General considerations and testing procedures, *Cutis*, 26, 458, 1980.
2. **Fisher, A. A.,** *Contact Dermatitis*, 3rd ed., Lea & Febiger. Philadelphia, 1986.
3. **Nikitakis, J. M.,** *CTFA Cosmetic Ingredient Handbook*. 1st ed., The Cosmetic, Toiletry and Fragrance Association, Washington, D.C., 1988.
4. **Emmons, W. W. and Marks, J. G.,** Immediate and delayed reactions to cosmetic ingredients, *Contact Dermatitis*, 13, 258, 1985.
5. **Thune, P.,** Photosensitivity and allergy to cosmetics, *Contact Dermatitis*, 7, 54, 1981.
6. **Veien, N. K., Hattel, T., Justesen, O., and Norholm, A.,** Patch testing with perfume mixture, *Acta Dermat. Venereol.*, 62, 341, 1982.
7. **Larsen, W. G.,** Perfume dermatitis, *J. Am. Acad. Dermatol.*, 12, 1, 1985.
8. **Klaschka, F. and Ring, J.,** Systemically induced (hematogenous) contact eczema, *Sem. Dermatol.*, 9, 210, 1990.
9. **Fraki, J. E., Peltonen, L., and Hopsu-Havu, V. K.,** Allergy to various components of topical preparations in stasis dermatitis and leg ulcer, *Contact Dermatitis*, 5, 97, 1979.
10. **Bandmann, H. J., Calnan, C. D., Cronin, E., Fregert, S., Hjorth, N., Magnusson, B., Maibach, H., Malten, K. E., Meneghini, C. L., Pirila, V., and Wilkinson, D. S.,** Dermatitis from applied medicaments, *Arch. Dermatol.*, 106, 335, 1972.
11. **Adams, R. M. and Maibach, H. I.,** A five-year study of cosmetic reactions, *J. Am. Acad. Dermatol.*, 13, 1062, 1985.
12. **de Groot, A. C., Liem, D. H., Nater, J. P., and van Ketel, W. G.,** Patch tests with fragrance materials and preservatives, *Contact Dermatitis*. 12, 87, 1985.
13. **Malten, K. E., van Ketel, W. G., Nater, J. P., and Liem, D. H.,** Reactions in selected patients to 22 fragrance materials, *Contact Dermatitis*, 11, 1, 1984.
14. **Kuiters, G. R. R., Smitt, J. H., Cohen, E. B., and Bos, J. D.,** Allergic contact dermatitis in children and young adults, *Arch. Dermatol.*, 125, 1531, 1989.

PETROLATUM

I. REGULATORY CLASSIFICATION

Petrolatum is classified as an ointment base.

II. SYNONYMS

A. Yellow Petrolatum

> Yellow soft paraffin
> Yellow petroleum jelly
> Yellow paraffin jelly
> Petroleum jelly

Yellow petrolatum is a pale yellow or amber, translucent purified semisolid hydrocarbon mixture derived from steam or vacuum distillation of petroleum. Polycyclic aromatic hydrocarbons may be present as an impurity. Yellow Ointment USP contains yellow petrolatum 95% and yellow beeswax 5%.

B. White Petrolatum

> White soft paraffin
> White petroleum jelly
> Vaseline officinale

White petrolatum is bleached yellow petrolatum, prepared by extended purification of yellow petrolatum. Official preparations containing white petrolatum are Hydrophilic Ointment USP, Hydrophilic Petrolatum USP, Petrolatum Gauze USP, and White Ointment USP.

III. AVAILABLE FORMULATIONS

A. Constituents

There are at least four different types of petrolatum. Natural petrolatum is derived from residues of petroleum distillation. Artificial petrolatum is a mixture of natural solid hydrocarbons, such as paraffin wax, and mineral oil. Gatsch or slack wax petrolatum is derived from byproducts of lubricating oil production and is blended with mineral oil. Synthetic petrolatum is made by ethylene polymerization or cohydrogenation. The first three types are obtained from natural sources and are purified using catalytic hydrogenation, followed by filtration with bleaching earth.[1]

The constituents of natural petrolatum are primarily branched-chain paraffins and cycloparaffins, along with some saturated n-paraffins, olefins, and aromatic hydrocarbons. The relative percentage of these hydrocarbon compounds varies according to the geographical origin of the petrolatum.[1]

Stabilizers and antioxidants may be added to natural petrolatum. Polyisobutylene and polyacrylic acid esters may be added to synthetic petrolatum to improve lubricity.[2]

B. Foods

Petrolatum may be used as a direct or indirect food additive. The petrolatum used in foods must meet more stringent UV absorbance requirements than those specified in the USP or NF for pharmaceuticals.[1]

C. Drugs

Petrolatum is used as a skin protectant in over-the-counter pharmaceuticals and as an ointment base in topical prescription pharmaceuticals.

D. Cosmetics

Petrolatum is used as an occlusive skin and hair conditioner. Types of products include eye makeup, hair tonics and grooming aids, facial makeup products, creams, and lotions.[3]

IV. ANIMAL TOXICITY DATA

A. Comedogenicity

Pure petrolatum was found to be totally noncomedogenic in rabbit ear assay. Hydrophilic Ointment USP contains 1% sodium lauryl sulfate, which is strongly comedogenic in this model.[4]

V. HUMAN TOXICITY DATA

A. Allergic Contact Dermatitis

Petrolatum is commonly used as a vehicle for most substances used in patch testing due to its nonirritating, nonsensitizing, and occlusive properties.[5] Occasional patients have been documented to have allergic or irritant contact dermatitis to petrolatum.

A 30-year-old man with psoriasis developed an immediate eczematous and exudative reaction after application of a petrolatum-based salicylic acid ointment. Patch testing was positive with both white and yellow petrolatum.[6]

In another case report of a 24-year-old black male, a chronic scaling dermatitis with hyperpigmentation was described at the site of application of petrolatum, which had been done twice daily for 11 years. Although patch testing failed to confirm an immunologic basis for this reaction, the dermatitis cleared after discontinuation of the ointment and recurred when rechallenged with petrolatum. This was probably an irritant reaction, but allergy to a minor component of the petrolatum is possible.[7]

Two patients with chronic leg ulcers tested positive to all substances in the test battery diluted in yellow petrolatum. Subsequent testing with several brands of yellow petrolatum produced positive results with two brands. No reactions to related products such as white petrolatum, mineral oil, and hard paraffin, were observed.[2]

Allergy to both yellow and white petrolatum in the same patient was also reported by Fisher[5] and Lawrence and Smith, who described a chronic leg ulcer patient.[8] A patient, reported by Malten, reacted only to yellow petrolatum.[9] Thus, of six reported cases, three were sensitive to yellow petrolatum only and three were sensitive to both yellow and white petrolatum.

There is some evidence that the allergenic fraction of petrolatum is in the polycyclic aromatic portion. Testing "as is" produced positive tests for those brands containing higher amounts of polycyclic aromatic hydrocarbons, as measured indirectly by UV absorbance limits. Those products meeting strict European pharmacopoeial requirements with specified UV absorbance limits were not allergenic. White petrolatums were more likely to meet purity standards than most yellow petrolatum brands with a few exceptions. Yellow petrolatums manufactured by Chesebrough-Pond, Penreco, and Witco were less allergenic than most white petrolatums.[1]

The most likely allergens were determined to be phenanthrene derivatives with molecular weights of 230 and 244 after analysis of the allergenic fraction of yellow petrolatum.[10]

B. Acne

Although petrolatum is noncomedogenic, it provides an occlusive area that may promote bacterial growth leading to pustule formation. Pomade acne was reported to be widespread in East African children who applied petroleum jelly frequently to the face, representing up to 2% of patients attending outpatient dermatology clinics.[11]

Limited acne with comedones, discrete and confluent pimples, and pustules was reported on the right side of the face where white petrolatum had been applied nightly. The untreated left side of the face remained clear.[12] Two similar cases were described in two black women following frequent application of petroleum jelly or Chap Stick®, which consisted primarily of petrolatum to the lips. In both cases, the acne was limited to a single row of large, open comedones along the cutaneous margin of the upper lip.[13]

C. Myospherulosis

An unusual foreign body reaction, consisting of large fungal-like structures in cystic tissue spaces, was observed in 16 patients following surgery of the paranasal sinuses, nose, and middle ear. The feature in common with all of these patients was the postoperative gauze packing. Petrolatum-containing gauze or the addition of petrolatum-based nonsterile antibiotic ointments was used in 15 patients.

The lesions had a "Swiss-cheese" pattern of closely arranged holes within tissue fragments lined by histiocytes and foreign-body giant cells. Within the holes were large sac-like structures with numerous spherules, giving an appearance described as "partly filled bags of marbles". The characteristic lesions could be reproduced by applying a nonsterile tetracycline antibiotic ointment into the abdominal cavity of rats.[14]

Subsequent isolated reports have described myospherulosis in other anatomical sites of surgery, such as the jaws or cranium, involving the use of petrolatum-based antibiotic ointments.[15-17]

Many attempts were made to identify the nature of the spherule structures, which did not stain for fungi or bacteria and could not be cultured.[18] It was soon determined that the structures represent erythrocytes altered by the petrolatum ointments, following demonstration of positive stains for hemoglobin.[19]

The association with the petrolatum vehicle was confirmed after *in vitro* experiments showing reproducibility of the myospherule structures after incubation of human red blood cells with individual components of the antibiotic ointment.[20]

D. Lipogranulomas

Lesions consisting of clear spaces in conjunction with an inflammatory response, called paraffinomas, lipogranulomas, liponecrosis, oleofibroma, oleogranulomas, and oil granulomas, have been reported following contamination of various tissues with various petrolatum products. Rectal administration of an ointment containing paraffin and lanolin alcohol was associated with development of oleogranuloma in a 57-year-old man.[21]

E. Fire Hazard

Datta[22] reported a case of a flash fire in a patient undergoing electrocautery of a lower eyelid mole with isoflurane and nitrous oxide in oxygen anesthesia, presumably due to application of an ophthalmic ointment containing white petrolatum, mineral oil, and lanolin alcohol. Burns of the eyelashes and eyelid skin resulted. Experimentation showed reproduction of ignition of the ointment only occurred in the presence of the nitrous oxide-oxygen mixture.

Subsequent to this report, attempts were made to reproduce these results by the manufacturer, and the flash hazard could not be confirmed using a 3:2 nitrous-oxygen mixture.[23] An independent group of ophthalmologists could produce a brief ignition and flame under circumstances not likely to occur in actual operative situations with a 50:50 nitrous-oxygen mixture administered directly over the ointments and cautery placed in the ointment.[24]

Five black patients sustained face and scalp burns following application of a sheen-type hair grease immediately prior to lighting a cigarette or touching the hair to an ignition source, such as an electric burner or outdoor barbecue. Petrolatum was the only ingredient common to all of the hair grease products. The inhalation injury, which required intubation in two of four cases, was attributed to combustion products of petrolatum, such as acrolein, formic acid, and acetic acid.[25]

Another series of ten patients using hair grease who suffered burns after exposure to various ignition sources, all in enclosed spaces such as cars, telephone booths, and oven pilot lights, has been reported.[26]

VI. CLINICAL RELEVANCE

A. Contact Dermatitis

White petrolatum has an extremely low incidence of allergic contact dermatitis and is the preferred vehicle for patch testing in most cases. Rare cases of allergic dermatitis have been reported, mostly in patients with stasis ulcers of the leg and multiple concomitant hypersensitivities. The allergen appears to be a phenanthrene compound which is present only in low concentrations in highly purified brands of petrolatum. Allergenicity varies with the source of the raw materials and the extent of refinement done during manufacturing. Patients with positive patch tests to multiple substances in a petrolatum vehicle should be patch testing with the vehicle "as is".

REFERENCES

1. **Dooms-Goossens, A. and Degreef, H.,** Contact allergy to petrolatums (I). Sensitizing capacity of different brands of yellow and white petrolatums, *Contact Dermatitis,* 9, 175, 1983.
2. **Dooms-Goossens, A. and Degreef, H.,** Sensitization to yellow petrolatum used as a vehicle for patch testing, *Contact Dermatitis,* 6, 146, 1980.
3. **Nikitakis, J. M.,** *CTFA Cosmetic Ingredient Handbook.* 1st ed., The Cosmetic, Toiletry and Fragrance Association, Washington, D.C., 1988.
4. **Fulton, J. E., Jr., Pay, S. R., and Fulton, J. E., III,** Comedogenicity of current therapeutic products, cosmetics, and ingredients in the rabbit ear, *J. Am. Acad. Dermatol.,* 10, 96, 1984.
5. **Fisher, A. A.,** Cutaneous reactions to petrolatum, *Cutis,* 28, 23, 1981.
6. **Grimalt, F. and Romaguera, C.,** Sensitivity to petrolatum, *Contact Dermatitis,* 4, 376, 1978.
7. **Maibach, H.,** Chronic dermatitis and hyperpigmentation from petrolatum, *Contact Dermatitis,* 4, 62, 1978.
8. **Lawrence, C. M. and Smith, A. G.,** Ampliative medicament allergy: concomitant sensitivity to multiple medicaments including yellow soft paraffin, white soft paraffin, gentian violet and Span 20, *Contact Dermatitis,* 8, 240, 1982.
9. **Malten, K. E.,** A case of contact eczema to yellow soft paraffin, *Contact Dermatitis,* 5, 106, 1969.
10. **Dooms-Goossens, A. and Degreef, H.,** Contact allergy to petrolatums (II). Attempts to identify the nature of the allergens, *Contact Dermatitis,* 9, 247, 1983b.
11. **Verhagen, A. R.,** Pomade acne in black skin, *Arch. Dermatol.,* 110, 465, 1974.
12. **Frankel, E. B.,** Acne secondary to white petrolatum use, *Arch. Dermatol.,* 121, 589, 1985.
13. **Shelley, W. B. and Shelley, E. D.,** Chap Stick acne, *Cutis,* 37, 459, 1986.
14. **Kyriakos, M.,** Myosperulosis of the paranasal sinuses, nose and middle ear: a possible iatrogenic cause, *Am. J. Clin. Pathol.,* 67, 118, 1977.
15. **Dunlop, C. L. and Barker, B. F.,** Myospherulosis of the jaws, *Oral Surg.,* 50, 239, 1980.
16. **Mills, S. E. and Lininger, J. R.,** Intracranial myospherulosis, *Hum. Pathol.,* 13, 596, 1982,
17. **Ide, F., Saito, I., Otsuka, K., Hori, K. M., and Kudo, I.,** Myospherulosis of the mandible, *Int. J. Oral Surg.,* 13, 65, 1984.
18. **De Schryver-Kecskemeti, K. and Kyriakos, M.,** Myospherulosis: an electron-microscopic study of a human case, *Am. J. Clin. Pathol.,* 69, 555, 1977.
19. **Rosai, J.,** The nature of myospherulosis of the upper respiratory tract, *Am. J. Clin. Pathol.,* 69, 475, 1978.
20. **Wheeler, T. M., Sessions, R. B., and McGavran, M. H.,** Myospherulosis: a preventable iatrogenic nasal and paranasal entity, *Arch. Otolaryngol.,* 106, 272, 1980.
21. **Greaney, M. G. and Jackson, P. R.,** Oleogranuloma of the rectum produced by Lasonil ointment, *Br. Med. J.,* 2, 997, 1977.
22. **Datta, T. D.,** Flash fire hazard with eye ointment, *Anesth. Analg.,* 63, 698, 1984.
23. **Robins, D. S.,** Safety of Lacri-Lube, *Anesth. Analg.,* 65, 1091, 1986.
24. **Carpel, E. F., Rice, S. W., Lang, M., and Schochet, L. H.,** Fire risks with ophthalmic ointments, *Am. J. Ophthalmol.,* 100, 477, 1985.
25. **Bascom, R., Haponik, E. F., and Munster, A. M.,** Inhalation injury related to use of petrolatum-based hair grease, *J. Burn Care Rehabil.,* 5, 326, 1984.
26. **Boucher, J., Raglon, B., Valdez, S., et al.,** Possible role of chemical hair care products in 10 patients with face, scalp, ear, back, neck and extremity burns, *Burns,* 16, 146, 1990.

PHENOL

I. REGULATORY CLASSIFICATION

Phenol is classified as an antimicrobial preservative. It is primarily used in parenteral preparations and is not listed in ophthalmic or oral products. In topical products, it is usually listed as an active ingredient, depending on the concentration.

FIGURE 47. Phenol.

II. AVAILABLE FORMULATIONS

A. Drugs

Phenol is approved for use in over-the-counter topical products as an active ingredient in concentrations of 0.5 to 2%. When used as an inactive ingredient, phenol is present at concentrations of less than 0.5%.[1] Parenteral drug products may contain phenol in amounts of 0.2 to 5%.

B. Cosmetics

Phenol is used as an active ingredient in cosmetics, such as mouthwashes and deodorant soaps, and also as an inactive ingredient in products such as douches; face, body, and hand creams; lotions, moisturizers, and other skin care products.[2,3]

III. TABLE OF COMMON PRODUCTS

A. Parenteral Drug Products

Trade name	Manufacturer	Amount (mg/ml)
Antivenin crotalidae polyvalent	Wyeth-Ayerst	0.25
Antivenin micrurus fulvius	Wyeth-Ayerst	0.25
A.P.L.	Wyeth-Ayerst	0.2
Cholera vaccine	Wyeth-Ayerst	0.5
Enlon	Anaquest	0.45
Glucagon diluent	Lilly	0.2
H.P. Acthar gel	Rorer	0.5
Hydeltrasol	Merck Sharp & Dohme	5
Konakion 0.5 ml ampule	Roche	0.45
Levo-Dromoran multiple-dose vial	Roche	0.45
Mepergan	Wyeth-Ayerst	5
Metubine iodide	Lilly	0.5
Morphine sulfate	Winthrop	5
Morphine sulfate	Elkins-Sinn	2.5
Nebcin	Lilly	5
Phenergan	Wyeth-Ayerst	5
Pneumovax 23	Merck Sharp & Dohme	0.25
Prostigmin multiple-dose vial	ICN	0.45
Roferon-A	Roche	3
Sus-Phrine	Forest	5
Synkayvite	Roche	0.45
Tagamet	Smith Kline & French	5
Tensilon multiple-dose vial	ICN	0.45
Zantac	Glaxo	5

IV. HUMAN TOXICITY DATA

A. Parenteral Drug Exposure

Phenol is present in parenteral drug products in low concentrations that have not been associated with adverse reactions when usual therapeutic doses of the drug are administered. The concern over the presence of phenol becomes significant when large doses of the drug are used for atypical indications, resulting in an accompanying large dose of phenol. Examples include the use of high-dose glucagon in the treatment of beta-blocker overdose and myocardial contractile failure, and the use of high-dose epidural morphine infusions in cancer patients. Although the minimum toxic parenteral dose of phenol was suggested to be 50 mg by one author,[4] there is little evidence that this amount is harmful when used in patients with beta-blocker toxicity.[5]

In the only documented case of phenol toxicity, a daily epidural dose of 30 mg of phenol resulted in confusion and disorientation in a 50-year-old man with pancreatic carcinoma, which resolved when a nonpreserved morphine product was substituted, even though the daily morphine dose was increased.[6]

B. Topical Exposure

Phenol is a local anesthetic in concentrations of 0.5 to 2%. At concentrations of 5% and greater, tissue protein coagulation and necrosis occur.[7] Primary irritant reactions to 2% phenol patch testing were reported in three of eight black patients with contact dermatitis following use of carbolated vaseline, which contains 0.2% phenol and 0.5% para-chloro-meta-xylenol. Twenty additional patients developed chronic reactions, consisting of dusky erythema followed by hyperpigmentation and diffuse fine scaling, which were attributed to a primary irritant response to phenol.[8] Irritant reactions can occur with concentrations of less than 1% in some patients, as described in two adults using P & S liquid or Panscol lotion.[9]

Phenol has been rarely documented as a sensitizer with one case report of allergic contact dermatitis reported.[10] Another case of accidental exposure to 10% phenol in an epoxy resin stripper described primary irritation which progressed after a delay to irregular areas in a "skip area" pattern. This was suggestive of a delayed cell-mediated reaction, but patch testing was not performed.[7]

Systemic toxicity may occur following application of low concentrations of topical phenol, especially in infants with excoriated skin. Drowsiness, shallow breathing, and blue urine were noted in a 6-month-old infant treated for seborrheic eczema of the scalp, face, skin folds, and diaper area with Castellani's Paint containing 4% phenol.[11] This concentration is also a primary irritant and resulted in burning and irritation in a 38-year-old man following application to the inguinal area.[9]

C. Aerosol Products

Spontaneous reports of four cases of laryngeal or epiglottal edema, resulting in two deaths, were submitted to the Committee on Safety of Medicines in the U.K. following use of a throat spray containing 1.4% phenol.[12]

V. CLINICAL RELEVANCE

A. Parenteral Drug Exposure

Potential phenol toxicity following infusion of large doses of phenol-preserved drugs is avoidable. If glucagon is necessary in doses greater than 10 mg, it should be reconstituted in 10 ml of 5% dextrose in water or normal saline just before initiation of the infusion.[13]

A nonpreserved morphine product, Duramorph®, is preferred for high-dose epidural therapy, but the cost may be prohibitive. Alternatives include two products using chlorobutanol as a preservative. This has also been noted to cause toxicity in patients treated with high-dose morphine therapy. If doses greater than 25 mg are used, the nonpreserved solution or the IMS® morphine system, which uses a nontoxic concentration of sodium bisulfite (0.1%), may be used.[6]

REFERENCES

1. **Food and Drug Administration,** OTC drug products for the control of dandruff, seborrheic dermatitis, and psoriasis; establishment of a monograph. *Fed. Reg.,* 47, 54646, 1982.
2. **Nikitakis, J. M.,** *CTFA Cosmetic Ingredient Handbook,* 1st ed., The Cosmetic, Toiletry and Fragrance Association, Washington, D.C., 1988.
3. **Weiner, M. and Bernstein, I. L.,** *Adverse Reactions to Drug Formulation Agents,* Marcel Dekker, New York, 1989.
4. **Brancato, D. J.,** Recognizing potential toxicity of phenol, *Vet. Hum. Toxicol.,* 24, 29, 1982.

5. **Weinstein, R. S.,** Recognition and management of poisoning with beta-adrenergic blocking agents, *Ann. Emerg. Med.,* 80, 1123, 1984.
6. **Du Pen, S. L., Ramsey, D., and Chin, S.,** Chronic epidural morphine and preservative-induced injury, *Anesthesiology,* 67, 987, 1987.
7. **Schmidt, R. and Maibach, H.,** Immediate and delayed onset "skip area" dermatitis presumed secondary to topical phenol exposure, *Contact Dermatitis,* 7, 199, 1981.
8. **Rubin, M. B. and Pirozzi, D. J.,** Contact dermatitis from carbolated vaseline, *Cutis,* 12, 52, 1973.
9. **Fisher, A. A.,** Irritant and toxic reactions to phenol in topical medications, *Cutis,* 26, 363, 1980.
10. **Baer, R. L., Serri, G., and Weissenbach-Vial, C.,** Studies on allergic sensitization to certain topical therapeutic agents, *Arch. Dermatol.,* 71, 19, 1955.
11. **Rogers, S. C. F., Burrows, D., and Neill, D.,** Percutaneous absorption of phenol and methyl alcohol in Magenta Paint B. P. C., *Br. J. Dermatol.,* 98, 559, 1978.
12. Committee on Safety of Medicines, *WHO Drug Inf.,* 4, 67, 1990.
13. **Mofenson, H. C., Caraccio, T. R., and Laudano, J.,** Glucagon for propranolol overdose, *JAMA,* 255, 2025, 1986.

PHENYLMERCURIC SALTS

I. REGULATORY CLASSIFICATION

Phenylmercuric acetate, phenylmercuric borate, and phenylmercuric nitrate are classified as antimicrobial preservatives. They are bactericidal against many Gram-positive and Gram-negative species. Fungicidal activity has been demonstrated against *Candida albicans* and *Aspergillus niger*.

II. SYNONYMS

A. Phenylmercuric Acetate

Acetoxyphenylmercury
Mercury, (acetato) phenyl

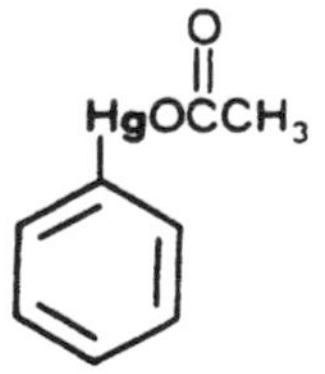

FIGURE 48. Phenylmercuric acetate.

B. Phenylmercuric Borate

Phenomerborum
Phenylmercuriborate
Orthoborato(1-)-O-phenylmercury
(Dihydrogen borato) phenylmercury

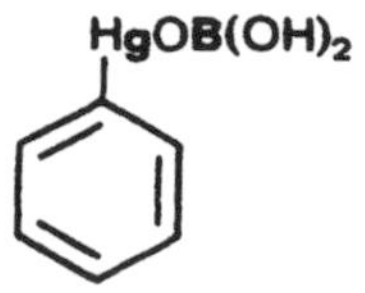

$$HgOB(OH)_2$$

FIGURE 49. Phenylmercuric borate.

C. Phenylmercuric Nitrate

Merphenyl nitrate
Nitratophenylmercury
PMN

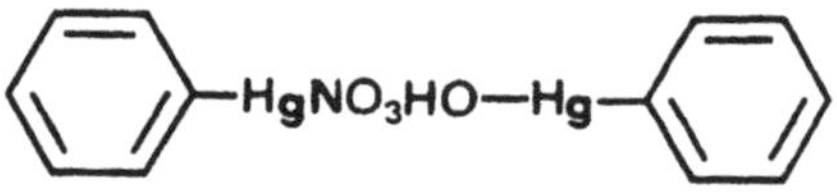

$$-HgNO_3HO—Hg-$$

FIGURE 50. Phenylmercuric nitrate.

III. AVAILABLE FORMULATIONS

A. Drugs

Phenylmercuric salts are used in parenteral products in concentrations of 0.001%, in ophthalmic products in concentrations of 0.002 to 0.004%, and in vaginal products in concentrations of 0.02%. Despite the lack of human epidemiological data demonstrating an increased risk of malformations, the FDA banned the use of phenylmercuric salts in vaginal contraceptives based on animal studies.[1]

B. Cosmetics

Phenylmercuric salts are used as preservatives in cosmetics, particularly in eye products such as eye shadow, mascara, eyeliner, eye makeup, and makeup removers.[2] Phenylmercuric acetate was present in 144 of 20,183 cosmetics listed with the FDA in 1982.[3]

IV. TABLE OF COMMON PRODUCTS

A. Parenteral Drug Products

Trade name	Manufacturer
Antivenin (Crotalidae) Polyvalent	Wyeth-Ayerst
Estradurin injection	Wyeth-Ayerst

B. Ophthalmic Drug Products

Trade name	Manufacturer
Gantrisin ophthalmic solution	Roche
Gantrisin ophthalmic ointment	Roche
Bleph 10 ophthalmic ointment	Allergan
Blephamide ophthalmic ointment	Allergan
FML ophthalmic ointment	Allergan

C. Nasal Drug Products

Trade name	Manufacturer
Afrin nasal spray, drops	Schering
Afrin chilren's strength drops	Schering
Coricidin nasal mist	Schering
Duration nasal pump, spray	Plough
4-Way Long acting spray	Bristol-Myers

V. HUMAN TOXICITY DATA

A. Immediate Hypersensitivity Reactions

Occupational exposure to hospital linen laundered with phenylmercuric propionate resulted in recurrent asthma and contact urticaria in a 34-year-old physician. Challenge testing, histamine release assay, and passive transfer confirmed immediate sensitivity to several phenylmercuric salts.[4]

B. Primary Irritant Reactions

One group of patients tested with 0.1% phenylmercuric acetate in petrolatum developed consistent primary irritant reactions, but this has not been the experience of other dermatologists. If irritant reactions are observed, a 0.05% concentration is advised.[5]

C. Delayed Hypersensitivity Reactions

Allergic contact dermatitis has been reported following occupational exposure to phenylmercuric salts in a crab grass killer, use of a medicated ointment, and use of a contraceptive jelly.[6]

D. Ocular Effects

Chronic use of phenylmercuric-preserved ophthalmic preparations has been reported to cause mecurialentis, a brown pigmentation of the anterior lens capsule, in over 140 glaucoma patients treated for more than 3 years. The pigmentation is not usually associated with visual impairment, and is estimated to occur in 6% of patients treated for more than 6 years.[7-10] The opacities are due to deposits of mercury and may resolve after topical treatment with EDTA.[9]

Occasional patients with chronic ophthalmic exposure display an atypical band keratopathy, beginning centrally and gradually extending toward the corneal periphery with corneal opacity. Symptoms may include reduced visual acuity, photophobia, lacrimation, visual haloes, and recurrent epithelial erosions.[11]

VI. CLINICAL RELEVANCE

Phenylmercuric salts may occasionally cause immediate and delayed hypersensitivity reactions. Most currently available ocular formulations containing phenylmercuric salts as preservatives are designed for short-term use and should not present a significant problem. Chronic use of these products should be avoided.

REFERENCES

1. **Goyan, J. E.,** Vaginal contraceptive drug products for over-the-counter human use: establishment of monograph: proposed rulemaking, *Fed. Reg.,* 45, 82014, 1980.
2. **Nikitakis, J. M.,** *CTFA Cosmetic Ingredient Handbook,* 1st ed., The Cosmetic, Toiletry and Fragrance Association, Washington, D.C., 1988.
3. **Decker, R. L. and Wenninger, J. A.,** Frequency of preservative use in cosmetic formulas as disclosed to FDA-1982 update, *Cosmet. Toilet.,* 97, 57, 1982.
4. **Mathews, K. P.,** Immediate type hypersensitivity to phenylmercuric compounds, *Am. J. Med.,* 44, 310, 1968.
5. **Fisher, A. A.,** Phenylmercuric acetate as primary irritant, *Arch. Dermatol.,* 106, 129, 1972.
6. **Morris, G. E.,** Dermatoses from phenylmercuric salts, *Arch. Environ. Health,* 1, 65, 1960.
7. **Abrams, J. D.,** Iatrogenic mercurialentis, *Trans. Am. Ophthalmol. Soc.,* 83, 263, 1963.
8. **Kennedy, R. E., Roca, P. D., and Landers, P. H.,** Atypical band keratopathy in glaucoma patients, *Trans. Am. Ophthalmol. Soc.,* 69, 124, 1971.
9. **Kennedy, R. E., Roca, P. D., and Platt, D. S.,** Further observations on atypical band keratopathy in glaucoma patients, *Trans. Am. Ophthalmol. Soc.,* 72, 107, 1974.
10. **Garron, L. K., Wood, I. S., Spencer, W. H., and Hayes, T. L.,** A clinical pathologic study of mercurialentis medicamentosus, *Trans. Am. Ophthalmol. Soc.,* 74, 295, 1967.
11. **Brazier, D. J. and Hitchings, R. A.,** Atypical band keratopathy following long-term pilocarpine treatment, *Br. J. Ophthalmol.,* 73, 294, 1989.

Polyethoxylated Castor Oil

I. REGULATORY CLASSIFICATION

Cremophor® is used as an emulsifying, wetting, and/or solubilizing agent.

II. SYNONYMS

Cremophor EL®
Cremophor RH 40®
Cremophor RH 60®
Emulphor EL®
Glycerol polyethyleneglycol riciinoleate
Glycerol polyethyleneglycol oxystearate
PEG 40 hydrogenated castor oil
PEG 60 hydrogenated castor oil
Polyoxyl 35 castor oil
Polyoxyl 40 hydrogenated castor oil
Polyoxyethylene 40 castor oil
Polyoxyethylene 60 castor oil
Polyethoxylated hydrogenated castor oil
Polyethoxylated vegetable oil

III. AVAILABLE FORMULATIONS

A. Constituents
Cremophor EL® is a mixture of a hydrophobic portion containing ricinoleic acid esters, glycerol and polyglycol ethers, and castor oil and a hydrophilic portion containing polyethylene glycols and ethoxylated glycerol.

B. Drugs
Cremophor EL® is used as an emulsifier and solubilizer in pharmaceuticals containing volatile oils, fat-soluble vitamins, and other hydrophobic substances, such as diazepam, propanidid, alfaxolone/alfadolone acetate, miconazole, methotrimeprazine, thiopental, and

glycerin suppositories. When used in oral liquid medications, the taste must be masked with flavors, such as banana. Cremophor RH 40® is almost tasteless and is preferred for oral liquid preparations.[1]

C. Cosmetics

Polyethoxylated castor oils are used in cosmetics as a perfume, in alcoholic vehicles as a volatile oil solubilizer, and in hand lotions as a substitute for castor oil. PEG hydrogenated castor oils are used in aftershave lotions, cleansing products, skin fresheners, colognes, hair tonics, and other cosmetics with alcoholic vehicles.[2]

IV. TABLE OF COMMON PRODUCTS

A. Injectable Drug Products

Trade name	Amount of Cremophor®	Manufacturer
AquaMephyton	70 mg/ml	Merck Sharp & Dohme
Monistat i.v.	0.115 mg/ml	Janssen
Sandimmune i.v.	928 mg/ml	Sandoz

V. ANIMAL TOXICITY DATA

A. Renal Dysfunction

The nephrotoxic potential of polyethoxylated castor oil (Cremophor®) was evaluated in three groups of five rats given 0.7 mg/kg/min of cyclosporine in Cremophor®, Cremophor® alone, or normal saline alone. Renal blood flow and glomerular filtration rate were decreased by more than 50% within 2 h of intravenous administration in the two groups receiving Cremophor®. These doses are approximately 300 times higher than those used in humans, thus the clinical significance of this finding is unclear.[3]

B. Anaphylactoid Reactions

Studies in the miniature pig demonstrated induction of anaphylactoid reactions after intravenous administration of Cremophor®-containing medications 1 to 3 weeks after a priming dose. The reactions could not be induced when the solvent was replaced with a propylene glycol/alcohol mixture.[4]

Continuous infusion of 20% Cremophor EL® at a rate of 30 ml/h in anesthetized beagle dogs resulted in severe cardiac collapse with decreased systemic arterial pressure peaking in 6 to 8 min. Six dogs developed cutaneous erythema and edema of the paws and muzzle. Platelet counts decreased sharply at the end of infusion, returning to baseline within 150 min. Plasma histamine increased sharply, peaking at 10 min and remaining at double the baseline level at 150 min.[5]

VI. HUMAN TOXICITY DATA

A. Immediate Anaphylactoid Reactions
1. Vitamin K

Anaphylactoid reactions, with precipitous hypotension, chest pain, dyspnea, facial flushing, abdominal pain, and asystole, have been reported within 2 to 5 min after bolus intravenous injections of phytonadione (Vitamin K1). In three of five cases, the rate of infusion was 1 mg/min; in one case it was 5 mg/min; and in another case 0.6 mg/min.[6-9] Four of five cases occurred with bolus injection of undiluted drug; in one case the dose was diluted in 100 ml of D5W.[9]

In one case, the reaction was prevented by reinstituting the infusion at a slow rate over 1 h.[6] Administration at a rate of 0.5 mg/min for 20 min did not cause adverse reactions in 12 patients.[8]

Cardiovascular collapse occurred during the first dose in three cases[6-8] and during the second dose in two cases, which resulted in one death.[7,9] Often, a prodrome sensation of uneasiness occurred prior to the onset of cardiovascular collapse.[7,8]

These reactions are consistent with those reported with the Cremophor® solvent in the phytonadione preparation. None of the patients were challenged with the vehicle alone, nor underwent immunologic testing, therefore the association with the solvent remains speculative but probable.

2. Althesin

This drug is a combination of alphaxalone and alphadolone, formerly used as a general anesthetic. It was withdrawn from the market in 1983 due to the high incidence of anaphylactic reactions to the Cremophor EL® solvent.

3. Cyclosporine

Anaphylactoid reactions, consisting of bronchospasm, dyspnea, laryngeal spasm, band-like chest pain, flushing, pruritus, urticaria, hypotension, and generalized maculopapular rash, have been reported to occur within 2 to 5 min after initiation of cyclosporine infusion, generally reversing within 30 min.

All but one of these patients have tolerated the oral cyclosporine solution, which contains an olive oil/polyoxy-5-oleate/ethanol-based vehicle, when it has been substituted. The patient who developed hypersensitivity reactions 8 h after a dose of oral solution and 1 h after intravenous dosing had a strongly positive intracutaneous skin test to the polyoxy-5-oleate solvent and a slight reaction to the Cremophor® solvent. Subsequent administration of cyclosporine in corn oil in an oral capsule was well tolerated.[10] Positive skin tests suggested an immunologic mechanism in another case.[11]

Most of these cases occurred after administration of multiple doses,[11-13] but have occurred with the first dose of cyclosporine in four cases. In two of these cases, previous exposure to Cremophor EL® as a solvent in a different drug (phytonadione and Althesin) was documented.[14-16] Premedication with antihistamines and corticosteroids failed to prevent a serious reaction in one case.[12] Premedication with diphenhydramine and slowing the rate of infusion to 6 h enabled tolerance of the drug in one case.[14]

A severe reaction, manifested primarily by cardiopulmonary arrest, occurred 5 min after the first dose of intravenous cyclosporine in a 38-year-old woman 8 h prior to a scheduled liver transplantation. Corticosteroid therapy had been discontinued in this patient for several months. She was also receiving propranolol, which has been shown to be a risk factor for severe anaphylactic reactions.[15]

Mild shortness of breath and a diffuse pruritic urticarial reaction were noted in one healthy volunteer, who was dropped from the study, 4 h after the end of an infusion of 190 mg over 2 h. A repeat dose 2 weeks later in another subject, who had no reaction to the first dose, resulted in immediate onset of erythema, pharyngeal irritation, urticaria, and abdominal pain. Three other subjects developed milder reactions after the second dose, consisting of facial and truncal erythema. Unlike the renal and liver transplant patients previously described to rarely experience these reactions, these subjects were not receiving chronic oral steroid therapy, which may explain the high incidence of reactions.[17]

4. Mechanism

Anaphylactoid reactions have been attributed to a peripheral vasodilatory effect of polyethoxylated castor oil, secondary to the release of histamine by nonimmunologically mediated degranulation of mast cells.[6] While histamine release has been readily documented in patients receiving drugs containing Cremophor® as a solvent, administration of Cremophor® alone produced no increase in histamine and clinical erythema in only one subject of eight tested.[18]

Complement-mediated reactions have been a major feature of anaphylactoid reactions to Althesin. Both nonimmune activation of the alternate complement pathway after the first dose and immune activation of complement after the second dose have been documented in human volunteers. It was suggested that Cremophor® enhances the immunogenicity of concurrently administered drugs, since the solvent alone did not induce immune responses.[19] Complement titers have also been decreased in patients treated with other drugs containing Cremophor EL® solvent, such as disoprofol,[20] and diazepam.[21]

A case report of anaphylaxis secondary to Althesin administration in a 16-year-old girl described the onset of cyanosis and generalized erythema. Complement titers were normal, but evidence of histamine release was documented. Immunological evaluation demonstrated a positive skin test to both the Althesin preparation and Cremophor EL®, a negative human basophil degranulation test, ruling out an IgE-mediated reaction, and a positive passive transfer test with heated serum. The mechanism in this case was presumed to be anti-Cremophor® IgG4 antibodies.[22]

Immediate hypersensitivity with positive intracutaneous skin tests has been documented in two other cases.[10,11]

The diverse mechanisms proposed to explain the pathogenesis of Cremophor®-related anaphylactoid reactions may explain differences in presentation and history in reported cases. Patients reacting to the first dose have generally milder reactions, which have been attributed to direct histamine release or alternate complement pathway activation. Patients reacting to subsequent exposures have more severe clinical reactions, which are immunologically related, via an IgG4 aggregate anaphylaxis, IgE-mediated Type I anaphylaxis, Type II cytotoxicity, or Type III immune complex hypersensitivity reaction.[23]

A 10-year survey of 118 patients with reactions to Althesin attributed 1% to IgE-mediated Type I hypersensitivity, 36% to immune complement-mediated reactions involving other antibodies, 40% to alternate pathway complement activation, and 23% to mixed reactions.[24]

B. Fat Embolism

It has been suggested that the encephalopathy associated with cyclosporine administration is related to fat embolism. Fat droplets were found in 20 of 40 samples of prepared cyclosporine intravenous solution. An 18-year-old girl with cyclosporine-induced encephalopathy following bone marrow transplantation had clinical findings consistent with fat embolism.[25] The manufacturer investigated the formation of droplets and concluded that fine droplets were formed only if the stoppers and solvent containers were contaminated with silicone oil and fatty substances, which are extracted by the Cremophor® solvent.[26]

Pulmonary fat embolism has been found at necropsy in the lungs of patients given bone marrow transplantation and cyclosporine therapy.[27]

C. Hyperlipidemia

Hyperlipidemia has been attributed to the Cremophor EL® solvent in intravenous miconazole preparations.[28-30] Serum lipoprotein electrophoretic patterns consistently demonstrated large beta, prebeta, and gamma-2 bands, with total serum triglyceride increases of more than tenfold. *In vitro* studies with the vehicle added to normal serum showed similar changes, which were only demonstrated with the polyethoxylated castor oil component.[30,31] The mechanism was shown to be selective delipidation of high-density lipoproteins.[31]

In one case, the peripheral blood smear was abnormal, with contraction of the blood on the slide, rouleaux formation, and distorted cellular morphology. These findings were attributed to a Cremophor®-related surface active phenomenon.[29]

Impairment of leukocyte function, specifically cell adherence and leukotaxis, have been documented with the miconazole vehicle.[32]

D. Bioavailability

One important aspect to be considered in evaluating the toxicity of excipients is the effect of the excipient on the bioavailability of the active ingredient, whether by an increased or decreased effect. Cremophor EL® has potent emulsifying properties that can increase the bioavailability of fat-soluble active ingredients. While this can be a beneficial attribute, reformulation of drugs, thereby altering their bioavailability, can result in adverse consequences.

When an injectable vitamin A product was changed from an oil-based vehicle to a water-based vehicle emulsified with Cremophor RH®, bioavailability of the vitamin increased from 1 to 50%, resulting in a cluster of reports of adverse local reactions. Pain, swelling, peeling at the site of injection, local pruritus, erythema, and long-lasting hyperpigmentation at the site of injection were described.[33]

E. Fluid Inbalance

Polyethoxylated castor oils have a surfactant effect that decreases the drop size of parenteral solutions administered by intravenous drip. The failure to account for this change in drop size resulted in fluid deprivation, with a deficit of 50% between calculated and measured fluid volumes. Administration via a volumetric pump will minimize this type of error.[34]

F. Pharmacokinetics

Polyethoxylated castor oil has a plasma half-life of about 22 h. Binding to albumin and other plasma proteins occurs initially. Once released from the protein site, urinary excretion occurs with a half-life of 13 min. Prolonged infusion of drugs containing Cremophor EL® solvents has been associated with various effects attributed to the persistence in plasma of this surfactant, including inability to spread blood on a slide, decreased plasma albumin levels, and decreased serum sodium levels, possibly due to hemodilution.[35]

VII. CLINICAL RELEVANCE

A. Anaphylactoid Reactions

Intravenous bolus of undiluted drugs with Cremophor® solvents should be avoided. Dilution of vitamin K1 in 100 to 125 ml of D5W or other appropriate diluent is recommended when the intravenous route is indicated. The rate of infusion should not exceed 0.5 mg/min. Administration via the intravenous route should be avoided in patients treated previously with drugs containing this solvent.

Patients treated with intravenous cyclosporine are generally receiving concomitant corticosteroid therapy, which may minimize the incidence of anaphylactoid reactions in this population. Patients receiving concomitant beta blocker therapy or who have a history of previous exposure to IV cyclosporine or other Cremophor®-containing drugs should be considered at increased risk of developing severe adverse reactions. Pretreatment of high risk patients with antihistamines and corticosteroids may be beneficial, but does not preclude development of serious toxicity. Alternate solvents for parenteral agents should continue to be investigated.

The incidence of anaphylactoid reactions in patients receiving a diazepam injectable preparation containing Cremophor EL® was 1 per 1040 doses.[36] Comparable rates have been published for other Cremophor®-containing parenterals, such as Althesin, with ranges from 1 in 930 to 1 in 433.[37,38]

REFERENCES

1. Technical leaflet, *Cremophor EL*, BASF, West Germany, 1981.
2. Nikitakis, J. M., *CTFA Cosmetic Ingredient Handbook*, 1st ed., The Cosmetic, Toiletry and Fragrance Association, Washington, D.C., 1988.
3. Thiel, G., Hermle, M., and Brunner, F. P., Acutely impaired renal function during intravenous administration of cyclosporine A: a cremophore side-effect, *Clin. Nephrol.*, 25 (Suppl. 1), S40, 1986.
4. Glen, J. B., Davies, G. E., Thompson, D. S., Scarth, S. C., and Thompson, A. V., An animal model for the investigation of adverse responses to i.v. anaesthetic agents and their solvents, *Br. J. Anaesth.*, 51, 819, 1979.
5. Gaudy, J. H., Sicard, J. F., Lhoste, F., and Boitier, *Can. J. Anaesth.*, 34, 122, 1987.
6. Barash, P., Kitahata, L. M., and Mandel, S., Acute cardiovascular collapse after intravenous phytonadione, *Anesth. Analg.*, 55, 304, 1976.
7. Rich, E. C. and Drage, C. W., Severe complications of intravenous phytonadione therapy, *Postgrad. Med.*, 72, 303, 1982.
8. Lefrere, J. J. and Girot, R., Acute cardiovascular collapse during intravenous vitamin K1 injection, *Thromb. Haemostasis*, 58, 790, 1987.
9. de la Rubia, J., Grau, E., Montserrat, I., Zuazu, I., and Paya, A., Anaphylactic shock and vitamin K1, *Ann. Intern. Med.*, 110, 943.
10. Van Hooff, J. P., Bessems, P., Beuman, G. H., and Leunissen, K. M. L., Absence of allergic reaction to cyclosporin capsules in patient allergic to standard oral and intravenous solution of cyclosporin, *Lancet*, 2, 1456, 1987.
11. Chapuis, B., Helg, C., Jeannet, M., Zulian, G., Huber, P., and Gumovski, P., Anaphylactic reaction to intravenous cyclosporine, *N. Engl. J. Med.*, 312, 1259, 1985.
12. Kahan, B. D., Wideman, C. A., Flechner, S., and Van Buren, C. T., Anaphylactic reaction to intravenous cyclosporin, *Lancet*, 1, 52, 1984.
13. Leunissen, K. M. L., Waterval, P. W. G., and Van Hooff, J. P., Anaphylactic reaction to intravenous cyclosporin, *Lancet*, 1, 636, 1985.
14. Howrie, D. L., Ptachcinski, R. J., Griffith, B. P., Hardesty, R. J., Rosenthal, J. T., Burckart, G. J., and Venkataramanan, R., Anaphylactoid reactions associated with parenteral cyclosporine use: possible role of Cremophor EL, *Drug Intell. Clin. Pharm.*, 19, 425, 1985.
15. Friedman, L. S., Dienstag, J. L., Nelson, P. W., Russell, P. S., and Cosimi, A. B., Anaphylactic reaction and cardiopulmonary arrest following intravenous cyclosporine, *Am. J. Med.*, 78, 343, 1985.
16. Magalini, S. C., Nanni, G., Agnes, S., Citterio, F., and Castagneto, M., Anaphylactic reaction to first exposure to cyclosporine, *Transplantation*, 42, 443, 1986.
17. Ptachcinski, R. J., Gray, J., Venkataramanan, J. R., Burckart, G. J., Van Thiel, D. H., and Rosenthal, J. T., Anaphylactic reaction to intravenous cyclosporin, *Lancet*, 1, 636, 1985.
18. Doenicke, A., Lorenz, W., Beigl, R., Bezecny, G., Uhlig, L., Kalmar, Praetorius, B., and Mann, G., Histamine release after intravenous application of short-acting hypnotics, *Br. J. Anaesth.*, 45, 1097, 1973.

19. **Watkins, J., Clark, A., Appleyard, T. N., and Padfield,** Immune-mediated reactions to althesin (alphaxalone), *Br. J. Anaesth.*, 48, 881, 1976.

20. **Briggs, L. P., Clarke, R. S. J., and Watkins, J.,** An adverse reaction to the administration of disoprofol (Diprivan), *Anaesthesia*, 37, 1099, 1982.

21. **Huttel, M. S., Schou Olesen, A., and Stoffersen, E.,** Complement-mediated reactions to diazepam with Cremophor as solvent (Stesolid MR), *Br. J. Anaesth.*, 52, 77, 1980.

22. **Moneret-Vautrin, D. A., Laxenaire, M. C., and Viry-Babel, F.,** Anaphylaxis caused by anti-Cremophor EL IgG STS antibodies in a case of reaction to Althesin, *Br. J. Anaesth.*, 55, 469, 1983.

23. **Radford, S. G., Lockyer, J. A., and Simpson, P. J.,** Immunological aspects of adverse reactions to althesin, *Br. J. Anaesth.*, 54, 859, 1982.

24. **Watkins, J.,** The allergic reaction to intravenous induction agents, *Br. J. Hosp. Med.*, 36, 45, 1986.

25. **Hoefnagels, W. A. J., Gerritsen, E. J. A., Brouwer, O. F., and Souverijn, J. H. M.,** Cyclosporin encephalopathy associated with fat embolism induced by the drug's solvent, *Lancet*, 2, 901, 1988.

26. **Krupp, P., Busch, M., Cockburn, I., and Schreiber, B.,** Encephalopathy associated with fat embolism induced by solvent for cyclosporin, *Lancet*, 1, 168, 1989.

27. **Paradinas, F. J., Sloane, J. P., Depledge, M. H., et al.,** Pulmonary fat embolisation after bone marrow transplantation, *Lancet*, 1, 715, 1983.

28. **Sung, J. P. and Grendahl, J. G.,** Side effects of miconazole for systemic mycosis, *N. Engl. J. Med.*, 297, 786, 1977.

29. **Niell, H. B.,** Miconazole carrier solution, hyperlipemia and hematological problems, *N. Engl. J. Med.*, 296, 1479, 1977.

30. **Bagnarello, A. G., Lewis, L. A., McHenry, M. C., Weinstein, A. J., Naito, H. K., McCullough, A. J., Lederman, R. J., and Gavan, T. L.,** Unusual serum lipoprotein abnormality induced by the vehicle of miconazole, *N. Engl. J. Med.*, 296, 497, 1977.

31. **Naito, H. K., McHenry, M. C., and Lewis, L. A.,** Drug-induced dyslipoproteinemia: a report of two cases, *Clin. Chem.*, 26, 163, 1980.

32. **Lee, C. and Maderazo, E. G.,** Interference of granulocyte function by the vehicle of miconazole, *Antimicrob. Agents Chemother.*, 13, 548, 1978.

33. **McCormick, P. A., Hughes, J. E., Burroughs, A. K., and McIntyre, N.,** Reformulation of injectable vitamin A: potential problems, *Br. Med. J.*, 301, 924, 1990.

34. **Wraight, W. J. and Cox, D. J.,** Fluid deprivation due to Althesin solution affecting drop size, *Br. Med. J.*, 1, 904, 1980.

35. **Knell, A. J., Turner, P., and Chalmers, E. P. D.,** Potential hazard of steroid anaesthesia for prolonged sedation, *Lancet*, 1, 526, 1983.

36. **Schou Olesen, A. and Huttel, M. S.,** Circulatory collapse following intravenous administration of Stesolid MR, *Ugeskr. Laeg.*, 140, 2644, 1978.

37. **Evans, J. M. and Keogh, J. A. M.,** Adverse reactions to intravenous anaesthetic induction agents, *Br. Med. J.*, 2, 735, 1977.

38. **Beamish, D.,** Adverse response to Althesin, *Anaesthesia*, 34, 683, 1979.

POLYETHYLENE GLYCOL

I. REGULATORY CLASSIFICATION

Polyethylene glycol (PEG) is classified as coating agent, ointment base, plasticizer, solvent, and tablet or capsule lubricant.

$$H(OCH_2CH_2)_nOH$$

FIGURE 51. Polyethylene glycol.

II. SYNONYMS

Carbowax
Macrogol
PEG

III. AVAILABLE FORMULATIONS

A. Description

Polyethylene glycols are designated by numbers correlating with the average molecular weight. PEG 100 to 700 are liquids. PEG 1000 to 10,000 are solids. The most commonly used liquid PEG 400 has a molecular weight ranging from 380 to 420. Carbowax 1500 is a mixture of PEG 300 and PEG 1540, resulting in a solid compound. Polyethylene glycol ointment USP is a mixture containing 60% PEG 400 and 40% PEG 3350.

B. Drugs

PEG 200 to 600 are used in soft gelatin capsules, oral liquids, and parenteral medications as solvents for the active ingredients. Solid derivatives are found in ointment bases, tablet binders and film coatings, and lubricants. Several colonic lavage solutions are available which contain PEG 3350 to PEG 4000 as active ingredients.

C. Cosmetics

Polyethylene glycols used in cosmetics contain from 4 to 115,000 oxyethylene groups (named PEG 4 to PEG 115M). Liquid derivatives are used as solvents and humectants in bath oils, fragrances, shampoos, hair conditioners, facial makeup, creams, lotions, suntan products, dentifrices, cleansing products, and mud packs. Higher molecular weight solid derivatives are used as binders, emulsion stabilizers, or viscosity-increasing agents in eyeliners, hair products, aftershave lotions, skin fresheners, mascara, makeup foundations, creams, and lotions.[1]

D. Foods

Polyethylene glycols with mean molecular weights of 200 to 9500 may be used in foods, with PEG 3350 to 4000 most common.[2]

IV. TABLE OF COMMON PRODUCTS

A. Topical Drug Products

Trade name	Manufacturer
Bactroban ointment	Beecham
Cortaid cream with aloe	Upjohn
Cyclocort lotion	Lederle
Furacin soluble dressing	Norwich Eaton
Halog ointment	Westwood-Squibb
Halog solution	Westwood-Squibb
Halotex cream	Westwood-Squibb
Lotrimin topical solution	Schering
Retin A liquid	Ortho
Tinactin cream	Schering
Tinactin solution	Schering

B. Parenteral Drug Products

Trade name	PEG type	Amount of PEG (mg/ml)	Manufacturer
Aristocort	3350	30	Fujisawa
Ativan	400	0.18	Wyeth-Ayerst
Depo-Medrol	3350	28–30	Upjohn

V. ANIMAL TOXICITY DATA

A. Topical Application to Burn Wounds

Following an observation of renal failure, hypercalcemia, acidosis, and high osmolal gaps in human burn patients, a rabbit model was employed to confirm the relationship of these findings to polyethylene glycol. Application of 20 g of the PEG-containing cream or the PEG vehicle was done every 12 h with dressing changes to surgically induced full-thickness injuries. Control animals received dressing changes without topical medication. Death resulted within 7 d in seven of eight animals in the PEG/antimicrobial group and in three of four animals in the PEG vehicle group. All control animals survived. PEG-treated animals had significantly increased total serum calcium levels, serum osmolal gaps, anion gaps, BUN, and creatinine levels compared to the control group.[3]

B. Sperm Motility

The spermicidal activity of a vaginal lubrication product containing polyethylene glycols and glycerin was demonstrated in an *in vitro* study of semen samples from normospermic donors. A marked dose-dependent decrease in motility and velocity was found, which lasted at least 180 min after exposure to the mixture. Since previous studies demonstrated no spermicidal activity for glycerin, this activity can be attributed to the PEGs.[4]

C. Cardiovascular Toxicity

Polyethylene glycol (PEG 500) was found to be devoid of significant arrhythmogenic activity in a study of cats pretreated with deslanoside. A significant increase in blood pressure, averaging 30 to 40 mmHg, was observed with the first two doses of 0.4 ml intravenously. The solvent also appeared to interfere with the pharmacologic activity of diazepam and chlordiazepoxide when mixed with these drugs, possibly due to precipitation of the drugs *in vivo*.[5]

VI. HUMAN TOXICITY DATA

A. Topical Application in Burn Patients

Topical application of an antimicrobial burn cream containing 63% PEG 300, 5% PEG 100, 32% PEG 4000, and trace amounts of ethylene glycol (0.01%) was associated with a triad of high anion gap acidosis, hypercalcemia, and hyperosmolality in three patients with burns covering 20 to 56% of the total body surface area. While total serum calcium was increased in two of the three patients, ionized calcium were decreased in all patients. Osmolal gaps (difference between measurement by freezing point depression and calculated value from sodium, glucose, and urea) ranged from +20 to +53. Serum ethylene glycol levels were measured in two patients with values of 1.3 mmol/l (8 mg/dl) in the patient with a 56% burn surface area and 0.4 mmol/l (2.5 mg/dl) in the patient with a 20% burn surface area. All patients had a fatal outcome with evidence of renal proximal tubular necrosis at autopsy.[6]

Acidosis was attributed to hydroxyacid and diacid metabolites of polyethylene glycol. Two of the diacid metabolites, 3-oxapentane-1,5-dicarboxylic and 3,6-dioxaoctane-1,8-dicarboxylic acid, bind avidly to calcium and explained the decrease in ionized calcium observed in all three patients. These metabolites were measured and found in the serum and urine of samples from these patients. The increase in total calcium levels was suggested to occur from a negative feedback mechanism in response to the fall in ionized calcium, producing increased parathyroid hormone levels.[6]

These investigators then performed a retrospective review of 40 patients with 20 to 75% body surface burns treated with this antimicrobial cream. Nine of ten fatalities were associated with renal failure, high anion gap metabolic acidosis, and high osmolal gaps. Serum polyethylene glycol and its metabolites were measured in the serum by mass spectrometry. Autopsy findings included severe proximal renal tubular necrosis and hydropic degeneration, with oxalate crystals in two patients. These lesions were not found in comparable patients not treated with the PEG-based cream.[7]

B. Immediate Hypersensitivity Reactions

One case of anaphylaxis has been described after ingestion of a multivitamin tablet containing PEG 8000 and PEG 20000. A severe reaction with hypotension, unconsciousness, and a grand mal seizure resulted on one occasion and a limited urticarial rash on taking the same medication 1 week later. Four previous episodes of unexplained anaphylaxis and a history of urticaria from perfumes and aftershave lotions were described. Skin prick tests with both PEG components elicited a positive response.[8]

Immediate contact urticaria was related to the use of two different antifungal products containing PEG 400 as solvents in a 50-year-old man. Application of the solvent into the normal skin of the forearm produced urticaria. Five control subjects had negative results. The incidence of urticaria to PEG 400 was estimated to be 1 in 1000 by manufacturer.[9]

A second case of contact urticaria was described by the same investigator 1 year later. This patient was a 35-year-old woman who used an otic solution containing PEG 300 and presented with exacerbation of her chronic otitis externa. Open skin testing with PEG 300 produced an immediate urticarial reaction within 20 min.[10]

C. Allergic Contact Dermatitis

One of 200 subjects became sensitized to polyethylene glycol 300 during a human Draize test of an experimental bar soap. After the final elicitation, a strong spreading reaction occurred. Positive patch tests were only found for 1 and 3% PEG 300 in petrolatum. Subsequent tests revealed cross-sensitization to PEG 600, 1000, 4000, and 6000. Despite evidence of delayed hypersensitivity, an open use test with 3% PEG 300 in petrolatum did not result in clinical dermatitis.[11]

Delayed allergic eczematous contact dermatitis was described in a 39-year-old woman following application of Furacin soluble dressing to a leg burn. The base of this product contained PEG 300, 1000, and 4000. Patch tests were strongly positive to PEG 300 and 400 and negative to PEG 1000 and 4000. A similar case of a 64-year-old man who developed a severe, edematous, vesicular, and crusted contact dermatitis following the use of Furacin solution on a radiation burn was described. Patch test results were identical to the first patient.[10]

A 53-year-old man with recurrent stasis dermatitis developed allergic contact dermatitis following application of Bactroban ointment, containing PEG 400 and PEG 3350. Patch testing to PEG ointment USP was positive in concentrations ranging from 5 to 100%. Evidence of sensitization to both PEG ingredients was demonstrated.[12]

The incidence of delayed contact dermatitis to PEG 300 was 4% in a series of 92 dermatologic patients with contact allergies. Cross-sensitivity was observed between polyethylene glycols of similar molecular weight. In 12 subjects sensitized to PEG 300, 5 cross-reacted with PEG 400, and only 1 reacted to PEG 1500 and PEG 6000.[9]

In another series of 180 patients with suspected topical medicament, footwear, or plant sensitivity, 8 of 120 with medicament-related dermatitis (6.7%) had positive patch tests to polyethylene glycols. Most patients (seven) tested positive to PEG 400, which was isolated to this molecular weight in four cases. Cross-reactivity to PEG 1500 was demonstrated in three cases and to PEG 3000 in two cases. An evaluation of 467 cases of medicament-related dermatitis over a 5-year period found sensitivity to PEG 400 in 25 cases (5.3%).[13]

D. Colonic Lavage Solutions

Administration of colonic lavage solutions containing polyethylene glycol as an active ingredient represent the extremes of oral exposure to these agents. The high molecular weight PEGs are not well absorbed orally, with an estimated 0.06% of the total dose of PEG 3350 absorbed in normal subjects[14] and up to 0.09% in inflammatory bowel subjects.[15] These solutions are tolerated well by infants, children, and young healthy adults in doses delivering 60 g of PEG per liter.[16]

Elderly patients with underlying cardiac disease developed asymptomatic ventricular ectopy in 12 of 24 procedures done with continuous electrocardiographic monitoring.[17] Ventricular bigeminy, paroxysmal supraventricular tachycardia, and multifocal PVCs were found during lavage of a 67-year-old man with previous left bundle branch block and atrial fibrillation.[18] A Mallory-Weiss tear was reported in an elderly woman after ingestion of 4 l of colonic lavage solution.[19]

The inadvertent substitution of PEG 400 for PEG 4000 in a colonic solution resulted in coma and metabolic acidosis after ingestion of 2 l in a 59-year-old woman. Oral absorption of the liquid PEG and subsequent metabolism to diacids, hydroxyglycolic acid, and diglycolics were thought to be responsible for the acidosis. Ethylene glycol was not found in the blood or urine, excluding metabolism to this monomer.[20]

E. Intravenous Drug Products

Intravenous doses of 40 ml/d of PEG 300, as a vehicle in nitrofurantoin, was associated with renal proximal tubular necrosis and metabolic acidosis in 7 of 32 patients treated with this antibiotic alone, or commonly in combination with other antibiotics, for chronic pyelonephritis. Oxalic acid could not be identified in the urine.[21]

Two adult patients with severe extensive poliomyelitic paralysis and resistant urinary tract infections developed severe reactions following administration of intravenous nitrofurantoin. Profound metabolic acidosis was the major finding in both patients, resulting in death in one case. The manufacturer revealed that 25 similar cases had been reported, resulting in recall of the product.[22]

F. Intrathecal Drug Products

Adhesive arachnoiditis was observed in 2 of 23 multiple sclerosis patients following administration of methylprednisolone acetate intrathecally in a vehicle containing 3% polyethylene glycol. The first case described a 28-year-old woman who had received a total of 23 injections over a 6-year period before discovery of a complete block at T6-7 with evidence of extensive arachnoiditis. The second case was a 27-year-old man who received a total of 5 injections over 1 year. Five months after the final injection he was hospitalized with adhesive arachnoiditis.[23]

Subsequently, an *in vivo* rodent study confirmed an association between this vehicle and sciatic nerve degeneration.[24] An *in vitro* study of desheathed rabbit nerves found a reversible neurotoxic effect with exposure to PEG 3350 at concentrations of 20% or greater, but could not document an adverse effect with a 3% solution. The neurotoxic effects described in the earlier study may have been related to another inactive component of the vehicle, which contains PEG 3%, sodium chloride 0.9%, and myristyl-gamma-picolinium chloride 0.02%. Adhesive arachnoiditis may be related to a nonspecific cellular reaction to repeated intrathecal injections.[25]

In 1980, the manufacturer of this product included a warning against the subarachnoid use of methylprednisolone. Although this route of administration was generally abandoned, replacement with the epidural route resulted in additional reports of arachnoiditis.[26-28] The continued use of intraspinal steroids is controversial.[29]

VII. CLINICAL RELEVANCE

A. Burn Patients

Polyethylene glycol-based topical antimicrobial agents should be avoided in patients with burn surface areas of 20% or greater and in those with impaired renal function. When indicated, application should be limited to 5 d or less. The recent finding of efficacy of mupirocin in methicillin-resistant *Staphylococcus aureus* burn wound infections may provoke more widespread use of this polyethylene glycol-based product.[30]

B. Dermatologic Reactions

Polyethylene glycol has rarely caused dermatitis in North American patients. The incidence of delayed hypersensitivity was over 5% in a series of Indian patients with medicament dermatitis. Both immediate allergic contact urticaria and delayed eczematous contact dermatitis have been reported. Patients sensitized to polyethylene glycol of one weight are likely to cross-react to PEGs of a similar weight and, occasionally, to higher molecular weight derivatives.

C. Colonic Lavage

Elderly patients with a history of cardiac disease are at risk for adverse reactions during administration of polyethylene glycol-electrolyte colonic lavage solutions. Ventricular ectopy may occur in up to 50% of these patients and may be more frequent than ectopy induced by the colonoscopy procedure itself. Cardiac monitoring should be done continuously during the lavage procedure, as well as during colonoscopy in this patient population.

REFERENCES

1. **Nikitakis, J. M.,** *CTFA Cosmetic Ingredient Handbook,* 1st ed., The Cosmetic, Toiletry and Fragrance Association, Washington, D.C., 1988.
2. **Meadows, J. O.,** On the safety of Golytely, *Gastrointest. Endosc.,* 31, 108, 1985.
3. **Herold, D. A., Rodeheaver, G. T., Bellamy, W. T., Fitton, L. A., Bruns, D. E., and Edlich, R. F.,** *Toxicol. Appl. Pharmacol.,* 65, 329, 1982.
4. **Boyers, S. P., Corrales, M. D., Huszar, G., and DeCherney, A. H.,** The effects of Lubrin on sperm motility in vitro, *Fertil. Steril.,* 47, 882, 1987.
5. **Pearl, D. S., Quest, J. A., and Gillis, R. A.,** Use of various solvents to study the effect of diazepam on cardiac rhythm, *Toxicol. Appl. Pharmacol.,* 44, 653, 1978.
6. **Bruns, D. E., Herold, D. A., Rodeheaver, G. T., and Edlich, R. F.,** Polyethylene glycol intoxication in burn patients, *Burns,* 9, 49, 1982.
7. **Sturgill, B. C., Herold, D. A., and Bruns, D. E.,** Renal tubular necrosis in burn patients treated with topical polyethylene glycol, *Lab. Invest.,* 46, 81A, 1982.
8. **Kwee, Y. N. and Dolovich, J.,** Anaphylaxis to polyethylene glycol (PEG) in a multivitamin tablet, *J. Allergy Clin. Immunol.,* 69, 138, 1982.
9. **Fisher, A. A.,** Contact urticaria due to polyethylene glycol, *Cutis,* 19, 409, 1977.
10. **Fisher, A. A.,** Immediate and delayed allergic contact reactions to polyethylene glycol, *Contact Dermatitis,* 4, 135, 1978.
11. **Maibach, H. I.,** Polyethyleneglycol: allergic contact dermatitis potential, *Contact Dermatitis,* 1, 247, 1975.
12. **Daly, B. M.,** Bactroban allergy due to polyethylene glycol, *Contact Dermatitis,* 17, 48, 1987.
13. **Bajaj, A. K., Gupta, S. C., Chatterjee, A. K., and Singh, K. G.,** Contact sensitivity to polyethylene glycols, *Contact Dermatitis,* 22, 291, 1990.
14. **Dipiro, J. T., Michael, K. A., Clark, B. A., Dickson, P., Vallner, J. J., Bowden, T. A., Jr., and Tedesco, F. J.,** Absorption of polyethylene glycol after administration of a PEG-electrolyte lavage solution, *Clin. Pharm.,* 5, 153, 1986.

15. **Brady, C. E., DiPalma, J. A., Morawski, S. G., Santa Ana, C. A., and Fordtran, J. S.,** Urinary excretion of polyethylene glycol 3350 and sulfate after gut lavage with a polyethylene glycol electrolyte lavage solution, *Gastroenterology,* 90, 1914, 1986.

16. **Tuggle, D. W., Hoelzer, D. J., Tunell, W. P., and Smith, E. I.,** The safety and cost-effectiveness of polyethylene glycol electrolyte solution bowel preparation in infants and children, *J. Pediatr. Surg.,* 22, 513, 1987.

17. **Marsh, W. H., Bronner, M. H., Yantis, P. L., Kilgore, J. W., and Rickoff, M. I.,** Ventricular ectopy associated with peroral colonic lavage, *Gastrointest. Endosc.,* 32, 259, 1986.

18. **Gholson, C. F.,** Ventricular ectopy during colonic lavage, *Gastrointest. Endosc.,* 33, 334, 1987.

19. **Brinberg, D. E. and Stein, J.,** Mallory-Weiss tear with colonic lavage, *Ann. Intern. Med.,* 104, 894, 1986.

20. **Belaiche, J., Vesin, P., Cattan, D., Payen, D., Rapin, M., Ventura, M., and Astier, A.,** Coma acidosique apres preparation colique par du polyethylene glycol, *Gastroenterol. Clin. Biologique,* 7, 426, 1983.

21. **McCabe, W. R., Jackson, G. G., and Grieble, H. G.,** Treatment of chronic pyelonephritis. II. Short-term intravenous administration of single and multiple antibacterial agents; acidosis and toxic nephropathy from a preparation of intravenous nitrofurantoin, *Arch. Intern. Med.,* 104, 710, 1959.

22. **Sweet, A. Y.,** Fatality from intravenous nitrofurantoin, *Pediatrics,* 22, 1204, 1958.

23. **Nelson, D. A., Vates, T. S., Jr., and Thomas, R. B.,** Complications from intrathecal steroid therapy in patients with multiple sclerosis, *Acta Neurol. Scand.,* 49, 176, 1973.

24. **Wood, K. M., Arguelles, J., and Norenberg, M. D.,** Degenerative lesions in rat sciatic nerves after local injections of methylprednisolone in aqueous solution, *Reg. Anaesth.,* 5, 13, 1980.

25. **Benzon, H. T., Gissen, A. J., Strichartz, G. R., Avram, M. J., and Covino, B. G.,** The effect of polyethylene glycol on mammalian nerve impulses, *Anesth. Analg.,* 66, 553, 1987.

26. **Roche, J.,** Steroid-induced arachnoiditis, *Med. J. Aust.,* 140, 281, 1984.

27. **Sekel, R.,** Epidural Depo-Medrol revisited, *Med. J. Aust.,* 141, 688, 1984.

28. **Kepes, E. R. and Duncalf, D.,** Treatment of backache with spinal injections of local anesthetics, spinal and systemic steroids: a review, *Pain,* 22, 33, 1985.

29. **Nelson, D. A.,** Dangers from intraspinal steroid injections, *Arch. Neurol.,* 47, 255, 1990.

30. **Rode, H., De Wet, P. M., Cywes, S., and Millar, A. J. W.,** Mupirocin in a polyethylene glycol carrier base, *J. Antimicrob. Chemother.,* 24, 78, 1989.

POLYSORBATES

I. REGULATORY CLASSIFICATION

Polysorbates are classified as emulsifying and/or solubilizing agents.

II. SYNONYMS

Crillet®
Monitan®
Polyethylene oxide sorbitan esters
Polyoxyethylene 20 sorbitan monolaurate (polysorbate 20)
Polyoxyethylene 40 sorbitan monopalmitate (polysorbate 40)
Polyoxyethylene 60 sorbitan monostearate (polysorbate 60)
Polyoxyethylene 80 sorbitan monooleate (polysorbate 80)
Polyoxyethylene sorbitan fatty acid esters
Sorlate®
Tween 80®

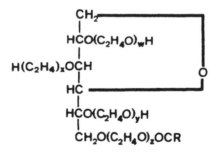

FIGURE 52. Polysorbate 20, 40, 80.

III. AVAILABLE FORMULATIONS

A. Drugs
Polysorbates are used as emulsifying agents in topical water-in-oil emulsions, in combination with hydrophilic emulsifiers in oil-in-water emulsions, and to increase water-holding properties

in ointments, in concentrations of 1 to 15%. As solubilizers, polysorbates may be a component of a lipophilic base for a water-insoluble drug in concentrations of 1 to 10%. Smaller amounts of polysorbates may be present as wetting agents in lipophilic bases.

Polysorbate 80 is approved for use as an active ingredient in ophthalmic demulcent preparations in concentrations of 0.2 to 1%. Polysorbate 20 and 80 may also be used in ophthalmic products and contact lens products in concentrations of up to 1%.[1]

B. Cosmetics

Polysorbates are widely used in cosmetics as oil-in-water emulsifiers, detergents, dispersing agents, solubilizers, and stabilizers. The total number of cosmetic products reported to contain polysorbates is summarized in the following table.[1]

Polysorbate type	Number of products
Polysorbate 20	702
Polysorbate 21	1
Polysorbate 40	69
Polysorbate 60	526
Polysorbate 61	10
Polysorbate 65	4
Polysorbate 80	203
Polysorbate 81	15
Polysorbate 85	30

Types of products include makeup bases and foundations; skin cleansers and fresheners; permanent waves; shampoos; hair conditioners; fragrance powders; face, body, and hand skin care products; eye makeup; shaving creams; colognes; and bath products.[1]

C. Foods

Polysorbates 20, 60, and 80 are approved as direct food additives as synthetic flavorings. Direct uses of polysorbates 60, 65, and 80 in foods include use as emulsifiers, solubilizers, dispersing agents, surfactants, wetting agents, opacifiers, defoaming agents, and dough conditioners. Limits depend on the food type and range from 10 ppm to 4.5%. Indirect food additive uses include adhesive components, emulsifiers, and surfactants.[1]

D. Impurities

Potential impurities in polysorbates include peroxides, isosorbide ethoxylates, free fatty acids, 1,4-dioxane, lead, and arsenic. Water-soluble impurities, such as 1,4-dioxane, are removed by steam stripping during the manufacturing process.[1]

IV. TABLE OF COMMON PRODUCTS

A. Parenteral Drug Products

Trade name	Manufacturer	Polysorbate type	Concentration	size
Kefzol	Lilly	Polysorbate 80	0.04%	500 mg/1g
M.V.I. vial 1	Rhone-Poulenc	Polysorbate 80	1.6%	5 ml
		Polysorbate 20	0.028%	5 ml

Trade name (cont'd)	Manufacturer (cont'd)	Polysorbate type (cont'd)	Concentration size (cont'd)	
M.V.I. Unit vial	Rhone-Poulenc	Polysorbate 80	0.8%	10 ml
		Polysorbate 20	0.014%	10 ml
M.V.I. Pediatric	Rhone-Poulenc	Polysorbate 80	50 mg	5 ml
		Polysorbate 20	0.8 mg	5 ml
Neupogen	Amgen	Polysorbate 80	0.004%	1 ml
Orthoclone OKT3	Ortho	Polysorbate 80	1 mg	5 ml

V. ANIMAL TOXICITY DATA

A. Neonates

Investigation of the cause of deaths in neonates due to a polysorbate-containing parenteral vitamin has been hampered by the inadequacies in animal models for the study of human neonatal toxicity. Preterm animals are difficult to maintain and involve excessive costs. Term neonatal animals are more mature than the premature human and provide an invalid model.[2]

Neonatal rats injected intraperitoneally with either polysorbate 80 or E-Ferol® developed a similar syndrome, with development of ascites, decreased weight gain, and increased mortality.[3]

A study of 63 term newborn rabbits found evidence of fatty liver changes following administration of the polysorbate vehicle in the E-Ferol® preparation. The combination of this vehicle and tocopherol or tocopherol acetate resulted in more pronounced toxicity, with cholestatic liver disease and elevated serum bilirubin levels. Tocopherol was not studied in an alternate vehicle, thus the relative contribution of each component to toxicity could not be determined. An interaction between polysorbate and tocopherol was implicated.[4]

B. Cardiovascular Effects

In dog and guinea pig models, intravenous administration of polysorbate 80, in doses of 10 mg/kg, has been associated with dose-dependent myocardial depressant properties, producing significant decreases in blood pressure within 10 min and lasting over 2 h. These effects were attributed to high levels of histamine measured in the animals, released from mast cells, peaking at 10 min after administration. No effect on the heart rate, respiration, or electrocardiogram was observed.[5-8] Histamine release has not been documented to occur in humans.

VI. HUMAN TOXICITY DATA

A. Immediate Hypersensitivity Reactions

An immediate, raised, whitish, intensely pruritic rash on the arms, chest, back, and head occurred after initiation of total parenteral nutrition with multivitamins in a 16-year-old boy with Burkitt's lymphoma. The rash resolved when the multivitamin was discontinued and recurred when resumed the following day. The vitamin vehicle contained polysorbates, BHA, and BHT. Immunologic testing was not done to determine which of the excipients was responsible. The antioxidants were not implicated since the boy had ingested foods known to contain these preservatives without adverse effects.[9]

B. Neonates

On November 8, 1983, O'Neal, Jones and Feldman (OJF) began distribution of a new parenteral aqueous vitamin E product (E-Ferol®) intended for the prevention of retrolental fibroplasia in premature newborns receiving supplemental oxygen. The excipients used to solubilize the vitamin, polysorbate 80 9% and polysorbate 20 1%, had a long history of safe use in food and drug products since the 1950s. Studies in adult animals indicated a very wide margin of safety. Thus, this product, containing an approved nutritional supplement and an approved excipient in a concentration ten times the usual amount, was marketed without submission of a new drug approval application.[10]

One month after the marketing of this product, the company received a report of two infants with weight gain, abdominal swelling, and clotting dysfunction believed to be related to administration of the vitamin. On January 24, the company received the second report of adverse reactions, involving kidney and liver dysfunction in four infants, resulting in three deaths. Distribution of the product was stopped for 11 d. After eight deaths had been reported, the FDA became involved, and a recall was issued on April 6, 1984. The final death toll was about 40 infants. On July 9, 1987, the distributor, manufacturer, the president and executive vice president of OJF, and the employee at Carter-Glogau Laboratories, who developed the formula, were indicted for continuing to market E-Ferol® without FDA approval. All three executives were found guilty and received suspended sentences with fines.[10]

The initial report of the syndrome observed by these two institutions was published by the Centers for Disease Control on April 13, 1984. Clinical signs and symptoms reported in 13 low-birth-weight (less than 1500 g) premature infants included ascites in all cases and hepatomegaly, splenomegaly, cholestatic jaundice, azotemia, and thrombocytopenia in some or all cases.[11]

A detailed report of the infants treated in one of the two institutions described eight premature infants with a mean birth weight of 934 g and mean gestational age of 26.8 weeks. The syndrome was found in 14% of infants treated with E-Ferol®. Thrombocytopenia was found in all infants at some point during deterioration of hepatic and pulmonary status. The average daily dose of E-Ferol® ranged from 26.6 to 50.5 U/kg. Five of the infants eventually died. *Cytomegalovirus* was isolated from the organs and urine at autopsy, but was dismissed as an insignificant finding.[12]

A review of clinical records and histologic material from infants reported from the other index hospital, along with records from two other institutions known to have used the preparation, included 50 patients administered doses of 7 to 136 U/kg/d. Autopsy materials from the 6 of the 7 deaths occurring in the index hospital and 14 deaths occurring in other participating hospitals were evaluated. Birth weights in the 21 infants who died ranged from 580 to 1431 g, with a gestational age of 25 to 35 weeks.[13]

Isolated thrombocytopenia was the earliest, most consistent sign heralding onset of deterioration, which was shortly followed by renal dysfunction, hypotension, and hyperbilirubinemia. Ascites was reported in ten cases. Initial signs of the syndrome appeared in 5 to 13 d after initiation of E-Ferol® in all but one case. In the index hospital, 11% of infants treated with E-Ferol® developed the severe syndrome, and 22% had probable evidence of the syndrome.[13]

Autopsy findings included intrahepatic cholestasis in 16 of 17 cases. A characteristic liver lesion was described in 15 infants, consisting of a progression from degeneration and exfoliation of hepatic sinusoidal lining cells, to panlobular sinusoidal dilatation with central accentuation, to focal obstruction of hepatic sinusoids and central veins by fibrous tissue, to a final stage involving extensive fibrosis, which obstructed sinusoidal and venular channels. These lesions were similar to those observed in *Cytomegalovirus* infection and hepatic veno-occlusive disease seen with certain herbal tea ingestions.[13]

Renal pathology revealed acute tubular necrosis in 9 of 15 patients. Three patients had substantial deposition of oxalate crystals in the distal renal tubules and collecting ducts; three others had tubular oxalate deposition without evidence of active tubular necrosis.[13]

An *in vitro* study found 37% inhibition of PHA-induced lymphyocyte transformation in the presence of E-Ferol®. Polysorbate 80 alone caused 44% inhibition, while tocopherol acetate enhanced the lymphocyte response. T11 cells were decreased after incubation of cells in culture with polysorbates. An immunosuppressive effect *in vivo* on T-helper cells was postulated.[14] Although no consistent evidence of overwhelming opportunistic infection was found in infants with the E-Ferol® syndrome,[15] the consistent presence of *Cytomegalovirus* in one series and generalized candidiasis in one case may indicate a contributory immunosuppressive effect of the polysorbate vehicle.[12,13]

A follow-up survey of neonatal intensive care units in which the drug had been used was conducted by the FDA. Results of this survey indicated that more than 1000 infants had been exposed, with about 40 deaths attributed to the product. An increased risk for death, hepatomegaly, ascites, and thrombocytopenia was confirmed in exposed infants. A dose relationship was found for the mean daily dose and cumulative dose and the risk of death. Decreasing birth weight was also a risk factor. Increased risks for the clinical syndrome were confined to the first 1 to 14 d of exposure.[16]

The mechanism of action of polysorbates in causing this tragedy has not been fully resolved. One theory postulated an inability of the premature newborn to metabolize, and therefore excrete the drug. Adults excrete about 90% of a dose in the urine within 24 h.[1] Metabolism by hydrolysis of the fatty acid chain from the polyoxyethylated moiety is essential for urinary excretion to occur in the neonate. High levels of polysorbate were found in some of the infants who died, indicating immature metabolism and accumulation of the excipient in body fluids.[17,18]

One possible, but far-fetched, metabolite of polysorbate is oxalic acid. For this to occur, the polyoxyethylene moiety would have to be cleaved and degraded to ethylene oxide, which has been documented to occur in rats.[19] The ethylene oxide would then react with water in the urine to form ethylene glycol, which is subsequently metabolized to oxalic acid.[20] Alternatively, hepatic microsomal oxidative dealkylation of the polyoxyethylene chains to form glycolaldehyde would result in eventual oxalate formation. Concurrent oxygen therapy would be expected to enhance these reactions. It was estimated that a 1% conversion of polysorbate to oxalate in these infants would result in urine oxalate excretion two to ten times higher than the normal excretion in adults.[21] Calcium oxalate crystals were found in the renal tissues in two infants with the E-Ferol® syndrome.[13,20]

There is some evidence that the toxicity observed was dose related. Infants with the syndrome received higher total doses over a longer period of time and had lower birth weights than unaffected infants receiving the drug.[13,18] One infant had a polysorbate 80 concentration of 100 mg/l in ascitic fluid.[18] The mean daily dose of alpha-tocopherol in 17 cases of E-Ferol® syndrome was 37 U/kg, compared to 25.8 U/kg in 51 nonaffected infants. No cases were observed with doses of less than 20 U/kg/d of E-Ferol®. Infants who received doses of greater than 40 U/kg/d had a 70% attack rate.[15] A dose of 1 U/kg/d of E-Ferol® corresponds to a dose of 3.6 mg/kg/d of polysorbate 80 and 0.4 mg/kg/d of polysorbate 20. The recommended dose of the drug, 1 to 2 ml daily, contained 90 to 180 mg of polysorbate 80 and 10 to 20 mg of polysorbate 20.[16]

Finally, a contaminant may have been responsible for some or all of the syndrome observed in these infants. Gas chromatography (GC) and GC-mass spectrometry results suggested that an unstripped grade of polysorbate may have been used in the E-Ferol® product. Contaminants, such as 1,4-dioxane and butyl rubber elastomeric compounds leached from vial closures, may have played some role in the toxicity of the product.[18]

C. Cardiovascular Toxicity

A study of 20 adults undergoing coronary angiography compared the cardiac effects of amiodarone in a lyophilized powder dissolved in a nonspecified "inert" diluent or a polysorbate 80 vehicle, containing 100 mg of polysorbate per milliliter. An intravenous injection of amiodarone 5 mg/kg was given over 3 min. Short-lived hemodynamic changes were attributed to the polysorbate vehicle, including hypotension and tachycardia. A negative inotropic effect was not demonstrated. Vasodilation secondary to histamine release was the proposed mechanism of action of polysorbate 80.[22]

D. Dermatologic Reactions

Topical application of 10% polysorbate 80 produces slight erythema on healthy human skin.[5]

VII. CLINICAL RELEVANCE

A. Neonates

The administration of polysorbates is unquestionably associated with a dose-related syndrome of hepatomegaly, ascites, and thrombocytopenia in premature neonates with birth weights of less than 1500 g. The lowest dose associated with this syndrome delivered 72 mg/kg/d of polysorbate 80. The drug product associated with this syndrome contained alpha-tocopherol acetate. Increased bioavailability of this vitamin due to the polysorbate vehicle may have been a contributory factor.

Currently available multivitamins for addition to intravenous parenteral nutrition fluids for infants are estimated to deliver a total daily polysorbate dose of 10 to 24 mg/kg in infants weighing less than 1000 g, and 24 to 72 mg/d in infants weighing 1000 to 3000 g. There is no evidence that this amount is harmful.[13]

REFERENCES

1. **Moore, J.,** Final report on the safety assessment of polysorbates 20, 21, 40, 60, 61, 80, 81, and 85, *J. Am. Coll. Toxicol.,* 3, 1, 1984.
2. **Pesce, A. J.,** Toxic susceptibilities in the newborn with special consideration of polysorbate toxicity, *Ann. Clin. Lab. Sci.,* 19, 70, 1989.
3. **Farkus, W. R., Conover, B., Al Ansari, H., Lorch, V., Feld, N., and Sinha, S. N.,** Toxicity of polysorbate 80 in neonatal rats, *Pediatr. Res.,* 20, 227A, 1986.
4. **Rivera, A., Jr., Abdo, K. M., Bucher, J. R., Leininger, J. R., Montgomery, C. A., and Roberts, R. J.,** Toxicity studies of intravenous vitamin E in newborn rabbits, *Dev. Pharmacol. Ther.,* 14, 231, 1990.
5. **Varma, R. K., Kaushal, R., Junnarkar, A. Y., Thomas, G. P., Naidu, M. U. R., Singh, P. P., Tripathi, R. M., and Shridhar, D. R.,** Polysorbate 80: a pharmacological study, *Arzneim. Forsch.,* 35, 804, 1985.
6. **Masini, E., Planchenault, J., Pezziardi, F., Gautier, P., and Gagnol, J. P.,** Histamine-releasing properties of polysorbate 80 *in vitro* and *in vivo*: correlation with its hypotensive action in the dog, *Agents Actions,* 16, 470, 1985.

7. **Millard, R. W., Baig, H., and Vatner, S. F.,** Cardiovascular effects of radioactive microsphere suspensions and Tween 80 solutions, *Am. J. Physiol.,* 232, H331, 1977.

8. **Burnell, R. H. and Maxwell, G. M.,** General and coronary haemodynamic effects of Tween 20, *Aust. J. Exp. Med. Sci.,* 52, 151, 1974.

9. **Levy, M. and Dupuis, L. L.,** Parenteral nutrition hypersensitivity, *JPEN,* 14, 213, 1990.

10. **Carey, C.,** Three jailed for selling drug that killed 38 babies, *FDA Consumer,* July-August, 33, 1989.

11. **Centers for Disease Control,** Unusual syndrome with fatalities among premature infants: association with a new intravenous vitamin E product, *MMWR,* 33, 198, 1984.

12. **Lorch, V., Murphy, M. D., Hoersten, L. R., Harris, E., Fitzgerald, J., and Sinha, S. N.,** Unusual syndrome among premature infants: association with a new intravenous vitamin E product, *Pediatrics,* 75, 598, 1985.

13. **Bove, K. E., Kosmetatos, N., Wedig, K. E., Frank, D. J., Whitlatch, S., Saldivar, V., Haas, J., Bodenstein, C., and Balistreri, W. F.,** Vasculopathic hepatotoxicity associated with E-Ferol syndrome in low-birth-weight infants, *JAMA,* 254, 2422, 1985.

14. **Alade, S. L., Brown, R. E., and Paquet, A.,** Polysorbate 80 and E-Ferol toxicity, *Pediatrics,* 77, 593, 1986.

15. **Martone, W. J., Williams, W. W., Mortensen, M. L., Gaynes, R. P., White, J. W., Lorch, V., Murphy, D., Sinha, S. N., Frank, D. J., Kosmetatos, N., Bodenstein, C. J., and Roberts, R. J.,** Illness with fatalities in premature infants: association with an intravenous vitamin E preparation, E-Ferol, *Pediatrics,* 78, 591, 1986.

16. **Arrowsmith, J. B., Faich, G. A., Tomita, D. K., Kuritsky, J. N., and Rosa, F. W.,** Morbidity and mortality among low birth weight infants exposed to an intravenous vitamin E product, E-Ferol, *Pediatrics,* 83, 244, 1989.

17. **McKean, D. L. and Pesce, A. J.,** Determination of polysorbate in ascites fluid from a premature infant, *J. Anal. Toxicol.,* 9, 174, 1985.

18. **Balistreri, W. F., Farrell, M. K., and Bove, K. E.,** Lessons from the E-Ferol tragedy, *Pediatrics,* 78, 503, 1986.

19. **Nelson, M. F., Poulos, T. A., Gongwer, L. E., et al.,** Preparations of carbon-14-labeled polyoxyethylene (20) sorbitan monolaurate and their metabolic fate in rats, *J. Food Sci.,* 31, 253, 1966.

20. **Brown, R. E., Alade, S. L., and Krouse, M. A.,** Polysorbates and renal oxalate crystals in the E-Ferol syndrome, *JAMA,* 255, 2445.

21. **Conyers, R. A. J., Bais, R., and Rofe, A. M.,** Oxalosis and the E-Ferol toxicity syndrome, *JAMA,* 256, 2678, 1986.

22. **Munoz, A., Karila, P., Gallay, P., Zettelmeier, F., Messner, P., Mery, M., and Grolleau, R.,** A randomized hemodynamic comparison of intravenous amiodarone with and without Tween 80, *Eur. Heart J.,* 9, 142, 1988.

POVIDONE

I. REGULATORY CLASSIFICATION

Povidone is classified as a suspending and/or viscosity-increasing agent and tablet binder. Crospovidone is classified as a tablet disintegrant.

II. SYNONYMS

Kollidon®
Plasdone®
Polyvidone
Polyvinylpyrrolidone
PVP
Vinylpyrrolidinone polymer

Povidone is a mixture of linear polymers of 1-vinylpyrrolidine-2-one with varying chain lengths. The average molecular weight is indicated by the K-value. Povidone K15 has an average molecular weight of about 10,000, K30 an average of 40,000, K60 an average of 160,000, and K90 an average of 360,000. Crospovidone is a synthetic cross-linked povidone homopolymer.

$$\left[\begin{array}{c} CH_2-CH- \\ | \\ N \\ \end{array} \right]_n$$

FIGURE 53. Povidone.

III. AVAILABLE FORMULATIONS

A. Drugs
Povidone is present in 12 injectable drug products listed with the FDA, containing 0.03 to 0.9%. There are over 600 oral dosage forms containing 1 to 50 mg of povidone, including

303

capsules, oral powders for reconstitution, oral suspensions, and tablets. Crospovidone is less widely used and is present in 222 oral dosage forms and 5 transdermal delivery systems.[1]

IV. TABLE OF COMMON PRODUCTS

A. Injectable Drug Products

Trade name	Amount (%)	Manufacturer
Bicillin-CR	0.55	Wyeth-Ayerst
Bicillin-LA	0.6	Wyeth-Ayerst
Wycillin	0.5	Wyeth-Ayerst

V. HUMAN TOXICITY DATA

A. Panniculitis

Chronic intramuscular administration of procaine suspended in povidone weekly for 7 years was associated with the development of multiple subcutaneous nodules and periodic fever and malaise in a 60-year-old woman. The total amount of the compound administered over this time period was more than 3 l. The injection site was intermittently red, swollen, and painful at intervals of approximately every 3 weeks. Histological examination of the lesions revealed large oval or round granular globules within histiocytes, which stained gray-blue with hematoxylin-eosin. The reactions responded to systemic corticosteroids.[2]

B. Granulomas

The discovery of tissue storage of povidone was a consequence of massive use of povidone as a plasma expander in over 500,000 German soldiers during World War II.[3]

Povidone excipients were first associated with the development of cutaneous granulomas in the early 1960s. The first cases were reported in patients receiving chronic daily subcutaneous injections of posterior pituitary-povidone for the treatment of diabetes insipidus. The presentation varied from an onset of a few months to 15 years and a total dosage of povidone from 100 g to 2.5 kg. The lesions occurred at sites remote from the sites of injection and consisted of hundreds of reddish-brown flat or slightly raised papules primarily on the neck, upper chest, and around the waist. Povidone was readily identified within tissue macrophages by staining with Chlorazol Fast Pink or Congo Red.[4,5]

A series of four cases of patients with povidone storage after injection of vasopressin delivering total povidone doses of 250 to 900 g reported unusual sites of storage, such as the ovary and kidney. A renal cell clear cell carcinoma containing povidone-filled macrophages was found in one patient.[6]

The histologic changes associated with povidone storage have been mistaken for tuberculosis, paraamyloidosis, inborn errors of metabolism, and sarcoma.[6,7]

In one case of a relatively minimal exposure to a total of 45 g of povidone injected subcutaneously over a 5-year-period as an excipient in a kallidinogenase antihypertensive medication, a sarcoma-like induration of the buttock was found. The material extracted from the pseudotumor corresponded with povidone K30 when analyzed by infrared spectrophotometry.[7]

VI. CLINICAL RELEVANCE

Chronic administration of high-molecular weight povidone polymers (40,000 or higher) results in accumulation of globular deposits within macrophages in the liver, spleen, lymph nodes, bone marrow, and skin. Lower molecular weight compounds are excreted in the urine. Povidone is stored indefinitely in these tissues and has been found many years after a single injection.[4] The clinical significance of these deposits is debatable; rare cases of an inflammatory response with panniculitis, pancytopenia, or asymptomatic hepatosplenomegaly and lymphadenopathy have been described after long-term intramuscular injection. The amounts of povidone present in currently available injectable products (less than 1%) is much less than the amounts associated with these storage conditions (10 to 25%).

Povidone is not well absorbed after oral ingestion, and crospovidone is completely nonabsorbed. These do not present a measurable hazard in foods or oral pharmaceuticals. The temporary acceptable daily intake of povidone in foods established by the WHO is 25 mg/kg.[8]

REFERENCES

1. **Food and Drug Administration,** *Inactive Ingredients Guide,* FDA, Washington, D.C., March 1990.
2. **Kossard, S., Ecker, R. I., and Dicken, C. H.,** Povidone panniculitis, *Arch. Dermatol.,* 116, 704, 1980.
3. **Wessel, W., Schoog, M., and Winkler, E.,** Polyvinylpyrrolidone (PVP), its diagnostic, therapeutic and technical application and consequences thereof, *Arzneim. Forsch.,* 21, 1468, 1971.
4. **Dupont, A. and Lachapelle, J. M.,** The fate of foreign macromolecules, *Br. J. Dermatol.,* 80, 543, 1968.
5. **Thivolet, J., Leung, T. K., Duverne, J., Leung, J., and Volle, H.,** Ultrastructural morphology and histochemistry (acid phosphatase) of the cutaneous infiltration by polyvinylpyrrolidone, *Br. J. Dermatol.,* 83, 661, 1970.
6. **Christensen, M., Johansen, P., and Hau, C.,** Storage of polyvinylpyrrolidone (PVP) in tissues following long-term treatment with a PVP-containing vasopressin preparation, *Acta Med. Scand.,* 204, 295, 1978.
7. **Hizawa, K., Otsuka, H., Inaba, H., Izumi, K., and Nakanishi, S.,** Subcutaneous pseudosarcomatous polyvinylpyrrolidone granuloma, *Am. J. Surg. Pathol.,* 8, 393, 1984.
8. **World Health Organization,** Twenty-seventh report of the FAO/WHO expert committee on food additives, Tech. Rep. Ser. No. 696, WHO, Geneva, 1983.

PROPYLENE GLYCOL

I. REGULATORY CLASSIFICATION

Propylene glycol is classified as a humectant, plasticizer, and solvent.

II. SYNONYMS

1,2-Propanediol
Propane-1,2-diol
Methyl glycol
Methyl ethylene glycol
1,2-dihydroxypropane

III. AVAILABLE FORMULATIONS

A. Foods

Propylene glycol is used in foods as a solvent for flavorings in baked goods and candies, as an emulsifier, and as a preservative. Foods implicated in causing hypersensitivity reactions include carbonated beverages, flavored popcorn, coconut, salad dressings, cake mixes, whipped topping mixes, cheesecake mixes, french fried onions, potato sticks, and frozen cakes and cheesecakes.[1]

B. Drugs

Propylene glycol is a clear, colorless, odorless viscous liquid with a sweet glycerin-like taste. It is used as a solvent or co-solvent in oral solutions in concentrations of 10 to 25% and in parenteral pharmaceuticals in concentrations of 10 to 80%. Topical products may contain 5 to 80% and topical aerosol solutions 10 to 30%. Propylene glycol is contained in approximately 55% of topical corticosteroid, emollients, antibacterial, and antifungal products.[2]

C. Cosmetics

Propylene glycol is used as a humectant, solvent, and viscosity-reducing agent in a wide variety of cosmetics, including lipsticks, creams and lotions, makeup, hair dyes, shampoos, eye makeup, deodorants, antiperspirants, and bubble baths.[3]

IV. TABLE OF COMMON PRODUCTS AND ALTERNATIVES

A. Oral Liquid Drugs

Trade name	Manufacturer
Bentyl syrup	Lakeside
Butisol sodium elixir	Wallace
Children's Panadol drops	Glenbrook
Children's Panadol liquid	Glenbrook
Colace drops	Mead Johnson
Delsym	McNeil Consumer
Dorcol Children's Cough syrup	Sandoz
Elixophyllin GG	Forest
Iberet 500 liquid	Abbott
Lanoxin elixir pediatric	Burroughs Wellcome
Lortab liquid	Rugby
Loxitane oral concentrate	Lederle
Minocin oral suspension	Lederle
Naldecon DX pediatric drops	Bristol
Naldecon EX pediatric drops	Bristol
Naldecon EX children's syrup	Bristol
Neoloid	Lederle
Nucofed syrup	Beecham
Pediacare cold formula	McNeil Consumer
Pediacare cough-cold formula	McNeil Consumer
Pedicare infant's oral decongestant drops	McNeil Consumer
Pediacare night rest cough-cold formula	McNeil Consumer
Pentuss	Fisons
Polaramine syrup	Schering
Prelone syrup	Muro
Slo-Phyllin syrup	Rorer
Slo-Phyllin GG syrup	Rorer
Sodium polystyrene sulfonate suspension	Roxane
Tagamet liquid	Smith Kline & French
Tavist syrup	Sandoz
Tegretol suspension	Geigy
Theoclear-80 syrup	Central
Triaminic DM syrup	Sandoz
Triaminic Nite-Light	Sandoz
Triaminicol Multi-Symptom Relief	Sandoz
Tri-Vi-Flor 0.25 mg with iron and fluoride drops	Mead Johnson
Tussar 2	Rorer
Tussionex	Fisons
Tussi-Organidin liquid	Wallace
Tussi-Organidin DM liquid	Wallace
Tylenol children's liquid cold formula	McNeil Consumer

Trade name (cont'd)	Manufacturer (cont'd)
Tylenol drops	McNeil Consumer
Tylenol elixir	McNeil Consumer
Vicks Children's Nyquil	Richardson-Vicks
Vicks Pediatric Formula 44 Cough	Richardson-Vicks
Vicks Pediatric Formula 44 Cough & Congestion	Richardson-Vicks
Vicks Pediatric Formula 44 Cough & Cold	Richardson-Vicks
Vi-Daylin ADC Vitamin drops	Ross
Vi-Daylin F ADC drops with fluoride	Ross

B. Topical Drug Products

Trade name	Manufacturer
Aclovate cream	Schering
Aristocort cream	Fujisawa
Aristocort A ointment	Fujisawa
Benzac AC gel	Owen/Galderma
Benzac W gel	Owen/Galderma
Blemerase lotion	Young
Caldesene ointment	Schering
Carmol HC cream	Syntex
Cleocin T topical solution	Upjohn
Cortaid lotion	Upjohn
Cutivate cream	Glaxo
Dermolate cream	Schering
Desenex cream, ointment	Schering
Desowen cream	Owen/Galderma
Eldecort cream	ICN
Eldepaque forte cream	ICN
Elocon lotion	Schering
Erymax solution	Herbert
Eurax cream,lotion	Westwood-Squibb
Exelderm cream	Westwood-Squibb
Femstat vaginal cream	Syntex
Florone cream	Dermik
Florone E cream	Dermik
Fluonid solution	Herbert
Fluoroplex solution	Herbert
Fungizone cream, lotion	Squibb
Halog cream	Westwood-Squibb
Halog E cream	Westwood-Squibb
Hytone cream,lotion	Dermik
Kenalog cream,lotion	Westwood-Squibb
Kenalog H cream	Westwood-Squibb

Trade name (cont'd)	Manufacturer (cont'd)
Keralyt gel	Westwood-Squibb
Komed acne lotion	Sola/Barnes-Hind
Lac-Hydrin lotion	Westwood-Squibb
Lidex cream, gel, ointment, solution	Syntex
Lidex E cream	Syntex
Masse breast cream	Ortho
Maxiflor cream	Herbert
Medicone-Derma ointment	Medicone
Medicone dressing cream	Medicone
Mycolog II cream	Westwood-Squibb
Mycostatin cream	Westwood-Squibb
Neo-Synalar cream	Syntex
Nizoral cream	Janssen
Nutracort cream	Owen/Galderma
Oxistat cream	Glaxo
Penecort cream	Herbert
Pramosone cream	Ferndale
Psorcon ointment	Dermik
Rogaine solution	Upjohn
SSD cream	Boots
Sulfacet R acne lotion	Dermik
Sulfoxyl lotion	Stiefel
Sween cream	Sween
Synacort cream	Syntex
Synalar cream, solution	Syntex
Synalar HP cream	Syntex
Synemol cream	Syntex
Temovate cream, ointment	Glaxo
Tinactin cream	Schering
Tinactin jock itch cream	Schering
Tinver lotion	Sola/Barnes-Hind
Vanoxide HC acne lotion	Dermik
Vytone cream	Dermik
Westcort cream, lotion	Westwood-Squibb

C. Parenteral Drug Products

Trade name	Manufacturer	Dosage	Route	Amount of PG v/v (%)	Amount of PG w/v (mg/ml)	Alternative
Amidate	Abbott	2 mg/ml	i.v.	35	362.6	None
Apresoline	Ciba	20 mg/ml	i.m., i.v.	10	103.6	None
Ativan	Wyeth-Ayerst	2 mg/ml	i.m., i.v.	80	828.8	None
Ativan	Wyeth-Ayerst	4 mg/ml	i.m., i.v.	80	828.8	None

Trade name (cont'd)	Manufacturer (cont'd)	Dosage (cont'd)	Route (cont'd)	v/v (%) (cont'd)	w/v (mg/ml) (cont'd)	Alternative (cont'd)
				Amount of PG		

Trade name (cont'd)	Manufacturer (cont'd)	Dosage (cont'd)	Route (cont'd)	v/v (%)	w/v (mg/ml)	Alternative (cont'd)
Bactrim	Roche	TMP 16 mg/ml SMX 80 mg/ml	i.v.	40	414.4	None
Berocca PN	Roche	2 ml	i.v.	25	259	MVI Pediatric MVI-12 lyophilized
Brevibloc	DuPont	250 mg/ml	i.v.	25	259	None
Dilantin	Parke-Davis	50 mg/ml	i.m., i.v.	40	414.4	None
Dramamine	Searle	50 mg/ml	i.m., i.v.	50	518	None
Dramocen	Central	50 mg/ml	i.m., i.v.	50	518	None
Embolex	Sandoz	0.7 ml	s.c.	44	460	None
Konakion	Roche	10 mg/ml	i.m.	20	207	Aqua-Mephyton
Konakion	Roche	2 mg/ml	i.m.	20	207	Aqua-Mephyton
Lanoxin	Burroughs Wellcome	0.25 mg/ml	i.m., i.v.	40	414.4	None
Lanoxin Pediatric	Burroughs Wellcome	0.1 mg/ml	i.m., i.v.	40	414.4	None
Librium	Roche	50 mg/ml	i.m., i.v.	20	207	None
Loxitane	Lederle	50 mg/ml	i.m.	70	725.2	None
Luminal Sod	Winthrop	130 mg/ml	i.v.	67.8	702.4	Phenobarbital Sodium (Lilly)
MCV9 Plus	Lyphomed	10 ml	i.v.	30	310.8	MVI Pediatric MVI-12 lyophilized
MVI-12	Armour	10 ml	i.v.	30	310.8	MVI Pediatric MVI-12 lyophilized
Nitro-BID	Marion	5 mg/ml	i.v.	4.3	45	Nitrol
Nembutal	Abbott	50 mg/ml	i.m., i.v.	40	414.4	None
Nitrostat	Parke-Davis	5 mg/ml	i.v.	30	310.8	Nitrol
Nitroglycerin	Abbott	5 mg/ml	i.v.	50	518	Nitrol
Pentobarbital	Wyeth	50 mg/ml	i.m., i.v.	40	414.4	None
Phenobarbital	Elkin-Sinns	130 mg/ml	i.m., i.v.	67.8	702.4	Phenobarbital Sodium (Lilly)
Phenobarbital	Wyeth	130 mg/ml	i.m., i.v.	67.8	702.4	Phenobarbital Sodium (Lilly)

Trade name (cont'd)	Manufacturer (cont'd)	Dosage (cont'd)	Route (cont'd)	Amount of PG v/v (%) (cont'd)	Amount of PG w/w (mg/ml) (cont'd)	Alternative (cont'd)
Phenytoin	Lyphomed	50 mg/ml	i.m., i.v.	30	310.8	None
Phenytoin	Elkin-Sinns	50 mg/ml	i.m., i.v.	40	414.4	None
Septra	Burroughs Wellcome	TMP 16 mg/ml SMX 80 mg/ml	i.v.	40	414.4	None
Tridil	American Critical Care	0.5 mg/ml	i.v.	30	310.8	Nitrol
Valium	Roche	5 mg/ml	i.m., i.v.	40	414.4	None

V. ANIMAL TOXICITY DATA

A. Cardiovascular Toxicity

Much of our appreciation of the potential cardiovascular effects of propylene glycol comes from scant animal data. In a study of 20 anesthetized cats given rapid bolus injections of 40% propylene glycol in a dose of 0.5 ml/kg over 1 to 5 s, cardiorespiratory effects were observed, including hypotension, bradycardia, amplified QRS and T waves, transient ST elevation, and transient apnea. A larger dose, 1 ml/kg over 1 to 5 s, caused depressed AV conduction and multifocal ectopic ventricular arrhythmias, while doses of 1.5 to 2.5 ml/kg resulted in asystole. The arrhythmic effects were markedly decreased when the infusion rate was slowed to 1 ml/min or less or when phenytoin was given as pretreatment; premature ventricular beats were observed in some animals. Conversely, hypotension was worsened by phenytoin pretreatment. The arrhythmias and bradycardia were prevented by severing the vagus or pretreatment with atropine.[4]

In a subsequent study in 27 anesthetized adult cats, the cardiovascular effects of diazepam injection were compared to those of the propylene glycol vehicle. Rapid infusion of 2 mg/kg (0.8 to 1.6 ml) of diazepam over 1 s resulted in immediate transient apnea, hypotension, cardiac arrhythmias, and bradycardia. The same effects were noted after infusion of 40% propylene glycol, with an intensity approximately two thirds that of the diazepam-propylene glycol mixture.[5]

A similar study was done in unanesthetized calves to eliminate potential cardiac effects of the anesthetic. Rapid bolus infusion of a 79.2% propylene glycol solution, at a rate of 1 to 2 ml/s produced AV block, decreased pulmonary and renal arterial blood flow, asystole, hemolysis, and hemoglobinuria.[6]

A subsequent study in six intact awake calves compared the effects of two vehicles used in oxytetracycline preparations, propylene glycol 77.8% and polyvinylpyrrolidine 19%, and attempted to determine the mechanism of action. The drugs were given at a rate of 1 ml/s. Effects attributed to the propylene glycol vehicle included decreased heart rate, cardiac output, and stroke volume, and increased aortic pressure and pulmonary resistance. No effect on ventricular contractility was observed. The bradycardia was concluded to be secondary to a baroreceptor response, resulting in increased vagal tone. Pretreatment with antihistamines abolished the cardiovascular effects in three calves, leading to the conclusion that the mechanism involved release of histamine.[7]

Because deaths in humans attributed to propylene glycol toxicity have frequently occurred in patients receiving acute or chronic digitalis therapy, the effects of intravenous propylene glycol in anesthetized cats with deslanoside-induced ventricular arrhythmias were studied. In seven cats with deslanoside-induced persistent ventricular tachycardia, the diazepam solvent system, consisting of propylene glycol 40%, ethanol 10%, and sodium benzoate/benzoic acid 5%, was infused in a dose of 1 ml every 45 s for a total dose of 5 ml. The onset of ventricular fibrillation was significantly faster in cats receiving the solvent (mean 10.8 min) compared to cats not administered the solvent (mean 26.7 min). Solvent-treated animals also experienced transient decreases in arterial blood pressure and heart rate.[8]

B. Respiratory Depression

No significant effect of propylene glycol on central respiratory activity could be demonstrated in a study of 15 anesthetized artificially ventilated dogs as measured by phrenic nerve activity.[9] Previous studies using anesthetized spontaneously breathing animals did not control for the potential of interference by the anesthetic agent.[5]

C. Immunosuppression

In vitro concentrations of 1% propylene glycol were shown to significantly decrease natural killer cell cytotoxicity in blood collected from a healthy human volunteer. In the same model, neutrophil function was markedly depressed at concentrations of 0.5 and 1%, with some effect at 0.1%. These concentrations are comparable to those achievable by oral administration of 60 ml (predicted level of 0.4%) or intravenous administration of 15 g/d (predicted level of 0.12%), assuming distribution is limited to the extracellular space.[10]

VI. HUMAN TOXICITY DATA

A. Cardiovascular Toxicity

Human cases of propylene glycol-associated fatal cardiotoxicity have been complicated by the multiplicity of underlying conditions and concomitant medications and the advanced age of the patients. The rate of infusion has been cited as related to toxicity, but in 5 of 11 reported cases, phenytoin was administered at the recommended rate of 1 ml (50 mg) per minute or less. Six patients appeared to have been given excessive rates of 1.7 to 5 ml (83.3 to 250 mg) per minute. In 9 of 11 cases, phenytoin was given to reverse presumed or proven digoxin toxicity. Adverse effects attributed to propylene glycol in these cases, prior to death, included sinus arrhthymias, ventricular fibrillation, and asystole.[11-18]

The incidence of significant hypotension following phenytoin infusion correlated with patient age in a study of 139 adults, aged 17 to 94 years, treated for acute seizures. A total of 13 patients (24%) developed hypotension required slowing or temporary discontinuation of the infusion; all were 40 years old or greater.[19]

Propylene glycol-induced reversible cardiotoxicity can occur in normal young subjects given recommended rates of administration. Sudden onset of sinus bradycardia, hypotension, and syncope was reported in 4 of 15 young healthy volunteers given phenytoin intravenously in a total dose of 250 mg administered at a rate of 35 to 40 mg/min. The method of administration (i.e., i.v. push vs. piggyback) was not mentioned, thus the effects of concentration could not be evaluated.[20]

A prospective clinical trial in 200 patients treated with intravenous phenytoin for seizures reported reversible cardiovascular toxicity in 6 patients (3%). All patients with cardiac effects received phenytoin at administration rates of 40 to 50 mg/min, total doses of 800 mg or more, and infusion concentrations of 6.6 mg/ml or higher.[21] The effect of concentration on cardiac toxicity was underscored by two case reports of death in patients receiving phenytoin by manual intravenous push, a methodology that resulted in a calculated concentration of 17.5 mg/ml of phenytoin.[17]

Commercial intravenous nitroglycerin products may contain 4.5 to 96% propylene glycol. Experimental use of intracoronary nitroglycerin to reverse coronary vasospasm has prompted consideration of the effects of propylene glycol given by this route. The intracoronary infusion of 4 ml of a vehicle containing 0.045% w/v propylene glycol and 0.7% v/v ethanol did not produce consistent changes in heart rate or mean aortic pressure in four patients with myocardial ischemia.[22] The effects of higher concentrations have not been reported.

POSSIBLE PROPYLENE-GLYCOL (PG)/PHENYTOIN RELATED CARDIAC TOXICITY

Age (year)	Cardiac history	Phenytoin rate (mg/min)	PG dose mg/min	Clinical effect	Outcome	Ref.
26	No	40–50	331.5–414.4	Bradycardia, hypotension	Recovered	21
38	NS	40–50	331.5–414.4	SA block or arrest	Recovered	21
57	No	26.2	217	Atrial fibrillation	Recovered	19
65	Yes	400	3315	Apnea, hypotension	Recovered	23
66	Yes	40–50	331.5–414.4	Bradycardia, hypotension	Recovered	21
68	Yes	40–50	331.5–414.4	Bradycardia, hypotension	Recovered	21
NS	NS	40–50	331.5–414.4	PVCs	Recovered	21
NS	NS	40–50	331.5–414.4	PVCs	Recovered	21
44	No	35	290	Apnea, asystole	Died	17
60	No	100	828.8	Apnea, cardiac arrest	Died	17
67	Yes	50	414.4	Apnea, AV block, asystole	Died	13
67	Yes	125	1036	Cardiac arrest	Died	14
67	Yes	250	2072	Hypotension, apnea	Died	18
68	Yes	35	290	Cardiopulmonary arrest	Died	17
70	Yes	83	690	AV block, asystole	Died	14
76	Yes	83	690	Heart block, asystole	Died	15
76	Yes	33	276	Asystole	Died	16
82	Yes	100	828.8	Asystole	Died	12
85	Yes	50	414.4	Ventricular fibrillation	Died	11

NS = Not stated.

B. Neonates

In 1983, propylene glycol was identified as a cause of hyperosmolality in premature neonates, secondary to the presence of this solvent in a parenteral multivitamin preparation. The index case was a premature infant, birth weight 890 g, who developed acute renal failure on the 12th day of life, 9 d after initiation of parenteral nutrition, which included a multivitamin delivering 3 g/d of propylene glycol. The serum osmolality was elevated at 407 mOsm/kg which decreased when parenteral nutrition was discontinued.

Evaluation of ten additional infants who had received the multivitamin product for at least 5 d revealed hyperosmolality in four infants. The osmolal gap was found to be proportional to the serum propylene glycol concentration. The propylene glycol half-life ranged from 10.8 to 30.5 h.[24]

A subsequent study by these same investigators compared the effects of a parenteral multivitamin delivering a lower dose of propylene glycol (300 mg/d) with a preparation containing mannitol in neonates weighing less than 1500 g at birth. Hyperosmolality was present in 30% of a low weight subset (less than 999 g) in both groups of infants, but was maintained for 12 d in the propylene glycol group, while decreasing to an incidence of 10% after 2 d in the mannitol group. The serum osmolality correlated with the propylene glycol level in the low weight subset only.[25]

A retrospective study compared the effects of propylene glycol on the incidence of seizures, hyperbilirubinemia, and renal failure in low birth weight infants. Infants weighing less than 1500 g were divided into a high-dose propylene glycol group of 49 infants, who received 3 g/d, and a low-dose propylene glycol group of 78 infants, who received 300 mg/d. Significant increased incidence of the following parameters were found in the high-dose group: BUN, serum osmolality, seizures, and intraventricular hemorrhage. The serum bilirubin values were 50% higher in the high-dose group, but this was not statistically significant.[26]

The effect of propylene glycol on serum osmolality in normal weight infants is less clearly documented. The continuous infusion of a total of 13.2 g/d of propylene glycol, derived from enoximone and nitroglycerin infusions, resulted in hyperosmolality in a 3.4 kg infant with low cardiac output following open heart surgery. No other adverse effects were noted.[27]

C. Neurologic Toxicity

Propylene glycol has been demonstrated to have CNS depressant activities in animals,[28] which has subsequently been documented in humans.

Similar to the cardiovascular effects, the CNS properties may be related to the dose, but the rate and concentration may not be as important. Ataxia occurred in 10 of 200 patients receiving phenytoin for the treatment of seizures; all had received doses of 800 mg or greater. Administration rates and concentrations were low or moderate, a mean of 28.4 mg/min and 5.6 mg/ml, respectively. Other CNS effects included confusion, dizziness, and drowsiness.[21]

That the CNS symptoms may be related to the solvent is suggested by a study of 18 epileptic adult patients comparing the effects of oral administration of phenytoin in capsules vs. an oral 40% propylene-glycol-containing solution given in a daily dose of 150 ml. Despite attainment of similar phenytoin plasma concentrations, the incidence of mental symptoms (drowsiness, confusion, slurred speech, or lethargy) was 38% in the oral solution phase and 11% in the capsule phase.[29]

D. Children

The adverse effects of propylene glycol in children, other than in neonates, have not been well documented. The administration of 2 to 4 ml twice daily of propylene glycol, as a vehicle in an oral vitamin D preparation, was associated with development of seizures in an 11-year-old boy. The onset was 13 months after initiation of therapy, which lessens the probability of a cause-effect relationship. Seizures did not recur after discontinuation of the product.[30]

In another case report, a 15-month-old child was administered 7.5 ml of propylene glycol daily for 8 d as a vehicle in an oral vitamin C preparation. On the eighth day he developed an irregular heart rate and sinus arrhythmias. Two days later, he was unresponsive with tachypnea, tachycardia, and diaphoresis. The blood glucose level was 70 mg/dl. Several episodes of symptomatic hypoglycemia (nadir 42 mg/dl) occurred over the next 3 d. The symptoms abated following administration of i.v. dextrose and did not recur after discontinuing the vitamin C preparation.[31]

E. Lactic Acidosis

The proposed metabolic pathway for the metabolism of propylene glycol involves conversion to pyruvate and lactate.[32,33] In a case of overdose in an epileptic patient with an unknown amount of propylene glycol, the presenting signs and symptoms included metabolic acidosis, status epilepticus, apnea, and a high osmolar gap. The serum lactate was within normal limits, but was not measured until 16 h after admission.[34]

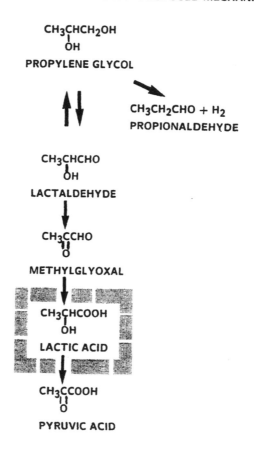

FIGURE 54. Propylene glycol metabolic pathway.

A prospective evaluation of adverse metabolic reactions to propylene glycol was performed in patients receiving a commercially available intravenous nitroglycerin preparation containing 50% propylene glycol. Total daily doses of propylene glycol were 30 to 400 ml. Six of 28 (21%) patients were found to have elevated lactate levels; however 5 had convincing evidence of other etiologies, such as hemodynamic instability and tissue hypoxia. Preexisting renal dysfunction, with a creatinine clearance of less than 30 ml/min, was strongly correlated with hyperosmolality.[35]

Lactic acidosis (5.1 mmol/l), hyperosmolality (441 mOsmol/kg), intravascular hemolysis, and progressive development of stupor, disorientation, and coma were described in an elderly woman with impaired renal function following administration of 322 ml of propylene glycol daily as a vehicle in an intravenous nitroglycerin product.[36]

Presumed inhalation exposure resulted in lactic acidosis (18 meq/l) and stupor in a 58-year-old man with chronic schizophrenia and azotemic renal disease.[37]

Elevated serum lactate levels were observed in five hospitalized patients with normal renal function, selected retrospectively on the basis of receiving medications containing propylene glycol and availability of biological specimens for analysis. A least-squares regression analysis suggested that serum propylene glycol concentrations of greater than 177 mg/l are necessary to increase lactate concentrations by 6 mg/l.[38]

The relationship between peak measured serum lactate levels and other clinical correlates to propylene glycol concentrations is summarized in the following table.

Case Reports of Propylene Glycol (PG) Lactic Acidemia

Age	Renal function (Cr (mg/dl) or CrCl (ml/min)	Time (h)	Peak serum PG (mg/l)	Peak serum lactate (meq/l)	PG dose (daily)	Symptoms PG	Half-life (h)	Ref.
45 years	NL	18	54	2.5	57 mg/kg	NS	NS	38
2 months	NL	120	711	24.1	267 mg/kg	NS	4.7	38
54 years	NL	NS	150	10	113 mg/kg	NS	NS	38
4 months	NL	24.5	173	13.7	771 mg/kg	NS	NS	38
4 months	NL	5	304	15.5	512 mg/kg	NS	NS	38
58 years	Cr 2.4	NS	700	18	Unknown	Stupor	NS	37
72 years	CrCl 39	8 days	9100	5.1	322 ml	Coma	3.1	36

NS = Not stated by author.

F. Hemolysis

Hemolysis is common *in vitro* with propylene glycol concentrations of 30% or greater. Concentrations higher than this cause direct cell injury, with crenation and agglutination. Administration of a 50% solution to rabbits resulted in decreased blood clotting time and thrombocytosis.[39] During a prospective study of the effects of intravenous administration of nitroglycerin in 50% propylene glycol vehicle, intravascular hemolysis was observed in three patients (10.5%); in one of these the nitroglycerin was given simultaneously with packed red cells.[35]

G. Topical Application in Burn Patients

Impressive hyperosmolality was first associated with the topical application of propylene glycol as a vehicle of silver sulfadiazine cream in two patients with second- and third-degree burns over 80 and 90% of the total body surface area. Both patients had been treated with massive amounts of the cream, totaling as much as 25 lb/d. The measured serum osmolality (by freezing point depression) peaked at 420 mOsm/kg on the 12th day of therapy in the first patient, and at 463 mOsm/kg on the 14th day of therapy in the second patient. Measurement of osmolality by the vapor pressure method, which reflects the contribution of propylene glycol and other volatile substances, showed a difference from the freezing point method of 20 and 35 mOsm/kg, respectively.[40]

These differences (osmolal gap) would correspond to a serum propylene glycol concentration of 240.6 and 357.6 mg/dl, respectively, using the formula published by Fligner et al. (PG level in milligrams per decaliter = 84.6 + (osmolal gap × 7.8).[41]

Following the discovery of these 2 patients, a retrospective study of 262 patients treated for flame burns at the same institution was performed. Nine of these patients (3.4%) had a serum osmolality measured by freezing point depression that differed from the calculated osmolality, using the formula 2(Na + K) + Glu/18 + BUN/2.8. The mean burn surface area was significantly greater in the patients with an osmolal gap (65%; range 35 to 90%) compared to patients without an osmolal gap (35%). None had a history of renal impairment. The osmolal gap in the hyperosmolar group ranged from 13 to 236 mOsm/kg (mean 84 mOsm/kg).[42]

This group of investigators subsequently performed a prospective study on 45 burn patients with measurement of serum propylene glycol levels by gas chromatography. Propylene glycol was detectable in 24 patients (53%) with a mean serum level of 45 mg/dl.[43]

The above-mentioned studies involved adult burn patients. Fligner et al.[41] reported an 8-month-old infant with second- and third-degree burns of the chest involving 8% of his total body surface area, which was complicated by development of toxic epidermal necrolysis involving 70% of his total body surface area. He received twice daily applications of silver sulfadiazine 600 to 800 g, corresponding to a total topical dose of propylene glycol of 9 g/kg/24 h.

The next day, he experienced a cardiorespiratory arrest 1 h after a dressing change, which was associated with a propylene glycol serum level of 369 mg/dl. An peak osmolal gap of 130 mOsm/kg was documented the following day. The serum propylene glycol level peaked at 1059 mg/dl. A mild lactic acidosis was also present, with a serum lactate of 5.8 meq/l. The apparent propylene glycol elimination half-life was 16.9 h.

Topical application of propylene glycol-containing creams to 12 adult patients with extensive psoriasis or other scaling disorders, in doses ranging from 1.5 to 6.1 gr/kg/24 h for 5 d, was not associated with hyperosmolality or changes in serum lactate concentrations.[44]

H. Contact Dermatitis

Propylene glycol is hypertonic in concentrations greater than 2% and is a moderate primary skin irritant in concentrations of 100%, expressed more measurably under occlusion and in cooler, less humid months.[1,45] The highest concentration documented to be nonirritating was 20%.[46] The irritant effects of propylene glycol, particularly in conjunction with occlusive conditions, may promote sensitization to other product ingredients.[47,48]

Due to concerns over possible irritant reactions, the recommended patch test concentration has been a matter of some debate and may account for the variable rates of positive tests observed in case series. Earlier studies used concentrations of 10 to 20%, while the current recommendation is 5% in aqueous solution.[49] Positive reactions are often not reproducible on later repeat tests, suggestive of a possible "angry back" reaction.[50]

Allergic reactions are rarely attributed to propylene glycol in foods.[51] One case report documents recurrence of propylene-glycol dermatitis at sites of previously affected areas and patch test areas associated with ingestion of a salad dressing.[52]

Oral and parenteral administration of propylene glycol may also cause dermatitis in sensitized individuals. A case series of 38 patients with positive patch tests to propylene glycol described the onset of dermatitis 3 to 16 h after oral ingestion of 2 to 15 ml in 15 patients (39%), which subsided within 24 to 48 h in all but one case.[50] An eczematous rash was also described in a patient given an intravenous dose of diazepam, containing a propylene glycol solvent.[53]

Topical pharmaceutical products reported to cause delayed-type contact hypersensitivity reactions have included corticosteroids,[51,48] antifungal products,[54] keratolytics, ECG electrode pastes,[51,54] silver sulfadiazine cream,[56] lubricating gels,[51] topical minoxidil lotion,[56] and external otic preparations.[51] An extemporaneous formulation of minoxidil, using glycerin in place of propylene glycol, has been successfully used in propylene-glycol sensitive patients.[57]

The incidence of positive patch tests to propylene glycol in series of patients with contact dermatitis is summarized in the following table.

Number of patients	Patch test concentration	% Positive tests	Country	Ref.
4097	2% aqueous	0.2	Finland	58
400	20% aqueous	1.5	Italy	59
3364	5% aqueous	0.8	Italy	60
167	10% aqueous	4.8	Belgium	48
487	Not stated	4.5	U.S.	61

I. CPK Elevation

Intramuscular administration of many pharmaceuticals is associated with a transient elevation in serum creatine phosphokinase (CPK) secondary to muscle injury. Chlordiazepoxide injection, containing propylene glycol 20%, polysorbate 4%, and benzyl alcohol 1.5%, resulted in comparable elevations in CPK when compared to the solvent system alone. Injection of a saline control did not produce significant changes in CPK.[62]

J. Thrombophlebitis

Injectable diazepam available in the U.S. is formulated in a solvent containing 40% propylene glycol, 10% ethanol, 5% sodium benzoate, and 1.5% benzyl alcohol. Ethanol has been reported to have a dose-related venous irritant effect, with an incidence of phlebitis of 8% in patients receiving rapid infusions of 5% w/v ethanol and 30% in patients receiving 10% w/v solutions,[63] and undoubtedly contributes to the high incidence of thrombophlebitis observed with diazepam. In evaluating studies on parenteral drug-induced thrombophlebitis, the sensitivity is increased with the duration of follow-up, since clinical onset may be delayed. In a study using 14 d of follow-up, the incidence of diazepam-related thrombophlebitis was 62%.[64]

Propylene glycol may also contribute to diazepam thrombophlebitis. Substitution of propylene glycol with a Cremophor® solvent reduced the incidence of phlebitis to 3.4 to 9%.[64,65] Studies comparing the incidence of phlebitis of etomidate in a propylene glycol or polyethylene glycol solvent found the former to be more irritating.[66]

K. Heparin Resistance

A dose-related resistance to the effects of heparin has been documented when high doses of intravenous nitroglycerin are given. The nitroglycerin preparation that was originally implicated in this interaction contained 93% propylene glycol in the vehicle. Testing of normal subjects showed inhibition of heparin activity with either the diluent alone (49% reduction in aPTT) or with the complete nitroglycerin preparation (70% reduction in aPTT).[67]

A subsequent report of seven patients compared the effects of a nitroglycerin product containing 30% propylene glycol to a nitroglycerin product containing 10% ethanol (with no propylene glycol). Heparin resistance was observed with both of these products, leading to the conclusion that propylene glycol was not involved in this interaction. Since the previous study did show an effect from the propylene glycol vehicle alone, and since this study did not involve a vehicle challenge, an alternate conclusion may be that while nitroglycerin is undoubtably responsible for heparin resistance, propylene glycol could contribute to this effect.[68]

In a pilot study of 18 patients receiving intravenous nitroglycerin (in 30% propylene glycol), heparin resistance correlated with a nitroglycerin dose of greater than 350 mcg/min.[69] A study using lower doses of nitroglycerin (67 to 333 mcg/min) in a propylene glycol vehicle failed to demonstrate heparin resistance.[70] A randomized double-blind cross-over study of nitroglycerin 83 mcg/min in a nonpropylene glycol vehicle also did not show an interaction with heparin.[71] Similar studies using a propylene-glycol vehicle control may clarify the role of propylene glycol in this interaction.

VII. CLINICAL RELEVANCE

A. Neonates

Propylene glycol is clearly a cause of serum hyperosmolality in very low birth weight infants, with doses of 3 g/d producing an incidence of about 30% in infants weighing less than 1000 g. The administration of lower doses, of 300 mg/d or less, has also been associated with hyperosmolality in the low birth weight group, but appears less likely to result in serious consequences, such as seizures.

Currently, it is recommended that very low birth weight infants should not be given parenteral doses of medications containing propylene glycol if this can be avoided. If administration of a propylene glycol-containing medication is unavoidable, the total daily dose should be kept to 300 mg/d or less, and serum osmolality should be monitored.

B. Cardiac Toxicity

Phenytoin is often given intravenously in small volumes of diluent to patients with seizure disorders. Administrations of concentrations of less than 5 mg/ml are associated with minimal pain on injection and did not cause cardiotoxicity in a series of 200 patients. It is recommended that this concentration not be exceeded unless increased intracranial pressure necessitates fluid restriction. The drug should be given by piggyback infusion rather than direct manual push to avoid inadvertent administration of excessive concentrations and to allow rate adjustment or discontinuation while maintaining the intravenous line. The rate of administration should be 40 mg/min or less. Elderly patients and those with known cardiac disease should receive decreased concentrations and rates of infusion and continuous cardiac monitoring during the infusion.[17,21]

C. Toxicity From Topical Exposure

Hyperosmolality has been associated with topical application of a burn cream containing 7% propylene glycol in adult patients with second- and third-degree burns covering a mean total body surface area of 65%. The multifactoral nature of complications in this patient population has compounded the difficulty in attributing causation of any of these complications to propylene glycol. Hyperosmolality is a common finding in burn patients and is more likely to be related to free water loss, components of tissue breakdown, or other osmotically active compounds in patients with extensive burn surface areas than to propylene glycol absorption.[42,43]

Infants and children have a larger surface area for weight and may be at increased risk for absorption and impaired metabolism of propylene glycol. One case of cardiorespiratory arrest from topical application to an 8-month-old burn victim may have been related to propylene glycol.[41]

Populations that are at increased risk for topical absorption and/or impaired elimination of propylene glycol include infants and children, patients with large burns, patients with epidermolyis bullosa or toxic epidermal necrolysis, and patients with hepatic or renal impairment. Consideration should be given to monitor the osmolal gap in these patients.[41] Topical application to large surface areas in adult patients with other skin disorders, such as psoriasis, does not appear to present a risk of hyperosmolality.[44]

D. Contact Dermatitis

Propylene glycol may be responsible for approximately 0.2 to 4.5% of cases of cosmetic or topical pharmaceutical-related contact dermatitis. Positive reactions are difficult to document, due to the inherent irritant effects of propylene glycol and the possibility of enhancement of other primary sensitizers. Positive reactions require careful confirmation with repeat patch tests, oral challenge tests, and usage tests.[50] Glycerin is the most acceptable alternative humectant in sensitive individuals.[51]

E. Lactic Acidosis

When given as a vehicle in pharmacologic doses, propylene glycol uncommonly causes unexplainable lactic acidosis. This is generally of minor clinical significance.[35] Patients with renal insufficiency may accumulate propylene glycol and are at increased risk for lactic acidosis and hyperosmolality.

REFERENCES

1. **Fisher, A. A.,** The management of propylene glycol-sensitive patients, *Cutis,* 25, 24, 1980.
2. **Cantanzaro, J. M. and Smith, J. G.,** Propylene glycol dermatitis, *J. Am. Acad. Dermatol.,* 24, 90, 1991.
3. **Nikitakis, J. M.,** *CTFA Cosmetic Ingredient Handbook,* 1st ed., The Cosmetic, Toiletry and Fragrance Association, Washington, D.C., 1988.
4. **Louis, S. and Kutt, H.,** The cardiocirculatory changes caused by intravenous Dilantin and its solvent, *Am. Heart J.,* 74, 523, 1967.
5. **Sharer, L. and Kutt, H.,** Intravenous administration of diazepam: effects on penicillin-induced focal seizures in the cat, *Arch. Neurol.,* 24, 169, 1971.
6. **Gross, D. R., Kitzman, J. V., and Adams, H. R.,** Cardiovascular effects of intravenous administration of propylene glycol and of oxytetracycline in propylene glycol in calves, *Am. J. Vet. Res.,* 40, 783, 1979.
7. **Gross, D. R., Dodd, K. T., Williams, J. D., and Adams, H. R.,** Adverse cardiovascular effects of oxytetracycline preparations and vehicles in intact awake calves, *Am. J. Vet. Res.,* 42, 1371, 1981.
8. **Pearl, D. S., Quest, J. A., and Gillis, R. A.,** Effect of diazepam on digitalis-induced ventricular arrhythmias in the cat, *Toxicol. Appl. Pharmacol.,* 44, 643, 1978.
9. **Al-Khudhairi, D., Whitwam, J. G., and Askitopoulou, H.,** Acute central respiratory effects of diazepam and its solvent and propylene glycol, *Br. J. Anaesth.,* 54, 959, 1982.
10. **Denning, D. W. and Webster, A. D. B.,** Detrimental effect of propylene glycol on natureal killer cell and neutrophil function, *J. Pharm. Pharmacol.,* 39, 236, 1987.
11. **Gellerman, G. L. and Martinez, C.,** Fatal ventricular fibrillation following intravenous sodium diphenylhydantoin therapy, *JAMA,* 200, 161, 1967.

12. **Goldschlager, A. W. and Karliner, J. S.,** Ventricular standstill after intravenous diphenylhydantoin, *Am. Heart J.,* 74, 410, 1967.

13. **Russell, M. A. and Bousvaros, G.,** Fatal results from diphenylhydantoin administered intravenously, *JAMA,* 206, 2118, 1968.

14. **Unger, A. H. and Sklaroff, H. J.,** Fatalities following intravenous use of sodium diphenylhydantoin for cardiac arrhythmias, *JAMA,* 200, 159, 1967.

15. **Voight, G. C.,** Death following intravenous sodium diphenylhydantoin, *John Hopkins Med. J.,* 123, 153, 1968.

16. **Zoneraich, S., Zoneraich, O., and Siegel, J.,** Sudden death following intravenous sodium diphenylhydantoin, *Am. Heart J.,* 91, 375, 1976.

17. **York, R. C. and Coleridge, S. T.,** Cardiopulmonary arrest following intravenous phenytoin loading, *Am. J. Emerg. Med.,* 6, 255, 1988.

18. **Karliner, J. S.,** Intravenous diphenylhydantoin sodium (Dilantin) in cardiac arrhythmias, *Dis. Chest,* 51, 256, 1967.

19. **Cranford, R. E., Leppik, I. E., Patrick, B., Anderson, C. B., and Kostick, B.,** Intravenous phenytoin: clinical and pharmacokinetic aspects, *Neurology,* 28, 874, 1978.

20. **Barron, S. A.,** Cardiac arrhythmias after small IV dose of phenytoin, *N. Engl. J. Med.,* 295, 678, 1976.

21. **Earnest, M. P., Marx, J. A., and Drury, L. R.,** Complications of intravenous phenytoin for acute treatment of seizures, *JAMA,* 249, 762, 1983.

22. **Feldman, R. L., Marx, J. D., Pepine, C. J., and Conti, C. R.,** Analysis of coronary responses to various doses of intracoronary nitroglycerin, *Circulation,* 66, 321, 1982.

23. **Wallis, W., Kutt, H., and McDowell, F.,** Intravenous diphenylhydantoin in treatment of acute repetitive seizures, *Neurology,* 18, 513, 1968.

24. **Glasgow, A. M., Boeckx, R. L., Miller, M. K., MacDonald, M. G., and August, G. P.,** Hyperosmolality in small infants due to propylene glycol, *Pediatrics,* 72, 353, 1983.

25. **MacDonald, M. G., Fletcher, A. B., Johnson, E. L., Boeckx, R. L., Getson, P. R., and Miller, M. K.,** The potential toxicity to neonates of multivitamin preparations used in parenteral nutrition, *J. Paren. Enteral. Nutr.,* 11, 169, 1987.

26. **MacDonald, M. G., Getson, P. R., Glasgow, A. M., Miller, M. K., Boeckx, R. L., and Johnson, E. L.,** Propylene glycol: increased incidence of seizures in low birth weight infants, *Pediatrics,* 79, 622, 1987a.

27. **Huggon, I., James, I., and Macrae, D.,** Hyperosmolality related to propylene glycol in an infant treated with enoximone infusion, *Br. Med. J.,* 301, 19, 1990.

28. **Zaroslinski, J. F., Browne, R. K., and Possley, L. H.,** Propylene glycol as a drug solvent in pharmacologic studies, *Toxicol. Appl. Pharmacol.,* 19, 573, 1971.

29. **Sawchuk, R. J., Pepin, S. M., Leppik, I. E., and Gumnit, R. J.,** Rapid and slow release phenytoin in epileptic patients at steady state:Comparative plasma levels and toxicity, *J. Pharmacokinet. Biopharm.,* 10, 365, 1982.

30. **Arulanantham, K. and Genel, M.,** Central nervous system toxicity associated with ingestion of propylene glycol, *J. Pediatr.,* 93, 515, 1978.

31. **Martin, G. and Finberg, L.,** Propylene glycol: a potentially toxic vehicle in liquid dosage form, *J. Pediatr.,* 77, 877, 1979.

32. **Miller, O. N. and Bazzano, G.,** Propanediol metabolism and its relation to lactic acid metabolism, *Ann. N.Y. Acad. Sci.,* 119, 957, 1965.

33. **Ruddick, J. A.,** Toxicology, metabolism, and biochemistry of 1, 2-propanediol, *Toxicol. Appl. Pharmacol.,* 21, 102, 1972.

34. **Lolin, Y., Francis, D. A., Flanagan, R. J., Little, P., and Lascelles, P. T.,** Cerebral depression due to propylene glycol in a patient with chronic epilepsy-the value of the plasma osmolal gap in diagnosis, *Postgrad. Med. J.,* 64, 610, 1988.

35. **Demey, H. E., Daelemans, R. A., Verpooten, G. A., De Broe, M. E., Van Campenhout, Ch. M., Lakiere, F. V., Schepens, P. J., and Bossaert, L. L.,** Propylene glycol-induced side effects during intravenous nitroglycerin therapy, *Intens. Care Med.,* 14, 221, 1988.

36. **Demey, H., Daelemans, R., De Broe, M. E., and Bossaert, L.,** Propyleneglycol intoxication due to intravenous nitroglycerin, *Lancet,* 1, 1360, 1984.

37. **Cate, J. C. and Hedrick, R.,** Propylene glycol intoxication and lactic acidosis, *N. Engl. J. Med.,* 303, 1237, 1980.

38. **Kelner, M. J. and Bailey, D. N.,** Propylene glycol as a cause of lactic acidosis, *J. Anal. Toxicol.,* 9, 40, 1985.

39. **Brittain, R. T. and D'Arcy, P. F.,** Hematologic effects following the intravenous injection of propylene glycol in the rabbit, *Toxicol. Appl. Pharmacol.,* 4, 738, 1962.

40. **Bekeris, L., Baker, C., Fenton, J., Kimball, D., and Bermes, E.,** Propylene glycol as a cause of an elevated serum osmolality, *Am. J. Clin. Pathol.,* 72, 633, 1979.

41. **Fligner, C. L., Jack, R., Twiggs, G. A., and Raisys, V. A.,** Hyperosmolality induced by propylene glycol: a complication of silver sulfadiazine therapy, *JAMA,* 253, 1606, 1985.

42. **Kulick, M. I., Lewis, N. S., Bansal, V., and Warpeha, R.,** Hyperosmolality in the burn patient: analysis of an osmolal discrepancy, *J. Trauma,* 20, 223, 1980.

43. **Kulick, M. I., Wong, R., Okarma, T. B., Falces, E., and Berkowitz, R. L.,** Prospective study of side effects associated with the use of silver sulfadiazine in severely burned patients, *Ann. Plastic Surg.,* 14, 407, 1985.

44. **Commens, C. A.,** Topical propylene glycol and hyperosmolality, *Br. J. Dermatol.,* 122, 77, 1990.

45. **Wahlberg, J. E. and Nilsson, G.,** Skin irritancy from propylene glycol, *Acta Derm. Venereol.,* 64, 286, 1984.

46. **Meneghini, C. L., Rantuccio, F., and Lomoto, M.,** Additives, vehicles and active drugs of topical medicaments as causes of delayed-type allergic dermatitis, *Dermatologia,* 143, 137, 1971.

47. **Agren-Jonsson, S. and Magnusson, B.,** Sensitization to propantheline bromide, trichlorocarbanilide and propylene glycol in an antiperspirant, *Contact Dermatitis,* 2, 79, 1976.

48. **Oleffe, J. A., Blondeel, A., and de Coninck, A.,** Allergy to chlorocresol and propylene glycol in a steroid cream, *Contact Dermatitis,* 5, 1979.

49. **Fisher, A. A.,** *Contact Dermatitis,* 3rd ed., Lea & Febiger, Philadelphia, 1986.

50. **Hannuksela, M. and Forstrom, L.,** Reactions to peroral propylene glycol, *Contact Dermatitis,* 4, 41, 1978.

51. **Fisher, A. A.,** Reactions to popular cosmetic humectants. Part III. Glycerin, propylene glycol, and butylene glycol, *Cutis,* 26, 243, 1980.

52. **Fisher, A. A. and Brancaccio, R. R.,** Allergic contact sensitivity to propylene glycol in a lubricant jelly, *Arch. Dermatol.,* 115, 1451, 1979.

53. **Fisher, A. A.,** Contact dermatitis from topical medicaments, *Semin. Dermatol.,* 1, 49, 1982.

54. **Hee, Ch. E. and Yoo, Ch. K.,** Propylene glycol allergy from ketoconazole cream, *Contact Dermatitis,* 21, 274, 1989.

55. **Cochran, R. J. and Rosen, T.,** Contact dermatitis caused by ECG electrode paste, *South. Med. J.,* 73, 1667, 1980.

56. **Degreef, H. and Dooms-Goossens, A.,** Patch testing with silver sulfadiazine cream, *Contact Dermatitis,* 12, 33, 1985.

57. **Fisher, A. A.,** Use of glycerin in topical minoxidil solutions for patients allergic to propylene glycol, *Cutis,* 45, 81, 1990.

58. **Hannuksela, M., Kousa, V., and Pirila, V.,** Allergy to ingredients of vehicles, *Contact Dermatitis,* 2, 105, 1976.

59. **Angelini, G. and Meneghini, C. L.,** Contact allergy from propylene glycol, *Contact Dermatitis,* 7, 197, 1981.

60. **Angelini, G., Vena, G. A., and Meneghini, C. L.,** Allergic contact dermatitis to some medicaments, *Contact Dermatitis,* 12, 263, 1985.

61. **North American Contact Dermatitis Group,** Prospective study of cosmetic reactions 1977–1980, *J. Am. Acad. Dermatol.,* 6, 909, 1982.

62. **Greenblatt, D. J., Shader, R. I., and Koch-Weser, J.,** Serum creatine phosphokinase concentrations after intramuscular chlordiazepoxide and its solvent, *J. Clin. Pharmacol.,* 16, 118, 1976.

63. **Isaac, M. and Dundee, J. W.,** Clinical studies of induction agents XXX: venous sequelae following ethanol anaesthesis, *Br. J. Anaesth.,* 41, 1070, 1969.

64. **Mattila, M. A. K., Ruoppi, M., Korhonen, M., Larni, H. M., Valtonen, L, and Heikkinen, H.,** Prevention of diazepam-induced thrombophlebitis with cremophor as a solvent, *Br. J. Anaesth.,* 51, 891, 1979.

65. **Schou-Olesen, A. and Huttel, M. S.,** Local reactions to IV diazepam in three different formulations, *Br. J. Anaesth.,* 52, 609, 1980.

66. **Zacharias, M., Clarke, R. S. J., Dundee, J. W., and Johnston, S. B.,** Venous sequelae following etomidate, *Br. J. Anaesth.,* 51, 779, 1979.

67. **Col, J., Col-Debeys, C., Lavenne-Pardonge, E. L., Meert, P., Hericks, L., Broze, M. C., and Moriau, M.,** Propylene glycol-induced heparin resistance during nitroglycerin infusion, *Am. Heart J.,* 110, 171, 1985.

68. **Habbab, M. A. and Haft, J. I.,** Heparin resistance induced by intravenous nitroglycerin: a word of caution when both drugs are used concomitantly, *Arch. Intern. Med.,* 147, 857, 1987.

69. **Becker, R. C., Corrao, J. M., Bovill, E. G., Gore, J. M., Baker, S. P., Miller, M. L., Lucas, F. V., and Alpert, J. A.,** Intravenous nitroglycerin-induced heparin resistance: a qualitative antithrombin III abnormality, *Am. Heart J.,* 119, 1254, 1990.

70. **Lepor, N. E., Amin, D. K., Berberian, L., and Shah, P. K.,** Does nitroglycerin induce heparin resistance?, *Clin. Cardiol.,* 12, 432, 1989.

71. **Bode, V., Welzel, D., Franz, G., and Polensky, U.,** Absence of drug interaction between heparin and nitroglycerin, *Arch. Intern. Med.,* 150, 2117, 1990.

PROPYL GALLATE[1]

I. REGULATORY CLASSIFICATION

Propyl gallate is classified as an antioxidant.

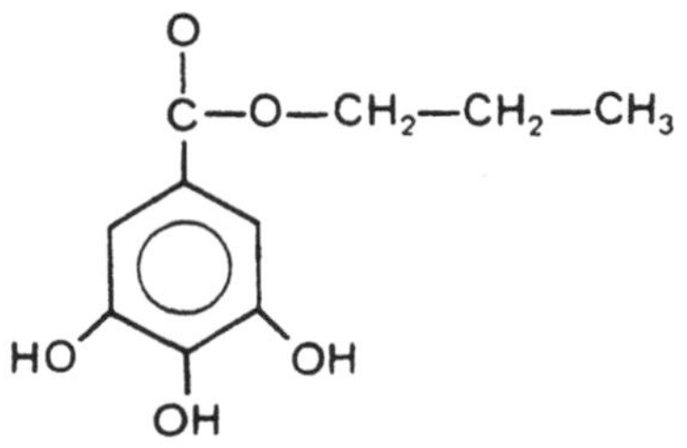

$$\underset{\text{HO}}{\bigcirc}\overset{\overset{\displaystyle O}{\parallel}}{C}-O-CH_2-CH_2-CH_3$$

FIGURE 55. Propyl gallate.

II. SYNONYMS

E310
Propyl 3,4,5-trihydroxybenzoate

III. AVAILABLE FORMULATIONS

A. Foods
Propyl gallate has been used as an antioxidant in foods since 1948 and has been on the GRAS list since 1973. In concentrations of up to 0.1% propyl gallate prevents oxidative rancidity of unsaturated oils. Types of food products include fats, oils, margarine, butter, meat products, snack foods, baked goods, nuts, grain products, frostings, chewing gum, soft candy, frozen dairy products, gelatins, puddings, alcoholic beverages, and nonalcoholic beverages.[1]

B. Drugs
Propyl gallate is used in topical and parenteral pharmaceuticals with an oily vehicle, often in combination with butylated hydroxytoluene. Rancidity is prevented with concentrations of 0.1%.

C. Cosmetics

Propyl gallate is commonly used in lipsticks, lip glosses, lip balms, cream, lotions, blushers, suntan oils and lotions, and bath oils.[2]

IV. TABLE OF COMMON PRODUCTS

A. Oral Drug Products

Trade name	Manufacturer
Bendadryl Plus Nightime liquid	Parke-Davis
Scott's Emulsion	Beecham
Serutan toasted granules	Beecham
Orajel mouth-aid	Commerce

B. Topical Drug Products

Trade name	Manufacturer
Aristocort cream	Fujisawa
Aristocort A ointment	Fujisawa
Bactine hydrocortisone skin care cream	·Miles
Cyclocort ointment	Lederle
Lidex gel	Syntex
Meclan cream	Ortho
Nutracort cream	Owen/Galderma

C. Parenteral Drug Products

Trade name	Manufacturer
Navane intramuscular injection	Roerig
Terramycin injection multidose vial	Roerig

V. ANIMAL TOXICITY DATA

Propyl gallate has demonstrated local anesthetic properties comparable to procaine hydrochloride in studies in frogs, rabbits, and guinea pigs. In extremely high doses, 100 to 200 mg/kg, CNS depression was observed in rats and mice.[3]

Guinea pigs fed 1.5 ml of a 10% propyl gallate solution daily for 7 d could not be sensitized to intradermal injections of propyl gallate. Animals that were not pretreated with the oral solution were readily sensitized, with reactivity that persisted for at least 3 months.[4]

VI. HUMAN TOXICITY DATA

A. Neonates

An outbreak of acquired methemoglobinemia in nine hospitalized infants aged 6 to 15 weeks was attributed to a change in a soybean-based infant formula to add an antioxidant mixture containing propyl gallate, BHA, and BHT. Methemoglobin levels of 7.7 to 35% were observed, which decreased with a change to cow's milk or an alternate soybean formula and increased when rechallenged with the original formula. Older infants, aged 6 months or greater, receiving the same formula had slightly elevated methemoglobin levels (3.8 to 5%), but did not develop clinical signs or symptoms. It was not determined which of the three antioxidants was responsible, but the structural similarity of propyl gallate to a known methemoglobin inducer, pyrogallol, implicated the gallate compound.[5]

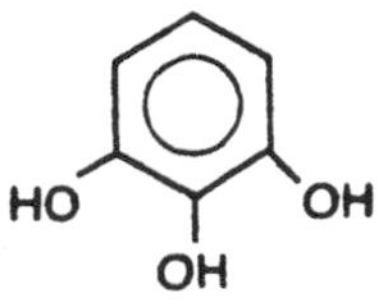

FIGURE 56. Pyrogallol.

B. Contact Dermatitis

The wide usage of propyl gallate in foods has been suggested to induce immunologic tolerance; this may explain the relatively few case reports of contact dermatitis and the high sensitization potential demonstrated in guinea pigs.[4] Prior ingestion of gallates in foods has also been suggested to sensitize individuals to topical products.[6]

Of 15 case reports of propyl gallate-induced allergic contact dermatitis published between 1975 and 1990, the source of exposure included deodorants,[7] lipsticks,[8,9] cosmetic creams,[10] antibiotic ointments,[11,12] body lotions,[13] moisturizing creams,[14] and Alphosyl® cream.[15]

Cross-reactivity or concomitant reactivity to octyl gallate and dodecyl gallate has been reported.[14] There is no cross-reactivity to the structurally related compound, pyrogallol, in guinea pigs.[4]

Studies of attempts to induce sensitization in humans have shown a limited ability with concentrations of greater than 10%; however an epidemic of contact dermatitis was observed in Italy following popularity of a topical antibiotic containing 8% propyl gallate.[11,12] Two of ten adult volunteers tested with 20% propyl gallate in 70% ethanol were sensitized after 14 d of repeat insult patch testing.[4] Similar testing with cosmetic products containing concentrations of less than 1% in 850 individuals failed to produce contact dermatitis.[16]

The threshold of sensitivity to propyl gallate is lower when applied under occlusion.[7] Of 15 case reports reviewed, one third have been reported following application to damaged skin, such as herpes zoster,[12] chapped lips,[9] stasis leg ulcers,[11] arm ulcers,[17] and psoriasis.[15]

VII. CLINICAL RELEVANCE

Propyl gallate is a strong contact sensitizer when applied in concentrations of greater than 1%. The fact that most cosmetics and topical pharmaceuticals contain 1% or less and the possibility that oral ingestion in foods induces tolerance may explain the relative scarcity of case reports of hypersensitivity. Patients with damaged skin or those occupationally exposed frequently to propyl gallate, such as bakers, may be at increased risk.

REFERENCES

1. **Lecos, C.,** Food preservatives: a fresh report, *FDA Consumer,* April 1984.
2. **Nikitakis, J. M.,** *CTFA Cosmetic Ingredient Handbook,* 1st ed., The Cosmetic, Toiletry and Fragrance Association, Washington, D.C., 1988.

3. **Modak, A. T. and Rao, M. R. R.,** Propyl gallate as a local anesthetic agent, *Ind. J. Med. Res.,* 59, 795, 1971.

4. **Kahn, G., Phanuphak, P., and Claman, H. N.,** Propyl gallate-contact sensitization and orally-induced tolerance, *Arch. Dermatol.,* 109, 506, 1974.

5. **Nitzan, M., Volovitz, B., and Topper, E.,** Infantile methemoglobinemia caused by food additives, *Clin. Toxicol.,* 15, 273, 1979.

6. **Postendorf, J.,** Oxidative deterioration in cosmetics and pharmaceuticals, *J. Soc. Cosmet. Chem.,* 16, 203, 1965.

7. **Kraus, A. L., Stotts, J., Altringer, L. A., and Allgood, G. S.,** Allergic contact dermatitis from propyl gallate: dose response comparison using various application methods, *Contact Dermatititis,* 22, 132, 1990.

8. **Cronin, E.,** Lipstick dermatitis due to propyl gallate, *Contact Dermatitis,* 1, 257, 1975.

9. **Wilson, A. G., White, I. R., and Kirby, J. D. T.,** Allergic contact dermatitis from propyl gallate in a lip balm, *Contact Dermatitis,* 20, 145, 1989.

10. **Bardazzi, F., Misciali, C., Borrello, P., and Capobianco, C.,** Contact dermatitis due to antioxidants, *Contact Dermatitis,* 19, 385, 1988.

11. **Cusano, F., Capozzi, M., and Errico, G.,** Safety of propyl gallate in topical products, *J. Am. Acad. Dermatol.,* 17, 308, 1987.

12. **Valsecchi, R. and Cainelli, T.,** Contact allergy to propyl gallate, *Contact Dermatitis,* 19, 380, 1988.

13. **Heine, A.,** Contact dermatitis from propyl gallate, *Contact Dermatitis,* 18, 313, 1988.

14. **Bojs, G., Nicklasson, B., and Svensson, A.,** Allergic contact dermatitis to propyl gallate, *Contact Dermatitis,* 17, 294, 1987.

15. **Liden, S.,** Alphosyl sensitivity and propyl gallate, *Contact Dermatitis,* 1, 257, 1975.

16. **Anon.,** Final report on the safety assessment of propyl gallate, *J. Am. Coll. Toxicol.,* 4, 23, 1985.

17. **Pigatto, P. D., Boneschi, V., Riva, F., and Altomare, G. F.,** Allergy to propylgallate, with unusual clinical and histological features, *Contact Dermatitis,* 11, 43, 1984.

QUATERNIUM 15

I. REGULATORY CLASSIFICATION

Quaternium 15 is a water-soluble antimicrobial preservative with good activity against bacteria and lesser activity against yeasts and molds.

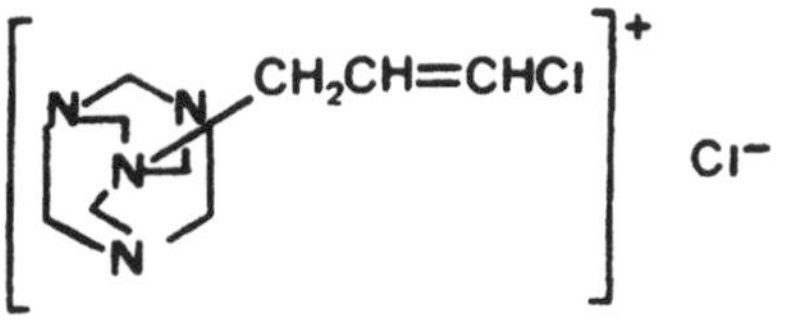

FIGURE 57. Quaternium-15.

II. SYNONYMS

Chloroallyl hexaminium chloride
Chloroallyl methenamine chloride
Dowicil 75®
Dowicil 100®
Dowicil 200®
N-(3-Chloroallyl)-hexaminium chloride
1-(3-Chloroallyl)-3,5,7-triaza-1-azonioadamatane chloride

III. AVAILABLE FORMULATIONS

A. Drugs
Quaternium 15 is infrequently used in topical pharmaceutical products. It was listed as an ingredient in two topical creams and one lotion in 1990 by the FDA. The concentration listed in these products is 0.02%.[1]

B. Cosmetics
Quaternium 15 was present in 1079 (5.3%) of 20,183 cosmetic formulations submitted to the FDA in 1982.[2] Commercial cosmetic products generally contain 0.1 to 0.2%

329

concentrations.[3] It is a highly water-soluble preservative, present in water-based products such as eye makeup, makeup foundations, hand and body lotions and creams, shampoos, conditioners, cleansing products, shaving products, and body powders.[4,5]

IV. TABLE OF COMMON PRODUCTS

A. Topical Drug Products

Trade name	Manufacturer
Atrac-tain lotion	Sween
Ionil Plus shampoo	Owen/Galderma
Ionil T Plus shampoo	Owen/Galderma
Keri lotion	Bristol Myers
Lac-Hydrin 12% lotion	Westwood-Squibb
Lubriderm cream	Warner-Lambert
Moisturel lotion	Westwood-Squibb
Solaquin forte cream 4%	ICN
Sween cream	Sween
Trans-Plantar dermal patch	Tsumura
Trans-Ver-Sal dermal patch	Tsumura

V. HUMAN TOXICITY DATA

A. Cosmetic Contact Dermatitis

Quaternium 15 is a formaldehyde releasing preservative. Concentrations used in cosmetics, 0.1 to 0.2%, contain about 100 to 200 ppm of free formaldehyde.[6] Patients with formaldehyde sensitivity will often react with a positive patch test to the commonly used 1 to 2% Quaternium 15 concentrations.[3] The incidence of positive patch tests to formaldehyde in patients sensitized to Quaternium 15 was 60% in one study.[7]

Sensitivity to Quaternium 15 is more frequent in countries with widespread use of this preservative. In a study of 713 North American patients with suspected cosmetic allergy, Quaternium 15 was the most commonly implicated preservative with positive patch tests in 65 (9.1%) patients.[8] It was also the most common allergen implicated in Great Britain with a prevalence of 3 to 4 % in women and 1% in men attending a dermatology clinic.[7]

In the Netherlands, where Quaternium is infrequently used, positive patch tests to 2% Quaternium 15 were found in 2.8% of 179 patients with suspected cosmetic allergy.[9] An unselected population of 501 patients with suspected contact dermatitis did not demonstrate any occurrences of sensitivity to Quaternium 15.[10]

Similarly, the incidence of allergic reaction in Italy was 0.22% in 4470 patients tested. Quaternium 15 is largely confined to rinse-off products in that country. Occupational contact dermatitis in hairdressers using rinse-off products containing Quaternium 15 accounted for two of these cases.[4]

B. Drugs

Quaternium 15 is infrequently used in drugs, thus there are few reports of sensitivity from this source. A case report of a 37-year-old man who developed severe, pruritic contact dermatitis of the face and eyelids 3 weeks after switching from a brand-name hydrocortisone product preserved with sorbic acid to a generic product preserved with Quaternium 15 indicates that reactions from pharmaceuticals may be more prevalent in the future.[11]

VI. CLINICAL RELEVANCE

The widespread use of Quaternium 15 in cosmetics sold in the U.S. has resulted in concomitant widespread sensitization to this preservative. Although dermatitis to topical pharmaceutical products has been minimal, due to the few number of products containing Quaternium 15, the increasing use in generic topical corticosteroid products is likely to provoke an increase in sensitization.[11]

Patients experiencing contact dermatitis from Quaternium 15 require careful evaluation. Primary sensitivity to formaldehyde or to Quaternium 15 may be present. Currently used patch test concentrations may release larger amounts of formaldehyde than those present in commercial formulations, leading to false positive reactions in the formaldehyde-sensitive individual. Usage tests should be used to confirm questionable reactions.

REFERENCES

1. **Food and Drug Administration,** *Inactive Ingredients Guide.* FDA, Washington, D.C., March 1990.
2. **Decker, R. L. and Wenninger, J. A.,** Frequency of preservative use in cosmetic formulas as disclosed to FDA-1982 update, *Cosmet. Toilet.,* 97, 57, 1982.
3. **Fisher, A. A.,** Cosmetic dermatitis. Part II: Reactions to some commonly used preservatives, *Cutis.* 26, 136, 1980.
4. **Tosti, A., Piraccini, B. M., and Bardazzi, F.,** Occupational contact dermatitis due to quaternium 15, *Contact Dermatitis,* 23, 41, 1990.
5. **Nikitakis, J. M.,** *CTFA Cosmetic Ingredient Handbook.* 1st ed., The Cosmetic, Toiletry and Fragrance Association, Washington, D.C., 1988.
6. **Jordan, W. P., Sherman, W. T., and King, S. E.,** Threshold responses in formaldehyde-sensitive subjects, *J. Am. Acad. Dermatol.,* 1, 44, 1979.
7. **Cronin, E.,** Allergy to cosmetics, *Acta Derm. Venereol. Stock.* (Suppl.) 134, 77, 1987.
8. **Adams, R. M. and Maibach, H. I.,** A five-year-study of cosmetic reactions, *J. Am. Acad. Dermatol.,* 13, 1062, 1985.
9. **De Groot, A. C., Liem, D. H., Nater, J. P., and van Ketel, W. G.,** Patch tests with fragrance materials and preservatives, *Contact Dermatitis.* 12, 87, 1985.
10. **De Groot, A. C., Bos, J. D., Jagtman, B. A., Bruynzel, D. P., Van Joost, T., and Weyland, J. W.,** Contact allergy to preservatives-II, *Contact Dermatitis,* 15, 218, 1986.
11. **Fisher, A. A.,** Preservative (Quaternium 15) dermatitis from the "generic equivalent" of a "brand name" hydrocortisone cream (Hytone® cream, Dermik), *Cutis.* 41. 153. 1988.

ROSIN

I. REGULATORY CLASSIFICATION

Rosin is classified as a stiffening agent in pharmaceuticals.

II. SYNONYMS

Colophane
Colophonium
Colophony BP
Pine resin
Resin
Resina pini
Resina terebinthinae

III. AVAILABLE FORMULATIONS

A. Constituents
There are three major types of unmodified rosin preparations. Gum rosin is tapped from living pine trees and is the product used for routine commercially available patch testing. It is a complex mixture of 90% resin acids, such as abietic acid. The remaining portion is a neutral fraction consisting of aldehydes and alcohols with the same basic structure as the resin acids. Wood rosin is derived from pine stumps or pine wood. Tall oil rosin is a by-product of paper pulp production.[1]

Modified rosins include maleic modified, pentaerythritol ester, polyterpene rosin, Abitol®, fumaric modified, glycerol ester, polymerized, terpene phenol, calcium resinate, zinc-calcium rosin, and alkylphenol rosin.[2]

B. Foods
Rosin is a common component of chewing gums[3] and is used as a polishing agent for roasted coffee beans.

C. Drugs

Rosin is the residue left after distillation of the volatile oil obtained from the *Pinus* species. It is a constituent of Flexible Collodion BP. Flexible Collodion USP does not contain rosin and is prepared with pyroxylin, a nitrated cellulose derived from wood pulp. Rosin is present in a few coated or sustained action tablets.[4]

D. Cosmetics

Rosin is used as a binder, film former, and plasticizer in products such as depilatories, hair tonics and dressings, makeup products, and mascara.[5]

E. Other Products

Rosin is present in the air sampled in pine forests. It may be a constituent of hot-melt glues, soft soldering fluxes, cutting oils, paints, furniture polish, soaps, and paper and is a permitted additive in cigarettes.[6] Rosin is also used in rubber manufacturing, protective coatings, defoaming agents in the manufacture of paper, closures in sealing gaskets for food containers, animal glues, wood preservatives, and reinforced waxes.[7]

IV. TABLE OF COMMON PRODUCTS

Trade name	Manufacturer
Boil-Ease	Commerce
Prid Salve	Walker Pharmacal

V. ANIMAL TOXICITY DATA

Experimental sensitization studies in guinea pigs demonstrated the resin acids, abietic acid, podocarpic acid, laevopimaric acid, and tetrahydroabietic acid to be weak sensitizers. The neutral fraction was comparable in strength with the two major resin acids.[8] When different commercial rosin preparations were compared in this model, the tall oil rosin was the most allergenic.[7]

VI. HUMAN TOXICITY DATA

A. Contact Dermatitis

Rosin is an infrequent cause of cosmetic dermatitis. It was implicated in 1 patient in a series of 119 Dutch patients with cosmetic dermatitis.[9] Rosin was not implicated in any of the 713 North American patients tested over a 5-year period.[10] Bilateral symmetrical eyelid dermatitis has been attributed to rosin in seven patients secondary to eyeshadows or mascaras.[11-13] Cross-reactions to dihydroabietyl alcohol, abietic acid, and abietate have been reported.[12]

Rosin may be a component of adhesives and is present in some salicylic acid plasters and gels. Three patients with dermatitis secondary to salicylic acid topical products have been attributed to rosin.[13-15]

In a series of 877 Polish patients patch tested to 20% rosin, positive reactions were demonstrated in 3.3%.[16] The incidence of positive reactions in a series of 4000 consecutive patients seen in five European clinics was 2.9% in males and 3.6% in females.[17] A later series of 5521 German patients demonstrated an incidence of 2.4% positive patch tests.[2]

While the resin acid components, especially abeitic acid, have been considered the primary allergens in rosin, some patients have reacted only to the neutral fraction.[18] The commonly used patch test concentration of 20% gum rosin in petrolatum was insufficient to detect patients allergic to the neutral fraction; a concentration of 60% in petrolatum was advised.[19]

A comparison of these two patch test concentrations done simultaneously in a series of patients concluded that there was no significant difference between the patch test results; four irritant reactions to the 60% preparation were observed. Serial dilution showed equivalent sensitivity with a 10% concentration.[1]

Modified rosins have been gradually dominating the world market. Some of these products are more allergenic than the unmodified rosin used in commercial patch test materials. Hydroabeitic alcohol (Abitol®) was the most frequently found sensitizer in a series of 75 patients tested with different modified rosin products. Other high sensitizers were maleic modified rosin, terpene phenol rosin, and polymerized rosin. The least sensitizing product was the polyterpene rosin.[2]

Occupational dermatitis due to the presence of rosin in cutting oil used by a file manufacturing company worker has been described.[20]

Hypersensitivity to rosin was demonstrated in an 8-year-old child who developed perioral dermatitis after frequent use of a rosin-containing chewing gum. Discontinuing the gum resulted in marked improvement, but not resolution, of the dermatitis. Other factors, such as concomitant allergy to perfumes, were implicated.[3]

B. Occupational Asthma

Asthma or peripheral airway reactions were described in 21 patients exposed to solder flux fumes containing rosin in the electronics industry; 5 had preexisting asthma. In most patients, a long period of exposure was experienced before recognition of the symptoms, with a median time of 6 years. The onset of reaction after arriving at the workplace was immediate in three patients, after 60 to 90 min in two patients, between 2 to 4 h in ten patients, and delayed for 5 to 8 h in the remaining six patients.[21]

Asthma was also described in a worker exposed to a hot-melt glue containing rosin, which was used as an adhesive for can labels.[22]

VII. CLINICAL RELEVANCE

European patient series have demonstrated a prevalence of hypersensitivity to rosin in 2.4 to 3.6% of the population in those countries. Some of the newer modified rosins are potentially more allergenic and may increase the incidence of sensitization. Cross-reactions do not necessarily occur between the unmodified and modified rosin preparations. Patients with a history of both adhesive and salicylic acid plaster dermatitis should be tested with rosin. The most commonly used patch test concentration is 10% of the unmodified commercial product in petrolatum. As the newer products become more prevalent, patch testing with a mixture of rosin products will become necessary.

REFERENCES

1. **Karlberg, A. and Liden, C.,** Comparison of colophony patch test preparations, *Contact Dermatitis*, 18, 158, 1988.
2. **Hausen, B. M. and Mohnert, J.,** Contact allergy due to colophony. V. Patch test results with different types of colophony and modified-colophony products, *Contact Dermatitis*, 20, 295, 1989c.

3. **Satyawan, I., Oranje, A. P., and van Joost, T.,** Perioral dermatitis in a child due to rosin in chewing gum, *Contact Dermatitis,* 22, 182, 1990.

4. **Food and Drug Administration,** *Inactive Ingredients Guide,* FDA, Washington, D.C., March 1990.

5. **Nikitakis, J. M.,** *CTFA Cosmetic Ingredient Handbook,* 1st ed., The Cosmetic, Toiletry and Fragrance Association, Washington, D.C., 1988.

6. **Burge, P. S.,** Colophony and asthma, *Lancet,* 2, 591, 1979.

7. **Hausen, B. M., Jensen, S., and Mohnert, J.,** Contact allergy to colophony. IV. The sensitizing potency of commercial products: an investigation of French and American modified colophony derivatives, *Contact Dermatitis,* 20, 133, 1989a.

8. **Hausen, B. M., Krueger, A., Mohnert, J., Hahn, H., and Konig, W. A.,** Contact allergy due to colophony. III. Sensitizing potency of resin acids and some related produts, *Contact Dermatitis,* 20, 41, 1989.

9. **De Groot, A. C., Bruynzeel, D. P., and Bos, J. D.,** The allergens in cosmetics, *Arch. Dermatol.,* 124, 37, 1988.

10. **Adams, R. M. and Maibach, H. I.,** A five-year-study of cosmetic reactions, *J. Am. Acad. Dermatol.,* 13, 1062, 1985.

11. **Calnan, C. D.,** Colophony in eyeshadow, *Contact Dermatitis Newletter,* 10, 235, 1971.

12. **Dooms-Goossens, A., Degreef, H., and Luytens, E.,** Dihydroabietyl alcohol (Abitol®): a sensitizer in mascara, *Contact Dermatitis,* 5, 340, 1979.

13. **Fisher, A. A.,** Allergic contact dermatitis due to rosin (colophony) in eyeshadow and mascara, *Cutis,* 42, 507, 1988.

14. **Rasmussen, J. E. and Fisher, A. A.,** Allergic contact dermatitis to a salicylic acid plaster, *Contact Dermatitis,* 2, 237, 1976.

15. **Veraldi, S. and Schianchi-Veraldi, R.,** Allergic contact dermatitis from colophony in a wart gel, *Contact Dermatitis,* 22, 184, 1990.

16. **Rudzki, E. and Kleniewska, D.,** The epidemiology of contact dermatitis in Poland, *Br. J. Dermatol.,* 83, 543, 1970.

17. **Bandmann, H.-J., Calnan, C. D., Cronin, E., Fregert, S., Hjorth, N., Magnusson, B., Maibach, H., Malten, K. E., Meneghini, C. L., Pirila, V., and Wilkinson, D. S.,** Dermatitis from applied medicaments, *Arch. Dermatol.,* 106, 335, 1972.

18. **Fregert, S. and Gurvberger, B.,** Patch testing with colophony, *Contact Dermatitis,* 11, 141, 1984.

19. **Anon.,** European Standard Series (ICDRG 1984), *Contact Dermatitis,* 11, 63, 1984.

20. **Matos, J., Mariano, A., Goncalo, S., Freitas, J. D., and Oliveira, J.,** Occupational dermatitis from colophony, *Contact Dermatitis,* 18, 53, 1988.

21. **Burge, P. S., Harries, M. G., O'Brien, I. M., and Pepys, J.,** Respiratory diseases in workers exposed to solder flux fumes containing colophony (pine resin), *Clin. Allergy,* 8, 1, 1978.

22. **Fawcett, I. W., Newman-Taylor, A. J., and Pepys, J.,** Asthma due to inhaled chemical agents-fumes from "Multicore" soldering flux and colophony resin, *Clin. Allergy,* 6, 577, 1976.

SACCHARIN

I. REGULATORY CLASSIFICATION

Saccharin, saccharin calcium, and saccharin sodium are listed as sweetening agents with the FDA.

In 1977, the FDA proposed a ban on the use of saccharin. A moratorium on imposition of this ban was instituted in 1983, pending more conclusive demonstration of risks to humans. This moratorium was extended in 1987 by President Reagan and is now scheduled to expire on May 1, 1992. Since 1977, saccharin-containing foods have been required to carry a label stating that "Use of this product may be hazardous to your health. This product contains saccharin which has been determined to cause cancer in laboratory animals."[1,2]

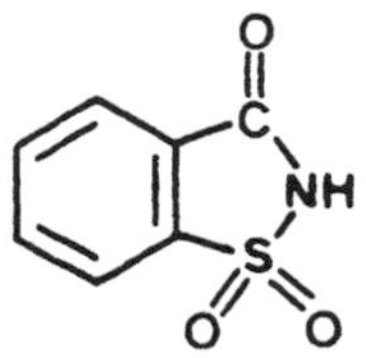

FIGURE 58. Saccharin.

II. SYNONYMS

Benzoic sulphimide
Saccharimide
Saccharinol
Saccharinose
Saccharol
Zaharina

III. AVAILABLE FORMULATIONS

A. Foods

Saccharin is widely used in foods and beverages, which account for approximately 70% of the total consumption of this non-nutritive sweetener. The estimated annual consumption in the U.S. was 3500 tons in 1976.[3] The use of saccharin has been steadily decreasing as a result of expanded use of other non-nutritive sweeteners, particularly aspartame.

Foods that may contain saccharin include powdered juices and drinks, other beverages, sauces and dressings, canned fruits, dessert toppings, cookies, gums, jams, candies, ice cream, and puddings.[3]

B. Drugs

Saccharin is used widely in prescription and over-the-counter pharmaceuticals, in such diverse products as intramuscular and intravenous injections, rectal tablets, sublingual tablets, chewable tablets, topical ointments, nasal solutions, oral solutions, oral suspensions, syrups, elixirs, and rectal solutions.[4]

C. Cosmetics

Cosmetics that may contain saccharin include lipsticks, dentifrices, aftershave lotions, skin cleansers, bubble baths, douches, mouthwashes, breath fresheners, moisturizing creams and lotions, hair tonics, colognes, and face powders.[3,5]

IV. TABLE OF COMMON PRODUCTS

A. Oral Prescription Drug Products

Trade name	Manufacturer
Alupent syrup	Boehringer Ingelheim
Amoxil chewable tablet 125, 250 mg	Beecham
Augmentin suspension 125, 250/5ml	Beecham
Bactrim suspension	Roche
Choledyl elixir	Parke-Davis
Kaochlor	Adria
Kaon elixir	Adria
Mysoline suspension	Wyeth-Ayerst
Septra oral suspension	Burroughs Wellcome

B. Oral Over-the-Counter Drug Products

Trade name	Manufacturer
Aspirin-free Excedrin extra strength caplet	Winthrop-Breon
Bonine tablets	Pfizer
Cerose DM syrup	Wyeth-Ayerst
Comtrex cough formula	Bristol-Myers
Congespirin chewable	Bristol-Myers
Creo-terpin cough suppressant	Medtech
Evac-Q-Mag	Adria
Liquid Hold syrup	Beecham
Riopan Plus suspension	Whitehall

Trade name (cont'd)	Manufacturer (cont'd)
Riopan Plus 2 chewable tablet	Whitehall
Riopan Plus 2 suspension	Whitehall
Serutan powder, granules	Beecham
Sominex liquid	Beecham
Sucrets Pain Relief cough formulas	Beecham
Sudafed Cough syrup	Burroughs Wellcome

C. Inhalation Drug Products

Trade name	Manufacturer
Bronkometer	Winthrop-Breon
Bronkometer 2	Winthrop-Breon
Isuprel inhalation solution	Winthrop-Breon

V. ANIMAL TOXICITY DATA

A. Carcinogenicity

Much of the controversy over the use of saccharin centers around concern over carcinogenicity in response to a series of studies done in rats exposed to diets containing 5 to 7.5% saccharin. These studies described a male-dominated increased incidence of bladder cancer.[6,7] Cancer promoting activity was also demonstrated in rats treated with most known bladder carcinogens and subsequently exposed to saccharin.[8-10]

A review of these issues done by the American Medical Association in 1985 concluded that saccharin induces species-specific bladder tissue changes with administration of large doses to rats; these effects are confined to the second generation of male rats. Changes included increased urine volume, increased bladder wall weight, increased accumulation of divalent cations in bladder tissue, and increased daily urinary excretion of indican, the main metabolite of indole. The no-effect level was 1% saccharin in the daily diet.[11] This amount is equivalent to 500 mg/kg/d.[3]

Several studies have been done since these reviews to determine the mechanism and further define the risk of exposure to saccharin. One theory suggests that impaired dietary saccharide metabolism in rats results in increased lower bowel volume, stimulating increased water consumption. The excess water excreted in the bladder causes hyperplasia and predisposes to tumor development. Increased water consumption and hyperplasia were confined to the sodium and potassium salts of saccharin, and these salts caused greater polydipsia and polyuria than calcium or free saccharin acid.[12]

Attempts have also been made to extrapolate the animal data to a lifetime bladder cancer risk in humans. One study estimated a maximum risk of 0.18 times the percentage of saccharin in the diet, assuming that saccharin only acts as a cancer promoter. For a person using 15 mg (one fourth grain) daily, this translates into an estimated risk of 3.8×10^{-7} to 1.8×10^{-4}. If a threshold dose exists for tumor promotion, the risk may be considerably less or nonexistent.[13]

VI. HUMAN TOXICITY DATA

A. Carcinogenicity

Case-control studies in humans in over 6400 patients have attempted to resolve the controversy generated by animal data suggesting a carcinogenic or promoting effect of saccharin.[14-20]

Only two of these studies, comprising 679 of these patients, showed a statistically significant increased risk of bladder cancer. The relative risk was less than that observed for cigarette smokers (3.3 vs. 5.6) in the smaller study.[21] In the larger study, a dose-response relationship was found with a relative risk of 1.6 in males who had ever used saccharin.[22]

An independent analysis of a large case-control study sponsored by the National Cancer Institute, involving 30,101 patients from ten geographic regions in the U.S., concluded that there was no association between saccharin use and bladder cancer.[23]

These human studies have a limited power to detect an increased risk of bladder cancer given the prevalence of this tumor in the general population. An increased risk of 2% over the basal risk has been estimated to occur with ingestion of saccharin in amounts exceeding 150 mg/d. Epidemiological case-control studies of the type that have been conducted can only detect increased risks of 200 to 300 times the basal rate, thus a small risk of human bladder cancer cannot be excluded from these data.[3]

B. Urticaria

Several cases of generalized urticaria with pruritus have been attributed to consumption of saccharin-sweetened beverages. One case reported involved a 50-year-old woman who developed hand and arm urticaria, generalized pruritus, and wheezing within 45 min after single-blind provocation with saccharin in solution.[24] Two cases reported by Gordon in 1975 displayed a variety of concomitant symptoms, including pruritus, cold sweats, skipped heart beats, a prickling sensation of the face, nausea, diarrhea, and tachycardia. One of these patients showed cross-sensitivity to sulfonamides.[25]

Other dermatological effects attributed to saccharin in a series of 42 patients included pruritus 33%, urticaria 26%, eczema 12%, and prurigo or other eruptions in 17%.[26] Fixed eruptions have also been described.[27]

C. Photosensitivity

At least 19 cases of photosensitivity reactions associated with saccharin have been reported. The majority of these cases have shown positive RAST tests for saccharin. These reactions have consisted of pruritic, eczematous eruptions on sun-exposed areas.[28] Photosensitization was observed in 12% of 42 patients with saccharin hypersensitivity in one series.[26]

D. Notalgia Paresthetica

Sensory neuropathy of the second through sixth dorsal nerves, known as notalgia paresthetica, is a condition often occurring in conjunction with postvaccinal neuritis or surgical procedures.[29] Fishman[30] reported an anecdotal case of a 69-year-old man with a chronic history of sporadic, persistent pruritus localized to the inferomedial border of the right scapula. A diagnosis of notalgia paresthetica was made. The patient's history revealed excessive use of saccharin and a hexachlorophene-containing soap. The symptoms resolved after discontinuation of these products and recurred after inadvertent exposure to beverages thought to contain aspartame, but which contained a combination of saccharin and aspartame.

E. Children

Adverse reactions were attributed to the accidental ingestion of 100 to 200 mg/kg of saccharin in a 5-year-old and a 3-year-old child. Symptoms included irritability, strabismus, crying, ataxia, depression, hallucinations, and dysmetria. These symptoms resolved within 24 h after discontinuation of the formula. Less severe symptoms were noted in four other infants who received lower doses in milk formulas; these included irritability, hypertonia, insomnia, opisthotonus, and strabismus.[31]

VII. CLINICAL RELEVANCE

The primary adverse reaction reported to saccharin appear to be immediate hypersensitivity reactions, although studies have not been done to confirm IgE mediation. Saccharin is an o-toluene sulfonamide derivative, and cross-reactions may occur in patients with sulfonamide sensitivity. Urticaria and pruritus are the most common features of these reactions. Photosensitivity reactions may also occur.

Animal studies have shown a weak carcinogenic effect, which may be related to promoter activity. These effects may be species-specific and occurred primarily in male rats exposed *in utero* and early in life in doses much higher than those likely to be consumed by humans. Nevertheless, currently available human epidemiological studies are insufficient to exclude a small contribution of saccharin to human bladder cancer.

It is generally recommended that saccharin be avoided in nondiabetic children, patients with sulfonamide allergy, pregnant women, and young women of childbearing age. Excessive use should be discouraged.

REFERENCES

1. **Hile, J. P.,** Saccharin and its salts: final guidelines, *Fed. Reg.,* 42, 62209, 1977.
2. **Wilms, G. H.,** *Talk Paper-Moratorium on Saccharin Ban Extended,* Food and Drug Administration, Washington, D.C., September 1, 1987.
3. **Food and Drug Administration,** Saccharin and its salts: proposed rule making, *Fed. Reg.,* 42, 199996, 1977.
4. **Food and Drug Administration,** *Inactive Ingredients Guide,* FDA, Washington, D.C., March 1990.
5. **Nikitakis, J. M.,** *CTFA Cosmetic Ingredient Handbook,* 1st ed., The Cosmetic, Toiletry and Fragrance Association, Washington, D.C., 1988.
6. **Arnold, D. L., Moodie, C. A., Grice, H. C., Charbonneau, S. M., Stavric, B., Collins, B. T., McGuire, P. F., Zawidzka, Z. Z., and Munro, I. C.,** Long-term toxicity of ortho-toluenesulfonamide and sodium saccharin in the rat, *Toxicol. Appl. Pharmacol.,* 52, 113, 1980.
7. **Arnold, D. L.,** Two-generation saccharin bioassays, *Environ. Health Perspect.,* 50, 27, 1983.
8. **Cohen, S. M., Arai, M., Jacobs, J. B., and Friedell, G. H.,** Promoting effect of saccharin and dl-tryptophan in urinary bladder carcinogenesis, *Cancer Res.,* 39, 1207, 1979.
9. **Nakanishi, K., Fukushima, S., Hagiwara, A., Tamano, S., and Ito, N.,** Organ-specific promoting effects of phenobarbital sodium and sodium saccharin the induction of liver and urinary bladder tumors in male F344 rats, *J. Natl. Cancer Inst.,* 68, 497, 1982.
10. **Nakanishi, K., Hagiwara, A., Shibata, M., Imalda, K., Tatematsu, M., and Ito, N.,** Dose response of saccharin in induction of urinary bladder hyperplasias in Fisher 344 rats pretreated with N-butyl-n-(4-hydroxybutyl)nitrosamine, *J. Natl. Cancer Inst.,* 65, 1005, 1980.
11. **Council on Scientific Affairs,** Saccharin: review of safety issues, *JAMA,* 254, 2622, 1985.
12. **Anderson, R. L., Lefever, F. R., and Maurer, J. K.,** The effect of various saccharin forms on gastrointestinal tract, urine and bladder of male rats, *Food Chem. Toxicol.,* 26, 665, 1988.
13. **Gaylor, D. W., Kadlubar, F. F., and West, R. W.,** Estimates of the risk of bladder tumor promotion by saccharin in rats, *Reg. Toxicol. Pharmacol.,* 8, 467, 1988.

14. **Moller-Jensen, O., Knudsen, J. B., Sorensen, B. L., and Clemmesen, J.,** Artificial sweeteners and absence of bladder cancer risk in Copenhagen, *Int. J. Cancer,* 32, 577, 1983.

15. **Morrison, A. S., Verhoek, W. G., Leck, I., Aoki, K., Ohno, Y., and Obata, K.,** Artificial sweeteners and bladder cancer in Manchester, U.K., and Nagoya, Japan, *Br. J. Cancer,* 45, 332, 1982.

16. **Hoover, R. N. and Strasser, P. H.,** Artificial sweeteners and human bladder cancer: preliminary results, *Lancet,* 1, 837, 1980.

17. **Kessler, I. I. and Clark, J. P.,** Saccharin, cyclamate, and human bladder cancer: no evidence of an association, *JAMA,* 240, 349, 1978.

18. **Morrison, A. S. and Buring, J. E.,** Artificial sweeteners and cancer of the lower urinary tract, *N. Engl. J. Med.,* 302, 537, 1980.

19. **Najem, G. R., Louria, D. B., Seebode, J. J., Thind, I. S., Prusakowski, J. M., Ambrose, R. B., and Fernicola, A. R.,** Life time occupation, smoking, caffeine, saccharine, hair dyes and bladder carcinogenesis, *Int. J. Epidemiol.,* 11, 212, 1982.

20. **Wynder, E. L. and Stellman, S. D.,** Artificial sweetener use and bladder cancer: a case-control study, *Science,* 207, 1214, 1980.

21. **Mommsen, S., Aagaard, J., and Sell, A.,** A case-control study of female bladder cancer, *Eur. J. Cancer Clin. Oncol.,* 19, 725, 1983.

22. **Howe, G. R., Burch, J. D., Miller, A. B., Morrison, B., Gordon, P., Weldon, L., Chambers, L. W., Fodor, G., and Winsor, G. M.,** Artificial sweeteners and human bladder cancer, *Lancet,* 2, 578, 1977.

23. **Walker, A. M., Dreyer, N. A., Friedlander, E., Loughlin, J., Rothman, K. J., and Kohn, H. I.,** An independent analysis of the National Cancer Institute study on non-nutritive sweeteners and bladder cancer, *Am. J. Public Health,* 72, 376, 1982.

24. **Miller, R., White, L. W., and Schwartz, H. J.,** A case of episodic urticaria due to saccharin ingestion, *J. Allergy Clin. Immunol.,* 53, 240, 1974.

25. **Gordon, H. H.,** Episodic urticaria due to saccharin ingestion, *J. Allergy Clin. Immunol.,* 56, 78, 1975.

26. **Birkbeck, J.,** Saccharin-induced skin rashes, *N.Z. Med. J.,* 102, 24, 1989.

27. **Domonkos, A. N., Arnold, J. L., and Odom, R. B.,** *Andrews' Diseases of the Skin: Clinical Dermatology,* 7th ed., W. B. Saunders, Philadelphia, 1982.

28. **Gordon, H. H.,** Photosensitivity to saccharin, *J. Am. Acad. Dermatol.,* 8, 565, 1983.

29. **Pleet, A. B. and Massey, E. W.,** Notalgia paresthetica, *Neurology,* 28, 1310, 1978.

30. **Fishman, H. C.,** Notalgia paresthetica, *J. Am. Acad. Dermatol.,* 15, 1304, 1986.

31. **Luis Sain, O. and Berman, J. M.,** Efectos adversos de edulcorantes en pediatria sacarina y ciclamato, *Arch. Arg. Pediatr.,* 82, 209, 1984.

SESAME OIL

I. REGULATORY CLASSIFICATION

Sesame oil is used as an oleaginous vehicle in pharmaceuticals.

II. SYNONYMS

Aceito de Ajonjoli
Benne oil
Gingelly oil
Gingilli oil
Oleum sesame
Sesame seed oil
Teel oil

III. AVAILABLE FORMULATIONS

A. Constituents
Sesame oil is the fixed oil obtained from the seeds of *Sesamum indicum*. Constituents that have been identified in the unsaponifiable fraction include sesamol, sesamin, and sesamolin. This fraction represents not more than 1.5% of the pharmaceutical grade oil.[1]

B. Drugs
Sesame oil is present in 50 parenteral pharmaceutical products registered with the FDA. There are also several oral capsules, concentrates, emulsions, tablets, and topical creams containing this excipient.[2]

C. Cosmetics
Sesame oil is used as a solvent and skin and hair conditioner in cosmetics. Types of products include eye makeup, lipsticks, makeup foundations, and hand and body creams and lotions.[3]

IV. TABLE OF COMMON PRODUCTS

A. Parenteral Drug Products

Trade name	Manufacturer
Haldol decanoate	McNeil
Prolixin decanoate	Princeton
Prolixin enanthate	Princeton
Solganal	Schering

V. HUMAN TOXICITY DATA

A. Immediate Hypersensitivity Reactions

Immediate urticarial reactions have been reported to sesame seed.[4] RAST testing was positive in three of four patients with a history of systemic anaphylaxis or angioedema-urticaria following ingestion of sesame seed or sesame oil-containing products.[5]

B. Allergic Contact Dermatitis

Contact hypersensitivity to sesame oil has mostly been reported in elderly patients with stasis leg ulcers[6] and was found in 14 of 81 patients in one series.[7] Specific allergens were demonstrated to be sesamolin and sesamin in 12 of 13 patients with contact allergy to sesame oil.[1] These two constituents were also shown to be the allergens responsible for dermatitis following the use of a burn ointment containing 60% sesame oil[8] and in a woman with lipstick cheilitis secondary to sesame oil.[9] Eight patients have been documented with positive patch tests to sesamol, another constituent reported to be found in crude and pharmaceutical grade sesame oil.[1] Other investigators have not been able to confirm the presence of sesamol in sesame oil.[9]

C. Bronchiolar Carcinoma

Alveolar carcinomata was reported in an elderly man who used sesame oil continuously to lubricate his tracheal cannula.[10]

VI. CLINICAL RELEVANCE

Sesame oil is a rare cause of allergic contact dermatitis and immediate hypersensitivity reactions. The primary inciting allergen is suspected to be sesamin, which is present in 100 times the concentration of the other allergenic component, sesamolin. Most cases have occurred in patients with chronically inflamed skin due to leg stasis ulcers who used frequent and prolonged applications of sesame oil.

REFERENCES

1. **Neering, H., Vitanyi, B. E., Malten, K. E., van Ketel, W. G., and van Dijk, E.,** *Acta Derm. Venereol.,* 55, 31, 1975.
2. **Food and Drug Administration,** *Inactive Ingredients Guide.* FDA, Washington, D.C., March 1990.
3. **Nikitakis, J. M.,** *CTFA Cosmetic Ingredient Handbook.* 1st ed., The Cosmetic, Toiletry and Fragrance Association, Washington, D.C., 1988.

4. **Tornsey, P. J.,** Hypersensitivity to sesame seed, *J. Allergy,* 35, 514, 1964.

5. **Malish, D., Glovsky, M. M., Hoffman, D. R., Ghekiere, L., and Hawkins, J. M.,** Anaphylaxis after sesame seed ingestion, *J. Allergy Clin. Immunol.,* 67, 35, 1981.

6. **van Dijk, E., Dijk, E., Neering, H., and Vitanyi, B. E.,** Contact hypersensitivity to sesame oil in patients with leg ulcers and eczema, *Acta Derm. Venereol.,* 53, 133, 1973.

7. **Maltern, K. E., Kuiper, J. P., and van der Staak, W. B. J. M.,** Contact allergic investigations in 100 patients with ulcus cruris, *Dermatologica,* 147, 241, 1973.

8. **Kubo, Y., Nonaka, S., and Yoshida, H.,** Contact sensitivity to unsaponifiable substances in sesame oil, *Contact Dermatitis,* 15, 215, 1986.

9. **Hayakawa, R., Matsunaga, K., Suzuki, M., Hosokawa, K., Arima, Y., Shin, C. S., and Yoshida, M.,** Is sesamol present in sesame oil?, *Contact Dermatitis,* 17, 133, 1987.

10. **Maesen, F. P., Lamers, J. H., and van den Tweel, J. G.,** Bronchiolo-alveolar carcinoma after inhalation of vegetable oil through a tracheal cannula, *Eur. J. Respir. Dis.,* 67, 136, 1985.

SHELLAC

I. REGULATORY CLASSIFICATION

Pharmaceutical glaze, which contains shellac, is classified as a coating agent.

II. SYNONYMS

Button Lac
Lac
Lacca
Leaf Lac

III. AVAILABLE FORMULATIONS

A. Constituents
Shellac is obtained from the purified resinous secretion of the insect *Laccifer lacca Kerr*. Orange shellac is purified by filtration or by hot solvent processing. Bleached or white shellac is treated with an alkaline sodium hypochlorite solution. Removal of the wax by filtration results in Refined Bleached shellac USP. Constituents include laccaic acid, erythrolaccin, aleurtic acid, and shellolic acid.

B. Drugs
Shellac is used as an enteric coating for solid oral dosage forms. It is present in 56 tablets listed as coated, film coated, enteric coated, or sustained action registered with the FDA.[1] Pharmaceutical glaze is a denatured alcohol solution containing 20 to 51% of anhydrous shellac.

C. Cosmetics
Bleached shellac is used in an alcoholic solution in hair sprays. It may also be present in eyeliners, mascara, and some creams and lotions.[2]

IV. TABLE OF COMMON PRODUCTS

A. Oral Drug Products

Trade name	Manufacturer
Atarax tablets	Roerig
AzoGantanol tablet	Roche
AzoGantrisin tablet	Roche
Donnatal extentabs	A.H. Robins
Donnazyme tablet	A.H. Robins
Entozyme tablet	A.H. Robins
Fastin capsule 30 mg	Beecham
Kaon Cl tablet	Adria
Larobec tablet	Roche
Ludiomil tablet	Ciba
Mexitil capsule	Boehringer Ingelheim
Miltown-200	Wallace
Motrin tablet 300, 400 mg	Upjohn
Norpace CR	Searle
Persantine tablet	Boehringer Ingelheim
PMB tablet 200, 400 mg	Wyeth-Ayerst
Prelu-2 capsule	Boehringer Ingelheim
Premarin tablets	Wyeth-Ayerst
Premarin with methyltesterone tablets	Wyeth-Ayerst
Quinidex tablets	A.H. Robins
Raudixin	Princeton
Rauzide	Princeton
Slow Fe tablet	Ciba
Sudafed tablet 60 mg	Burroughs Wellcome
Taractan tablet	Roche
Theo-24 capsule	Searle
Thiacide tablet	Beach
Uroqid acid tablet	Beach
Verelan capsule	Lederle
Voltaren tablet	Geigy

V. HUMAN TOXICITY DATA

A. Contact Dermatitis

Cheilitis secondary to the presence of shellac in a lipstick sealant was reported in an atopic 20-year-old woman who had been using the product for 6 months. She reacted to patch testing with 1% shellac in methyl ethyl ketone. Irritant reactions were excluded by negative patch tests to 100% shellac in 50 control subjects.[3]

One case of cosmetic-related contact dermatitis was attributed to shellac in a series of 713 patients tested over a 5-year study period in the U.S.[4]

B. Occupational Exposure

Pulmonary reactions have been associated with the use of shellac-containing hair sprays in hairdressers.[5]

VI. CLINICAL RELEVANCE

Shellac is nonirritating and is an extremely rare source of allergic dermatitis.

REFERENCES

1. **Food and Drug Administration,** *Inactive Ingredients Guide,* FDA, Washington, D.C., March 1990.
2. **Nikitakis, J. M.,** *CTFA Cosmetic Ingredient Handbook,* 1st ed., The Cosmetic, Toiletry and Fragrance Association, Washington, D.C., 1988.
3. **Rademaker, M., Kirby, J. D., and White, I. R.,** Contact cheilitis to shellac, Lanpol 5 and colophony, *Contact Dermatitis,* 15, 307, 1986.
4. **Adams, R. M. and Maibach, H. I.,** A five-year-study of cosmetic reactions, *J. Am. Acad. Dermatol.,* 13, 1062, 1985.
5. **McLaughlin, A. I., Bidstrup, P. L., and Konstam, K. M.,** The effects of hair lacquer sprays on the lungs, *Food Cosmet. Toxicol.,* 1, 171, 1963.

SODIUM BENZOATE

I. REGULATORY CLASSIFICATION

Sodium benzoate is classified as an antimicrobial preservative agent.

FIGURE 59. Sodium benzoate.

II. SYNONYMS

Benzoate of Soda
Benzoic acid, sodium salt
E211
Natrium Benzoicum
Sodii Benzoas

III. AVAILABLE FORMULATIONS

A. Foods
The GRAS status of sodium benzoate has been confirmed by the FDA. It may be added as a preservative in concentrations not exceeding 0.1% for combined benzoic acid and sodium benzoate.[1] Orange soft drinks contain a high amount of sodium benzoate. Information compiled from manufacturers of six different samples reported a range of 13.5 to 25 mg per 250 ml.[2]

B. Drugs
Sodium benzoate is a moderately effective bacteriostatic and fungistatic agent at a pH of 5 or less. It is more water soluble than benzoic acid, but antimicrobial activity is dependent on dissociation to the free acid . A concentration of 0.1% is used to preserve oral liquid and

parenteral products. Sodium benzoate is listed as an inactive ingredient in 2 dentifrice products in a concentration of 0.08%; 18 injectable products in concentrations of 4.75 to 5%; 147 oral concentrates, solutions, elixirs, suspensions, and syrups in concentrations of 0.02 to 0.5%; and in 67 oral capsules and tablets in amounts of 0.14 to 18 mg.[3]

C. Cosmetics

Sodium benzoate was among the top 25 preservatives used in cosmetics in 1982 and was present in 99 of 20,183 (0.5%) products surveyed by the FDA.[4] It is also used in cosmetics as a corrosion inhibitor. Types of products include shampoos, permanent waves, dentifrices, cleansing products, creams, lotions, face and dusting powders, bath products, and mascara.[5]

IV. TABLE OF COMMON PRODUCTS

A. Foods[1]

Alcoholic beverages
Baked goods
Cereals
Cheese
Condiments
Fats and oils
Frosting
Frozen dairy products
Fruit ices
Gelatins
Gravies
Hard candies
Imitation dairy products
Instant Coffee
Instant Tea
Jams
Jellies
Meat products
Milk products
Nonalcoholic beverages
Orange soft drinks
Processed fruit
Processed vegetables
Puddings
Relishes
Salted margarine
Seasonings
Soft candy
Sweet sauces

B. Oral Liquid Drug Products

Trade name	Manufacturer
Actifed with codeine cough syrup	Burroughs Wellcome
Advil children's suspension	Whitehall
Amoxil oral suspension	Beecham
Anacin 3 children's liquid, infant drops	Whitehall
Atarax syrup	Roerig
Cerose DM	Wyeth-Ayerst
Chloromycetin oral suspension	Parke-Davis
Compazine syrup	Smith Kline & French
Deconamine syrup	Berlex
Dilantin suspension	Parke-Davis
Dimetane DC, Dimetane DX	A.H. Robins
Donnagel PG	A.H. Robins
Entex liquid	Norwich Eaton
Gantanol suspension	Roche
Klorvess liquid	Sandoz
Marax syrup	Roerig
Naldecon syrup, pediatric syrup, drops	Bristol
Naldecon CX liquid	Bristol
Naldecon DX adult and children's, drops	Bristol
Naldecon EX children's syrup, drops	Bristol
Naldecon Senior DX and EX cough/cold	Bristol
Nucofed syrup	Beecham
Pediaprofen suspension	McNeil
Pediacare cough-cold liquid	McNeil
Pediacare Nightrest cough-cold liquid	McNeil
Pediacare Infant's oral decongestant drops	McNeil
Phenergan syrup, DM, VC syrup	Wyeth-Ayerst
Phenergan syrup and VC syrup with codeine	Wyeth-Ayerst
Prolixin oral concentrate, elixir	Princeton
Proventil syrup	Schering
Robitussin AC, DAC	A.H. Robins
Septra suspension	Burroughs Wellcome
Sumycin syrup	Squibb
Suprax suspension	Lederle
Temaril syrup	Herbert
Thorazine syrup	Smith Kline & French
Tussi-Organidin DM	Wallace
Tylenol cold nighttime medication liquid	McNeil
Tylenol with codeine elixir	McNeil
Tylenol Children's liquid	McNeil
Tylenol children's cold medication liquid	McNeil
Tylenol extra strength liquid	McNeil
Ultracef suspension	Bristol

V. ANIMAL TOXICITY DATA

Based on animal toxicology data, the WHO has recommended a maximum acceptable daily intake of up to 5 mg/kg as the sum of benzoic acid, potassium benzoate, and sodium benzoate.[6]

VI. HUMAN TOXICITY DATA

A. Anaphylactoid Reactions

A 42-year-old asthmatic woman undergoing surgery for removal of nasal and sinus polyps developed a severe anaphylactoid reaction, with bronchoconstriction and hypotension, despite premedication with antihistamines. The anesthetic regimen included diazepam injection, which the patient had received uneventfully the year before. Due to speculation that the benzoate preservative in the diazepam injection was responsible, she underwent oral challenge with sodium benzoate 100 mg, resulting in a decrease in peak expiratory flow rate of 50%. She had received 144.7 mg of sodium benzoate from the diazepam injection.[7]

B. Nonimmunologic Contact Urticaria

Topical exposure to sodium benzoate 5% in petrolatum induced nonimmunologic contact urticaria in 10% of 105 subjects tested.[8] The development of contact urticaria is thought to be related to formation of benzoic acid on skin contact. Conditions that maximize formation of benzoic acid are prolonged exposure and sweating, which lowers the skin pH.[9]

Patch testing of a series of three workers with complaints of urticaria after occupational exposure to sodium benzoate powder and three controls demonstrated positive immediate reactions to 0.5% sodium benzoate in saline (pH 7.3) in three of six subjects. Only one subject reacted to 10% sodium benzoate in petrolatum, and none reacted to 0.5% sodium benzoate in aqueous solution. In contrast, all six subjects reacted to 0.25% benzoic acid in aqueous solution. The workers experiencing reactions developed symptoms 15 to 30 min after exposure, which lasted 30 min to a few hours.[9]

C. Chronic Urticaria

A study of 75 patients with a history of recurrent urticaria and angioedema associated with food additives demonstrated positive oral provocation with sodium benzoate alone in two patients. Reactions to sodium benzoate and azo dyes were found in 7 subjects, to sodium benzoate and aspirin in 11 patients, and to all three substances in 24 patients. Thus 44 of 75 (58.5%) reacted to the benzoate compound.[10]

A placebo effect was suggested by Lahti and Hannuksela, who administered benzoic acid 500 mg in a double-blind placebo-controlled fashion (using lactose as the placebo) to 150 dermatological inpatients. Objective symptoms (rash, lip or throat edema, rhinitis, and urticaria) were seen in 3 to 7% of the benzoic acid group and in 6 to 14% in the placebo group.[11]

A double-blind placebo-controlled study of 34 children with a history of food additive intolerance demonstrated positive urticarial responses to 100 mg of sodium benzoate in 7 children (20.5%). Follow-up studies 1 to 5 years later indicated only one child with a persistent reaction, consisting of mild coughing and wheezing, after a higher dose than the original challenge. The urticaria found on the original challenge could not be reproduced.[12]

D. Orofacial Granulomatosis

Symptoms of recurrent upper lip and gum swelling and a fissured tongue "Melkersson-Rosenthal syndrome" were found to be triggered by sodium benzoate 50 mg and tartrazine 5 mg during oral double-blind challenges in a 34-year-old man. An elimination diet excluding these two additives resulted in complete remission, lasting at least 1 year.[13]

E. Asthma

Four of 14 asthmatic patients with a history of provocation of asthma by ingestion of orange soft drinks had positive challenge tests with sodium benzoate 20 to 100 mg (28.5%). Bronchodilators were withheld for 8 h prior to the challenge. The onset of the reactions were between 20 to 30 min.[2] Another single-blind study found only one of 504 unselected asthma or rhinitis patients with a positive challenge to sodium benzoate in a dose of 50 mg. Ten other positive challenges to doses of 200 mg or greater were reported (2.2% total).[14]

A double-blind study of 28 unselected patients with chronic asthma found 1 patient with a positive challenge to sodium benzoate (3.6%). This patient had a 29% fall in FEV_1 after 25 mg of sodium benzoate. Normal medications were continued during conduction of the study.[15] Another double-blind placebo-controlled study of 45 patients with severe perennial asthma found 1 confirmed positive reactor to a combination of sodium benzoate and para-hydroxybenzoic acid (2.2%). Sensitivity to both ingredients was demonstrated. Rechallenge 2 years later was negative.[16]

The mechanism of action of sodium benzoate was studied in an *in vitro* model using blood from healthy volunteers. The presence of 5 to 15 mM of sodium benzoate inhibited formation of thromboxane B2 by noradrenaline-activated platelets, to a degree about one half of that seen with 0.1 mM of aspirin.[17] These concentrations are comparable to those achieved in neonates receiving 500 mg/kg/d intravenously.[18] The inhibition of the cyclooxygenase pathway of arachidonic acid metabolism is predicted to cause preferential metabolism via the lipoxygenase pathway, resulting in formation of bronchospasm-inducing leukotrienes. This activity remains to be proven in serum levels achievable by excipient doses.

F. Neonates

An *in vitro* study demonstrated displacement of bilirubin from albumin binding sites in the presence of 3% sodium benzoate or from the vehicle used in injectable diazepam, which contains 5% sodium benzoate and benzoic acid. Concern was addressed over the potential for increasing the free bilirubin levels in hyperbilirubinemic newborns, leading to increased CNS penetration and development of kernicterus.[19]

There have been no documented case reports of diazepam-related kernicterus, probably due to the low concentration of sodium benzoate achieved. Intravenous administration of up to 500 mg/kg/d of sodium benzoate has been tolerated in term infants, even though this dose was calculated to increase free bilirubin levels by 4 to 25 times.[18] To exceed this amount of sodium benzoate, a 4 kg neonate would have to receive over 200 mg/d of diazepam. However, this calculation assumed first order elimination, which has been shown to be inappropriate. Neonates with immature liver metabolizing function may accumulate benzoate, since it has been shown to be eliminated by a saturable process. Most is metabolized to hippurate via conjugation with glycine. This pathway is limited by the availability of glycine.[20] Low-birth-weight premature infants and those with hyperbilirubinemia may be at increased risk.

VII. CLINICAL RELEVANCE

A. Urticaria

Nonimmunologic contact urticaria occurs in about 10% of normal subjects and is thought to be related to formation of benzoic acid on skin contact, which is enhanced by prolonged exposure and sweating, which lowers the skin pH. Symptoms of tingling, accompanied by a pronounced flare, appear within 30 min and generally last less than 2 h.

Exacerbation of chronic urticaria has also been described in patients who ingest foods containing this additive. This reaction has been reported in about 20% of children with a history of food additive sensitivity and may disappear within several years. A large placebo effect has been observed in adults, making the open challenge studies difficult to interpret.

B. Asthma

Unselected groups of asthmatic patients have demonstrated exacerbation of symptoms in about 2.2 to 3.6% following double-blind placebo-controlled challenges with sodium benzoate. The mechanism is unclear, but may be related to inhibition of cyclooxygenase.

C. Neonates

Term infants without hyperbilirubinemia do not appear to be at risk for adverse reactions from sodium benzoate present as a drug excipient. Low-birth-weight premature infants or term infants with significant hyperbilirubinemia may be at risk for accumulation of benzoate after repeated administration.

REFERENCES

1. **Lecos, C.,** Food preservatives: a fresh report, *FDA Consumer,* April 1984.
2. **Freedman, B. J.,** Asthma induced by sulphur dioxide, benzoate and tartrazine contained in orange drinks, *Clin. Allergy,* 7, 407, 1977.
3. **Food and Drug Administration,** *Inactive Ingredients Guide,* FDA, Washington, D.C., March 1990.
4. **Decker, R. L. and Wenninger, J. A.,** Frequency of preservative use in cosmetic formulas as disclosed to FDA-1982 update, *Cosmet. Toilet.,* 97, 57, 1982.
5. **Nikitakis, J. M.,** *CTFA Cosmetic Ingredient Handbook,* 1st ed., The Cosmetic, Toiletry and Fragrance Association, Washington, D.C., 1988.
6. **World Health Organization,** Seventeenth Report of the Joint FAO/WHO Expert Committee on Food Additives, Tech Rep Ser No. 539, SHO, Geneva, 1974.
7. **Moneret-Vautrin, D. A., Moeller, R., Malingrey, L., and Laxenaire, M. C.,** Anaphylactoid reaction to general anaesthesia: a case of intolerance to sodium benzoate, *Anaesth. Intens. Care,* 10, 156, 1982.
8. **Lahti, A.,** Non-immunologic contact urticaria, *Acta Derm. Venereol.,* 50 (Suppl. 91), 1, 1980.
9. **Nethercott, J. R., Lawrence, M. J., Roy, A., and Gibson, B. L.,** Airborne contact urticaria due to sodium benzoate in a pharmaceutical manufacturing plant, *J. Occup. Med.,* 26, 734, 1984.
10. **Ros, A., Juhlin, L., and Michaelsson, G.,** A follow-up study of patients with recurrent urticaria and hypersensitivity to aspirin, benzoates and azo dyes, *Br. J. Dermatol.,* 95, 19, 1976.
11. **Lahti, A. and Hannuksela, M.,** Is benzoic acid really harmful in cases of atopy and urticaria?, *Lancet,* 2, 1055, 1981.
12. **Pollock, I. and Warner, J. O.,** A follow-up study of childhood food additive intolerance, *J. R. Coll. Phys. London,* 21, 248, 1987.
13. **Pachor, M. L., Urbani, G., Cortina, P., Lunardi, C., Nicolis, F., Peroli, P., Coorrocher, R., and Gotte, P.,** Is the Melkersson-Rosenthal syndrome related to the exposure to food additives?, *Oral Surg. Oral Med. Oral Pathol.,* 67, 393, 1989.
14. **Rosenhall, L.,** Evaluation of intolerance to analgesics, preservatives and food colorants with challenge tests, *Eur. J. Respir. Dis.,* 63, 410, 1982.

15. **Tarlo, S. M. and Broder, I.,** Tartrazine and benzoate challenge and dietary avoidance in chronic asthma, *Clin. Allergy,* 12, 303, 1982.

16. **Weber, R. W., Hoffman, M., Raine, D. A., and Nelson, H. S.,** Incidence of bronchoconstriction due to aspirin, azo dyes, non-azo dyes and preservatives in a population of perennial asthmatics, *J. Allergy Clin. Immunol.,* 64, 32, 1979.

17. **Williams, W. R., Pawlowicz, A., and Davies, B. H.,** Aspirin-like effects of selected food additives and industrial sensitizing agents, *Clin. Exp. Allergy,* 19, 533.

18. **Green, T. P., Marchessault, R. P., and Freese, D. K.,** Disposition of sodium benzoate in newborn infants with hyperammonemia, *J. Pediatr.,* 102, 785, 1983.

19. **Schiff, D., Chan, G., and Stern, L.,** Fixed drug combinations and the displacement of bilirubin from albumin, *Pediatrics,* 48, 139, 1971.

20. **Oyanagi, K., Kuniya, Y., Tsuchiyama, A., Nakao, T., Owada, E., Sato, J., and Ito, K.,** Nonlinear elimination of benzoate in patients with congenital hyperammonemia, *J. Pediatr.,* 110, 634, 1987.

SODIUM LAURYL SULFATE

I. REGULATORY CLASSIFICATION

Sodium lauryl sulfate (SLS) is classified as an emulsifying, wetting, and/or solubilizing agent.

$$CH_3(CH_2)_{10}CH_2OSO_3Na$$

FIGURE 60. Sodium lauryl sulfate.

II. SYNONYMS

Dodecyl sodium sulfate
Lauryl sodium sulfate
Lauryl sulfate sodium salt
SLS
Sodium N-dodecyl sulfate
Sodium laurilsulfate

III. AVAILABLE FORMULATIONS

A. Constituents

Sodium lauryl sulfate is a mixture of sodium alkyl sulfates, primarily sodium lauryl sulfate, containing not more than a total of 8% of sodium sulfate and sodium chloride. Formaldehyde may be included as a preservative in concentrations of up to 0.1%.[1]

B. Foods

The addition of sodium lauryl sulfate to egg whites is permissible.[2]

C. Drugs

Sodium lauryl sulfate is present in 534 oral solid dosage forms approved by the FDA in amounts of 0.004 to 0.6 mg; in 11 oral liquid dosage forms in concentrations of 0.01 to 0.02%; and in 38 topical creams, lotions, ointments, a medicated sponge, and medicated shampoos in concentrations of 0.1 to 12.7%.[3]

D. Cosmetics

Sodium lauryl sulfate is widely used in cosmetics as an anionic surfactant with detergent, wetting, foaming, and emulsifying properties. In 1981, there were 703 cosmetic formulations listed with the FDA containing this ingredient. Types of products included hair shampoos (226); bubble baths (73); hair dyes and colors (61); dentifrices (28); skin cleansing products (28); face, body, and hand skin care products (27); skin moisturizers (23); hair bleaches (21); and sachets (20). Concentrations varied from less than 0.1% to more than 50%.[1]

IV. TABLE OF COMMON PRODUCTS

A. Topical Drug Products

Trade name	Manufacturer
Aqua Care lotion	Smith Kline Consumer
Aspercreme	Thompson Medical
Benoquin cream	ICN
Cetaphil cleanser	Owen/Galderma
Choromycetin cream	Parke-Davis
Cortizone-5 creme	Thompson Medical
Drithocreme	American Dermal
Dritho-scalp cream	American Dermal
Eldecort cream	ICN
Nutracort lotion	Owen/Galderma
Nutraderm lotion	Owen/Galderma
Oxy Clean facial scrub	Norcliff Thayer
Oxy Clean Night watch	Norcliff Thayer
Oxy Clean Medicated pads	Norcliff Thayer
Persa-gel	Ortho
Sebulex shampoo with conditioner	Westwood-Squibb
Selsun lotion	Abbott
Sween cream	Sween
Tridesilon cream	Miles
Tronolane cream	Ross
Tronthane cream	Abbott
Vioform-Hydrocortisone cream	Ciba
Westcort cream, ointment	Westwood-Squibb

V. ANIMAL TOXICITY DATA

A. Comedogenicity

Application of 1% sodium lauryl sulfate produced significant comedones (3 on a scale of 0 to 5) in the rabbit ear assay. A 5% solution produced a score of 4.[4]

VI. HUMAN TOXICITY DATA

A. Irritation

Sodium lauryl sulfate causes concentration-related epidermal damage. Thus, it is widely used as an experimental model for irritant contact dermatitis. Studies using human abdominal epidermis showed increased water permeability and some skin damage after soaking in a 1% solution for 22 h. Damage appeared within 2 to 6 h after application of a 5% solution and was more severe. Irritant reactions were observed in 52% of chronic eczema patients and 12% of controls following application of a 5% aqueous solution. Slight irritation occurs with a 1% concentration, consisting of faint or definite erythema. Products designed for prolonged skin contact should not contain greater than 1%.[1]

Increased susceptibility to SLS-induced irritation has been demonstrated for black and hispanic subjects,[5,6] while decreased susceptibility has been demonstrated in elderly subjects.[7]

Mild irritation (moderate erythema) occurred in normal subjects after a 24-h occlusive patch test with 2% sodium lauryl sulfate. Repeated patch tests 8 d later produced an augmented effect on 21 of 34 anatomic patch test sites, even though the skin had appeared completely normal for 4 to 5 d. Transepidermal water loss was a more sensitive indicator of continued functional skin impairment than visual observation.[8]

Sodium lauryl sulfate is present in a concentration of 1% in Hydrophilic Ointment USP and may produce irritant reactions when applied under occlusion for several days.[9]

B. Ocular Exposure

Allergic contact conjunctivitis, manifested by bilateral palpebral edema, conjunctival hyperemia and chemosis, follicular conjunctival hypertrophy, and a foreign body sensation described as burning, photophobia, and lacrimation, was reported in a 26-year-old woman occupationally exposed to hospital use disinfectants and detergents containing sodium lauryl sulfate. Marked ocular pruritus was produced after exposure to the product vapors. Patch testing was clearly positive to sodium lauryl sulfate 0.1%.[10]

C. Allergic Contact Dermatitis

Contact dermatitis to the formaldehyde preservative in a sodium lauryl sulfate solution was reported in a 37-year-old woman. Contact with the manufacturer revealed that formalin 0.1% was frequently used in sodium lauryl sulfate provided to shampoo manufacturers. Some manufacturers may not be aware of the addition of formaldehyde to the bulk solutions provided.[11] A cell-mediated hypersensitivity was documented by positive lymphocyte transformation in 10 of 12 patients with positive eczematous reactions to SLS. Only two of these subjects had positive patch tests to 0.1% SLS.[12]

One of 20 subjects developed sensitization to SLS after a maximization testing with a SLS-containing hair dye.[13]

The irritating effects of sodium lauryl sulfate were suggested to promote allergic sensitization to hydrocortisone in a topical ointment used to treat stasis dermatitis.[14]

VII. CLINICAL RELEVANCE

Sodium lauryl sulfate is a well-documented primary irritant displaying dose-related and time-related characteristics; race and age may also be factors in susceptibility. Concentrations of 2% and greater are unequivocal irritants, and a 1% aqueous solution is recommended for patch testing.[2]

Allergic contact dermatitis has been rarely documented. Sodium lauryl sulfate may alter skin function and enhance permeability, facilitating sensitization to active ingredients. The comedogenicity potential is high based on animal data.

REFERENCES

1. **Elder, R. L.,** Final report on the safety assessment of sodium lauryl sulfate and ammonium lauryl sulfate, *J. Am. Coll. Toxicol.,* 2, 127, 1983.
2. **Fisher, A. A.,** *Contact Dermatitis,* 3rd ed., Lea & Febiger, Philadelphia, 1986.
3. **Food and Drug Administration,** *Inactive Ingredients Guide,* FDA, Washington, D.C., March 1990.
4. **Fulton, J. E., Bradley, S., Aqueda, A., and Black, T.,** Noncomedogenic cosmetics, *Cutis,* 17, 344, 1976.
5. **Berardesca, E. and Maibach, H. I.,** Racial differences in sodium lauryl sulphate induced cutaneous irritation: black and white, *Contact Dermatitis,* 18, 65, 1988a.
6. **Berardesca, E. and Maibach, H. I.,** Sodium-lauryl-sulphate-induced cutaneous irritation: comparison of white and hispanic subjects, *Contact Dermatitis,* 19, 136, 1988b.
7. **Cua, A. B., Wilhelm, K. P., and Maibach, H. I.,** Cutaneous sodium lauryl sulphate irritation potential: age and regional variability, *Br. J. Dermatol.,* 123, 607, 1990.
8. **Freeman, S. and Maibach, H.,** Study of irritant contact dermatitis produced by repeat patch test with sodium lauryl sulfate and assessed by visual methods, transepidermal water loss, and laser Doppler velocimetry, *J. Am. Acad. Dermatol.,* 19, 496, 1988.
9. **Bergstresser, P. R. and Eagelstein, W. H.,** Irritation by hydrophilic ointment under occlusion, *Arch. Dermatol.,* 108, 218, 1973.
10. **Orlandini, A., Viotti, G., Martinoli, C., and Magno, L.,** Allergic contact conjunctivitis from synthetic detergents in a nurse, *Contact Dermatitis,* 23, 376, 1990.
11. **Fisher, A. A.,** Dermatitis due to the presence of formaldehyde in certain sodium lauryl sulfate (SLS) solutions, *Cutis,* 27, 360, 1981.
12. **Eubanks, S. and Patterson, J. W.,** Dermatitis from sodium lauryl sulfate in hydrocortisone cream, *Contact Dermatitis,* 11, 250, 1984.
13. **Foussereau, J., Petitjean, J., and Lants, J. P.,** Sodium lauryl sulphate, *Contact Dermatitis Newsletter,* 15, 460, 1974.
14. **Krook, G.,** Contact dermatitis due to Ficortril (hydrocortisone 1 percent ointment, Pfizer), *Contact Dermatitis Newsletter,* 15, 460, 1974.

SORBIC ACID/POTASSIUM SORBATE

I. REGULATORY CLASSIFICATION

Sorbic acid is an antimicrobial preservative with good activity against yeasts and molds and less activity against bacteria. Acceptable *in vitro* activity was demonstrated against most bacteria, with the exception of *Bacillus cereus*. Potassium sorbate has less antimicrobial activity, but is more water soluble. Most potassium sorbate-containing products contain some sorbic acid.[1,2]

Sorbic acid and the sodium and potassium salts are on the FDA's GRAS (generally recognized as safe) list.

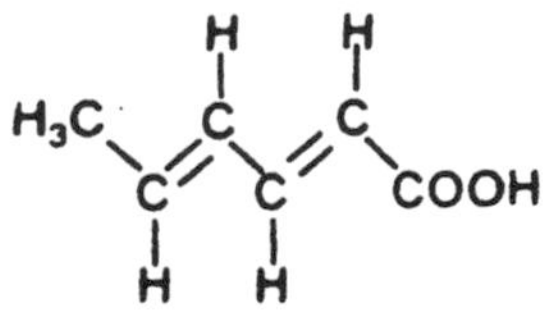

FIGURE 61. Sorbic acid.

II. SYNONYMS

2,4-Hexadienoic acid
2-Propenylacrylic acid

III. AVAILABLE FORMULATIONS

A. Foods

Sorbic acid occurs naturally in berries as the lactone, parasorbic acid. Commercial sorbic acid is obtained from mountain ash berries or synthesized. It is a white to yellow-white tasteless powder. Concentrations of 0.1 to 0.2% are commonly used in foods. Food products that may contain sorbic acid or potassium sorbate include baked goods, fats and oils, milk products, cheese, frozen dairy products, processed vegetables, juices, condiments, soft candy, jams, jellies, sweet sauces, nonalcoholic beverages, gravies, fruit ices, meat and fish products, snack foods, alcoholic beverages, and seasonings. Calcium sorbate is used in margarines.[3]

363

B. Drugs

Sorbic acid is commonly used in pharmaceutical syrup preparations for its antifungal activity in a concentration of 0.1%. Ophthalmic products may contain 0.1 to 0.13%. Oral enzyme products and gelatin capsules may contain up to 0.3%.[2]

C. Cosmetics

Sorbic acid is used in cosmetics in concentrations of 0.1 to 0.2%. In 1982 it was the seventh most commonly used preservative, present in 361 (1.8%) of 20,183 cosmetic formulations registered with the FDA.[4]

IV. TABLE OF COMMON PRODUCTS

A. Foods That May Contain Sorbic Acid[1]
1. Natural Sources

Cranberries
Currants
Fresh fruits and vegetables
Fruit juices
Strawberries

2. Present as Preservative

Baked goods
Cheese products
Meat and fish products
Pickles
Sauerkraut
Soft drinks
Wines

B. Topical Corticosteroid Products

Trade name	Manufacturer
Acticort lotion	Baker Cummins
Aristocort cream 0.025, 0.1, 0.5%	Fujisawa
Cortane lotion 1%	Blansett Pharmacal
Florone cream	Dermik
Hytone cream 1, 2.5%	Dermik
Hytone lotion 1, 2.5%	Dermik
Kenalog cream 0.025, 0.1, 0.5%	Westwood-Squibb
Maxiflor cream	Herbert
Mycolog II cream	Westwood-Squibb
Pramosone cream	Ferndale
Pramosone lotion	Ferndale
Vytone cream	Dermik

C. Ophthalmic Products

Trade name	Manufacturer
Clerz 2 lubricating/rewetting drops	CooperVision
Lens Clear soft lens cleaner	Allergan
Pliagel cleaning solution	CooperVision
SoftMate daily cleaning solution II	Barnes-Hind
SoftMate ps daily cleaning solution	Barnes-Hind
SoftMate lens drops	Barnes-Hind
SoftMate ps saline solution	Barnes-Hind
Sorbi-Care saline solution	Allergan

V. ANIMAL TOXICITY DATA

The acceptable daily intake of sorbic acid (as combined acid, calcium, potassium, and sodium salts) is up to 25 mg/kg.[5]

VI. HUMAN TOXICITY DATA

A. Immediate Nonspecific Contact Erythema

Sorbic acid is well-documented to be one of the causes of immediate contact erythema, which is most pronounced with water-in-oil emulsion bases. This occurs in 50 to 70% of patients applying ointments and creams containing this preservative, beginning 5 to 15 min after application, with erythema, slight pruritus, slight edema, and occasional stinging sensations reported. The reaction is transient and disappears completely within 1 to 2 h.[1] Contact urticaria to a shampoo has also been reported.[6]

Contact urticaria secondary to perioral contact from foods containing sorbic acid has also been described. This was observed in 18 of 20 children, aged 1 to 4 years, who smeared a mayonnaise and fruit salad around their mouths. Subsequent testing of ten healthy adults with the salad dressing and a 1% aqueous solution of sorbic acid applied in a closed 20-min patch test to the perioral region disclosed erythema and stinging after 10 to 30 min in four subjects. A dose-response curve was shown, with 20% reacting to 0.1%, 62% reacting to 1%, and 65% reacting to 5% concentrations.[7]

Curiously, an urticarial reaction could only be produced on challenges to the buccal mucosa with concentrations of 10% sorbic acid in a study of 11 healthy volunteers.[8]

Further investigation of this reaction, in an attempt to define a mechanism, involved application of varying concentrations of sorbic acid, vehicles, and sites to 15 to 17 healthy volunteers. The dose-response relationship was confirmed, with 100% of subjects reacting to 1% concentrations applied to the upper back, 76% of subjects reacting to 0.5%, and 17.6% reacting to 0.1%. Using the 1% concentration, the site with the greatest prevalence of erythema was the upper back (100%), followed by the cheek (86%), forehead (80%), forearm (80%), and deltoid (70%). The face showed significantly more reactions to the 0.1% concentration than were found on other body sites.

Most pretreatment modalities had little effect on the incidence and intensity of reactions, including systemic corticosteroids and antihistamines and intradermal lidocaine. Aspirin and topical corticosteroids, however, produced a striking decrease in erythema, with no effect on edema. No evidence of mast cell degranulation was found. It was suggested that sorbic acid penetrates the skin and induces prostaglandin formation, resulting in vasodilation.[9]

B. Allergic Contact Dermatitis

Sensitization has rarely been attributed to sorbic acid. The incidence of hypersensitivity in patients with suspected allergy to topical preservatives has ranged from 0.3 to 1.6%.[1,10] Patch tests are generally performed with a 5% concentration in petrolatum.[11]

Case reports of sorbic acid sensitivity include a patient with stasis eczema treated with a corticosteroid paste bandage,[12] an over-the-counter hydrocortisone cream,[13] and several cases of reactions to Unguentum Merck®, an emollient ointment used as a base for some corticosteroids.[14-16] These reactions have sometimes been generalized or exfoliative.[12,15,16]

C. Ocular Reactions

Sorbic acid-preserved soft contact lens solutions were reported to cause adverse ocular reactions in 15% of 135 patients. Symptoms resolved following substitution with nonpreserved saline.[17]

VII. CLINICAL RELEVANCE

Sorbic acid commonly causes immediate transient erythema, most frequently on the face and back. In concentrations found in topical cosmetics and pharmaceuticals, the face may be the only site involved. Facial reactors comprise about 5% of cases and include patients with "status cosmeticus" and patients who report flushing from niacin, wines, liquors, beta blockers, nitrates, and sulfites. Foods containing sorbic acid may rarely induce facial erythema.[18] This reaction may occur more frequently and with greater severity in patients with rosacea.[9]

Sorbic acid is a rare sensitizer. Patients with documented sensitivity to sorbic acid will also react to potassium sorbate. Foods containing these preservatives do not appear to be problematic, as sensitized patients have been challenged with sorbic acid-containing foods for up to 3 weeks with no exacerbation of dermatitis.[11]

REFERENCES

1. **Fisher, A. A.,** Cosmetic dermatitis. Part II. Reactions to some commonly used preservatives, *Cutis,* 26, 136, 1980.
2. **Reynolds, J. E. F.,** *Martindale: The Extra Pharmacopoeia,* (CD-ROM Version), Micromedex, Inc., Denver, CO, 1991.
3. **Lecos, C.,** Food preservatives: a fresh report, *FDA Consumer,* April 1984.
4. **Decker, R. L. and Wenninger, J. A.,** Frequency of preservative use in cosmetic formulas as disclosed to FDA-1982 update, *Cosmet. Toilet.,* 97, 57, 1982.
5. **World Health Organization,** Seventeenth Report of the Joint FAO/WHO Expert Committee on Food Additives, Tech. Rep. Ser. No. 539, WHO, Geneva, 1974.
6. **Rietschel, R. L.,** Contact urticaria from synthetic cassia oil and sorbic acid limited to the face, *Contact Dermatitis,* 4, 347, 1978.
7. **Clemmensen, O. and Hjorth, N.,** Perioral contact urticaria from sorbic acid and benzoic acid in a salad dressing, *Contact Dermatitis,* 8, 1, 1982.
8. **Clemmensen, O. J. and Schiodt, M.,** Patch test reaction of the buccal mucosa to sorbic acid, *Contact Dermatitis,* 8, 341, 1982a.
9. **Soschin, D. and Leyden, J. J.,** Sorbic acid-induced erythema and edema, *J. Am. Acad. Dermatol.,* 14, 234, 1986.

10. **Adams, R. M. and Maibach, H. I.,** A five-year-study of cosmetic reactions, *J. Am. Acad. Dermatol.,* 13, 1062, 1985.
11. **Fisher, A. A.,** Cutaneous reactions to sorbic acid and potassium sorbate, *Cutis,* 25, 350, 1980a.
12. **Simpson, J. B.,** Sorbic acid sensitivity from Cortacream bandages, *Contact Dermatitis Newsletter,* 10, 232, 1971.
13. **Fisher, A. A.,** Allergic reactions to the preservatives in over-the-counter hydrocortisone topical creams and lotions, *Cutis,* 32, 222, 1983.
14. **Saihan, E. M. and Harman, R. M.,** Contact sensitivity to sorbic acid in "Unguentum Merck", *Br. J. Dermatol.,* 99, 583, 1978.
15. **Brown, R.,** Another case of sorbic acid sensitivity, *Contact Dermatitis,* 5, 268, 1979.
16. **Coyle, H. E., Miller, E., and Chapman, R. S.,** Sorbic acid sensitivity from Unguentum Merck®, *Contact Dermatitis,* 7, 56, 1981.
17. **Josephson, J. E. and Caffery, B.,** Sorbic acid revisited, *J. Am. Optom. Assoc.,* 57, 188, 1986.
18. **Fisher, A. A.,** Erythema limited to the face due to sorbic acid, *Cutis,* 40, 395, 1987.

SORBITAN TRIOLEATE

I. REGULATORY CLASSIFICATION

Sorbitan trioleate is used as an emulsifying and/or solubilizing agent in pharmaceuticals.

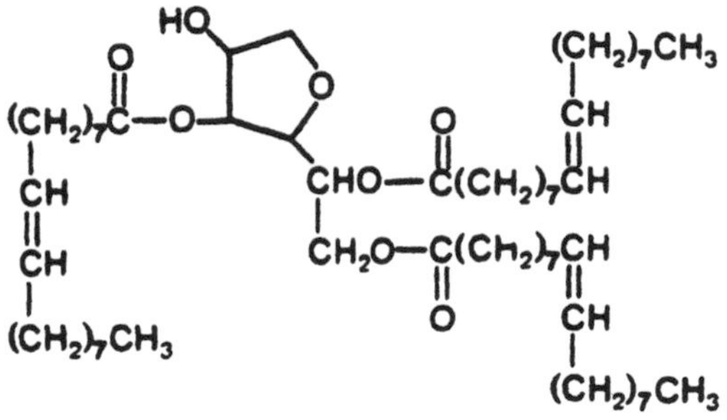

FIGURE 62. Sorbitan trioleate.

II. AVAILABLE FORMULATIONS

A. Cosmetics

Sorbitan trioleate is used as an emulsifying agent in moisturizing creams and lotions, personal cleanliness products, and eye shadows.[1]

III. TABLE OF COMMON PRODUCTS

A. Inhalation Drug Products

Trade name	Manufacturer
AeroBid	Forest
Alupent metered dose aerosol	Boehringer Ingelheim
Asthmahaler	Norcliff Thayer
Brethaire	Geigy
Bronitin Mist	Whitehall

Trade name (cont'd)	Manufacturer (cont'd)
Bronkaid Mist suspension	Winthrop
Intal inhaler	Fisons
Maxair	Riker
Metaprel metered dose aerosol	Sandoz
Primatene Mist suspension	Whitehall

IV. HUMAN TOXICITY DATA

A. Inhalation Drug Products

Significant maximal airway resistance changes and bronchoconstriction were observed in 5 of 13 subjects after inhalation of a placebo aerosol containing sorbitan trioleate and chlorofluorocarbons.[2]

An anaphylactoid or anaphylactic reaction consisting of bright facial erythema, diaphoresis, lip and periorbital angioedema, urticaria, sinus congestion, and profuse rhinorrhea was reported in a 31-year-old asthmatic, atopic patient within minutes following the second use of Alupent inhaler. Intradermal testing produced a positive reaction to sorbitan trioleate, an ingredient in the inhaler.[3]

V. CLINICAL RELEVANCE

Although an isolated immediate-type hypersensitivity reaction has been described to a sorbitan trioleate-containing inhaler, attempts to replace this excipient with soya lecithin resulted in a dramatic increase in adverse reactions, resulting in withdrawal of the reformulated product.[4] Bronchoconstriction, whether secondary to sorbitan trioleate or to the propellant mixture, is usually masked by administration of the active agent.

REFERENCES

1. **Nikitakis, J. M.,** *CTFA Cosmetic Ingredient Handbook.* 1st ed., The Cosmetic, Toiletry and Fragrance Association, Washington, D.C., 1988.
2. **Brooks, S. M., Mintz, S., and Weiss, E.,** Changes occurring after Freon inhalation, *Am. Rev. Respir. Dis.,* 105, 640, 1972.
3. **Malish, D. M.,** Possible allergic reactions to inert ingredient in Alupent metered dose inhaler, sorbitan trioleate, *Immunol. Allergy Pract.,* 7, 467, 1985.
4. **Wilms, H. G.,** Market removal of prescription asthma metered doseinhaler ... Alupent ... A brand of metaproterenol, Food and Drug Administration FAX to Poison Control Centers, October 27, 1989.

SORBITOL

I. REGULATORY CLASSIFICATION

Sorbitol is classified as a humectant, sweetening agent, and component of flavored and/ or sweetened vehicles in pharmaceuticals.

$$
\begin{array}{c}
CH_2OH \\
| \\
H—C—OH \\
| \\
HO—C—H \\
| \\
H—C—OH \\
| \\
H—C—OH \\
| \\
CH_2OH
\end{array}
$$

FIGURE 63. Sorbitol.

II. AVAILABLE FORMULATIONS

A. Foods

Dietetic foods are a common source of sorbitol, particularly for the diabetic patient. The amount of sorbitol in various natural and dietetic foods is summarized in the following table:[1-4]

Food	Amount of sorbitol
Natural foods	
Pears	4.6 g per 100 g dry weight
Pear juice	2.1 g per 100 g
Prunes	2.4 g per 100 g dry weight
Prune juice	12.7 g per 100 g
Peaches	1 g per 100 g dry weight
Peach juice	0.9 g per 100 g
Apple juice	0.3–0.9 g per 100 g of juice

Food (cont'd)	Amount of sorbitol (cont'd)
Sweet cherries	12.6 g per 100 g dry weight
Sweet cherry juice	1.4 g per 100 g
Plums	15.8 g per 100 g
Dietetic foods	
Orange marmalade	58 g per 100 g
Strawberry jam	60 g per 100 g
Chocolate	33 g per 100 g
Gum or candy	1.3 –2.2 g per piece
Bars	5–7 g per piece
Wafers	2.8–4.4 g per piece

B. Drugs

Sorbitol is a naturally occurring polyhydric alcohol containing 4 cal/g. It is used as a humectant and vehicle for oral and topical products and as a sweetener in many sugar-free pharmaceuticals. Labeling of pharmaceuticals may indicate sorbitol or sorbitol solution USP, a 70% w/w aqueous preparation. The FDA lists the available concentrations of sorbitol in NDA-approved products as 70 to 72% in suspensions, 6 to 35% in solutions, 5 to 25% in syrups, and 5 to 20% in elixirs.[5]

C. Cosmetics

Sorbitol is used as a humectant in toothpastes and as a humectant and skin conditioner in cosmetics. Types of products include colognes, sachets, shaving creams, lotions, moisturizing creams and lotions, mud packs, and skin fresheners.[6]

III. TABLE OF COMMON DRUG PRODUCTS

A. Oral Liquid Drug Products

Trade name	Manufacturer
Actifed with codeine	Burroughs Wellcome
Aludrox oral suspension	Winthrop
Alupent syrup	Boehringer Ingelheim
Alurate elixir	Roche
Ambenyl-D	Forest
Amicar syrup	Lederle
Amphojel	Wyeth-Ayerst
Artane elixir	Lederle
Asbron G elixir	Sandoz
Bactrim suspension	Roche
Bactrim pediatric suspension	Roche
Basaljel	Wyeth-Ayerst
Bayer Children's cough syrup	Glenbrook
Children's Advil suspension	Whitehall

Trade name (cont'd)	Manufacturer (cont'd)
Children's Panadol liquid	Glenbrook
Children's Tylenol liquid cold formula	McNeil
Children's Tylenol elixir	McNeil
Choledyl elixir	Parke-Davis
Cibalith-S syrup	Ciba
Coly-Mycin S oral suspension	Parke-Davis
Deconamine syrup	Berlex
Depakene syrup (contains 0.75g/5 ml)	Abbott
Dimetane DC cough syrup	Robins
Dimetane DX cough syrup	Robins
Ditropan syrup	Marion
Dorcol children's decongestant liquid	Sandoz
Dorcol children's liquid cold formula	Sandoz
Elixophyllin GG	Forest
Entex liquid	Norwich Eaton
Extra-Strength Tylenol adult liquid	McNeil
Gantanol suspension	Roche
Gaviscon liquid	Marion
Gaviscon Extra Strength liquid	Marion
Hycodan syrup	DuPont
Hycomine syrup	DuPont
Hycomine pediatric syrup	DuPont
Hycotuss expectorant	DuPont
Iberet liquid	Abbott
Iberet-500 liquid	Abbott
Indocin oral suspension	Merck Sharp & Dohme
Laniazid syrup	Lannett
Lasix oral solution	Hoechst-Roussel
Levsin elixir	Schwarz Pharma
Levsin oral drops	Schwarz Pharma
Lithium citrate syrup	Roxane
Lortab liquid	Rugby
Mellaril concentrate	Sandoz
Metaprel syrup	Sandoz
Methadone hydrochloride oral solution	Roxane
Minocin oral suspension	Lederle
Mintezol suspension	Merck Sharp & Dohme
Mylanta liquid	Stuart
Mylanta II liquid	Stuart
Naldecon syrup	Bristol
Naldecon pediatric syrup	Bristol
Naldecon pediatric drops	Bristol
Naldecon DX adult liquid	Bristol
Naldecon DX pediatric drops	Bristol

Trade name (cont'd)	Manufacturer (cont'd)
Naldecon EX children's syrup	Bristol
Naldecon Senior DX cough/cold liquid	Bristol
Naldecon Senior EX cough/cold liquid	Bristol
Naprosyn suspension	Syntex
Navane concentrate	Roerig
NegGram suspension	Winthrop
Neo-Calglucon syrup	Sandoz
Novahistine DH	Lakeside
Novahistine expectorant	Lakeside
Nucofed syrup	Beecham
Pamelor oral solution	Sandoz
Pediacare cold formula	McNeil
Pediacare cough-cold formula	McNeil
Pediacare night rest cough-cold formula	McNeil
Pediacare infant's oral decongestant drops	McNeil
Pediaflor drops	Ross
Peri-Colace syrup	Mead Johnson
Polaramine syrup	Schering
Prelone syrup 5 mg/5 ml	Muro
Reglan syrup	Robins
Riopan Plus 2 suspension	Whitehall
Robitussin AC	Robins
Robitussin DAC	Robins
Rondec oral drops	Ross
Rondec syrup	Ross
Rondec DM syrup	Ross
Rondec DM oral drops	Ross
Roxicodone oral solution	Roxane
Septra suspension	Burroughs Wellcome
Septra grape suspension	Burroughs Wellcome
Slo-Phyllin syrup	Rorer
Slo-Phyllin GG syrup	Rorer
Sodium polystyrene sulfonate suspension	Roxane
Sumycin syrup	Squibb
Tacaryl syrup	Westwood
Tagamet elixir	Smith Kline & French
Taractan concentrate	Roche
Tavist syrup	Sandoz
Tegretol suspension	Geigy
Theoclear-80 syrup	Central
Theolair liquid	3M Riker
Triaminic expectorant	Sandoz
Triaminic DH expectorant	Sandoz
Triaminic Nite Light	Sandoz

Trade name (cont'd)	Manufacturer (cont'd)
Triaminic oral infant drops	Sandoz
Triaminic syrup	Sandoz
Triaminic DM syrup	Sandoz
Triaminicol Multi-Symptom Relief	Sandoz
Trilafon concentrate	Schering
Tussar SF	Rorer
Tussi-Organidin liquid	Wallace
Tussi-Organidin DM liquid	Wallace
Tuss-Ornade liquid	Smith Kline & French
Tylenol cold medication liquid	McNeil
Vicks Pediatric Formula 44 cough	Richardson-Vicks
Vicks Pediatric Formula 44 cough/congestion	Richardson-Vicks
Vicks Pediatric Formula 44 cough/cold	Richardson-Vicks
Wingel liquid	Winthrop

IV. HUMAN TOXICITY DATA

A. Tube-Feedings

Sorbitol has become an increasingly reported cause of adverse gastrointestinal reactions in tube-fed patients. Both extemporaneously prepared and commercially available oral medications have been implicated.

Diarrhea was reported in a 5-year-old child receiving 9 g/d of sorbitol (equivalent to 0.36 g/kg/d) via jejunostomy tube as an excipient in a valproic acid syrup.[7] Intractable diarrhea was noted in a 16-year-old boy who received 60 g of sorbitol daily in a hydralazine preparation given via nasogastric tube.[8,9] Profuse diarrhea occurred in a 51-year-old tube-fed woman following administration of a baclofen liquid containing 1.9 g/5 ml of sorbitol. The total daily dose was 21.2 g.[10]

Osmotic diarrhea was associated with administration of a sugar-free theophylline elixir containing 30 g of sorbitol per 240 mg of theophylline in a series of 12 patients, 4 receiving tube feedings. In the index case, a 65-year-old man developed diarrhea exceeding 1.000 ml/24 h 5 d after institution of oral theophylline and tube feedings. The stool osmotic gap, calculated by the formula "stool osmotic gap = measured stool osmolality (by freezing point depression) – (2 × (stool sodium level + stool potassium level))", was 206 mmol/l. An osmotic gap of greater than 100 mmol/l is diagnostic of osmotic diarrhea.[11]

B. Normal Populations

Sorbitol intolerance may be more prevalent than commonly realized and masquerade as irritable bowel syndrome and other gastrointestinal disorders. Jain et al. demonstrated clinical sorbitol intolerance, defined as bloating, abdominal cramps, or diarrhea, in 48% of healthy volunteers given 10 g of sorbitol in solution.[12] In a subsequent study of 124 healthy adults, clinical intolerance was observed in 30 to 36% with no difference between ethnic groups.[13]

Diarrhea and abdominal cramps were reported in 8 of 12 children who ingested 3 to 16 pieces of dietetic candy containing 3 g per piece of sorbitol.[14]

The effect of orally administered sorbitol depends on the dosage form, with liquid formulations being much more potent than crystalline sorbitol. In a study of 12 healthy adult subjects, the laxative threshold dose was 50 g for the crystalline form and 20 g for the syrup.[15] In a study of 86 healthy adults, 5 subjects experienced severe diarrhea after doses of 12.5 g twice a day, given as a 50% aqueous solution.[16]

C. Foods

Malabsorption of sorbitol found in certain fruit juices, such as pear and apple juice, is associated with chronic nonspecific diarrhea in children. Restriction of these juices in children with this diagnosis had a beneficial effect in three of seven children studied.[17]

D. Fructose Intolerance

Sorbitol is metabolized in the liver to fructose and is hazardous in patients with fructose intolerance or fructose 1,6-diphosphatase deficiency. Fructose intolerance occurs in about 1 in 40,000 people. Injection of a sorbitol-containing actinomycin product resulted in severe liver and renal damage, culminating in death, in a woman with fructose intolerance.[18]

E. Peritoneal Dialysis Solutions

The addition of sorbitol 2 to 5% to peritoneal dialysis solutions, formerly used in place of dextrose, produced severe coma in four of ten uremic patients, presumably due to a direct toxic effect on the CNS. Delayed clearance of sorbitol was demonstrated in these cases compared to normal controls.[19]

F. Rectal Administration

Sorbitol is used as a vehicle in various enema preparations. The presence of sorbitol in a prednisone enema was associated with worsening clinical course and increased stool prostaglandin secretion in a patient with ulcerative colitis.[20]

Fatal colonic necrosis was associated temporally with administration of sodium polystyrene resin in sorbitol enemas in four uremic patients. A fifth patient with colon infarction survived after hemicolectomy. A subsequent experiment in both normal and uremic rats demonstrated transmural necrosis after sorbitol-containing enemas and not after normal saline or sodium polystyrene in saline controls.[21]

G. Activated Charcoal Products

There are four pediatric and three adult case reports of diarrhea and fluid or electrolyte inbalance associated with the use of sorbitol as a cathartic in commercially available activated charcoal-sorbitol combination products.

A 3-month-old infant was given 110 g of sorbitol (13.75 mg/kg) as a vehicle in an activated charcoal suspension for treatment of a theophylline overdose. He developed sudden onset of hypotension, dehydration, and increased serum osmolarity, which was attributed to the sorbitol.[22] Two children treated for chlorpromazine overdose with continuous nasogastric infusions of activated charcoal in 70% sorbitol developed dehydration, hypernatremia, and neurologic deterioration after receiving 100 g or greater of sorbitol.[23] A 2-year-old child developed dehydration and hypernatremia secondary to sorbitol administration in a multiple-dose activated charcoal regimen for treatment of imipramine overdose.[24]

Hypernatremia occurred in three adults given three, four, and six doses of 192 g of sorbitol in an activated charcoal preparation.[25]

V. CLINICAL RELEVANCE

A. Oral Drug Products

The minimum dose of sorbitol that is associated with diarrhea is 0.5 g/kg in children[26] and 10 g in adults. In a study of seven healthy adult subjects, a dose of 10 g resulted in mild effects such as bloating and flatulence, while cramps and diarrhea occurred after a dose of 20 g.[4] Similar effects were seen in a study of 42 healthy adults, with severe intolerance occurring in 32% of nonwhite subjects and 4% of white subjects.[12]

It is advisable that the total daily dose of sorbitol in children and tube-fed adults not exceed these amounts. As one study demonstrated, as much as 40% of diarrhea seen in tube-fed patients may be attributable to sorbitol in liquid medications.[10] However, it is practically impossible to estimate the amount of sorbitol in commercially available drug products, since the concentration is rarely listed by the manufacturer. As a conservative maximum estimate, liquid products listing sorbitol as an ingredient can be assumed to contain 70% w/w, equivalent to 4.5 g of sorbitol per teaspoonful. In general, pharmaceutical suspensions contain the most sorbitol, and elixirs the least.[5]

B. Diabetics

Although diabetics may not have a higher incidence of sorbitol intolerance than the general population, they are more likely to consume dietetic foods in amounts associated with gastrointestinal problems.

C. Functional Bowel Disease

Patients with functional bowel disorders may be more susceptible to the adverse gastrointestinal effects of sorbitol. In a study of 13 patients given 5 g of sorbitol in an aqueous solution, 7 reported mild to moderate gastrointestinal distress without diarrhea.[27]

D. Activated Charcoal/Sorbitol Mixtures

The recommended therapeutic dose of sorbitol, when used as a cathartic in combination with activated charcoal, is 0.5 to 1 g/kg for children and 1 to 2 g/kg for adults. However, there are no data available regarding safety or efficacy of lower doses in this situation. Since doses of sorbitol as low as 0.5 g/kg in children and 10 g in adults have been associated with diarrhea, the efficacy of lower doses in combination with activated charcoal should be explored. Activated charcoal is given at a dose of 15 to 30 g for children and 30 to 100 g for adults. Most commercial premixed combination products provide an excessive dose of sorbitol for small children. The amount of sorbitol delivered in commercial preparations is listed below:

Trade name	Charcoal/sorbitol (grams)	Sorbitol/15 g charcoal		
		10 kg child	20 kg child	30 kg child
Acta-Char	62/50	1.86 g/kg	0.93 g/kg	0.62 g/kg
Actidose	96/50	2.88 g/kg	1.44 g/kg	0.96 g/kg
Charcoaid	110/30	5.5 g/kg	2.75 g/kg	1.83 g/kg
Charcolex	40/18	0.675 g/kg	0.338 g/kg	0.225 g/kg

Therefore, it is recommended that children weighing less than 30 kg be given aqueous activated charcoal, along with a separate appropriate dose of sorbitol or other cathartic. Adults given multiple-dose charcoal regimens to enhance elimination of some poisons will also receive excessive doses of sorbitol with premixed products. The initial dose may be given as a premixed product, with subsequent doses as aqueous activated charcoal.

E. Foods

The American Academy of Pediatrics Committee on Nutrition has issued a policy statement which suggests taking a careful dietary history of sorbitol-containing fruit juice consumption in evaluating children with a history of chronic diarrhea, abdominal pain, or bloating. If this history is positive for significant consumption of these juices, a trial of curtailed intake of these products should be done to determine the relationship to symptoms. Juices that do not contain significant amounts of sorbitol include grape, strawberry, raspberry, blackberry, pineapple, and orange.[28]

REFERENCES

1. **Washuttl, J., Riederer, P., and Bancher, E.,** A qualitative and quantitative study of sugar-alcohols in several foods, *J. Food Sci.*, 38, 1262, 1973.
2. **Richmond, M. L., Brandao, S. C. C., Gray, J. I., et al.,** Analysis of simple sugars and sorbitol in fruit by high-performance liquid chromatography, *J. Agric. Food Chem.*, 29, 4, 1981.
3. **Bask, L. S., Griffin, D., and Prachuabpaibul, P.,** The determination of sorbitol in foodstuffs using thermometric titrimetry, *Analyst*, 101, 306, 1976.
4. **Hyams, J. S.,** Sorbitol intolerance: an unappreciated cause of functional gastrointestinal complaints, *Gastroenterology*, 84, 30, 1983.
5. **Food and Drug Administration,** *Inactive Ingredient Guide*, FDA, Washington, D.C., March 1990.
6. **Nikitakis, J. M.,** *CTFA Cosmetic Ingredient Handbook*, 1st ed., The Cosmetic, Toiletry and Fragrance Association, Washington, D.C., 1988.
7. **Veerman, M. W.,** Excipients in valproic acid syrup may cause diarrhea: a case report. *DICP*, 24, 832, 1990.
8. Subcommittee Report on Pediatric Extemporaneous Formulations of the American Society of Hospital Pharmacist Special Interest Group (SIG) on Pediatric Pharmacy Practice, *American Society of Hospital Pharmacist*, Washington, D.C., 1979.
9. **Charney, E. B. and Bodurtha, J. N.,** Brief clinical and laboratory observations: intractable diarrhea associated with the use of sorbitol, *J. Pediatr.*, 98, 157, 1981.
10. **Brown, A. M. and Masson, E.,** "Hidden" sorbitol in proprietary medicines-a cause for concern?, *Pharmaceutical J.*, 245, 211, 1990.
11. **Edes, T. E. and Walk, B. E.,** Nosocomial diarrhea: beware the medicinal elixir, *South. Med. J.*, 82, 1497, 1989.
12. **Jain, N. K., Rosenberg, D. B., Ulahannan, M. J., Glasser, M. J., and Pitchumoni, C. S.,** Sorbitol intolerance in adults, *Am. J. Gastroenterol.*, 80, 678, 1985.
13. **Jain, N. K., Vijaykumar, P. P., and Pitchumoni, C. S.,** Sorbitol intolerance in adults, *J. Clin. Gastroenterol.*, 9, 317, 1987.
14. **Lipin, R.,** Outbreak of diarrhea linked to dietetic candies — New Hampshire. *JAMA*, 252, 1672, 1984.
15. **Ellis, F. W. and Krantz, J. C.,** Sugar alcohols. XXII. Metabolism and toxicity studies with mannitol and sorbitol in man and animals, *J. Biol. Chem.*, 141, 147, 1941.
16. **Peters, R. and Lock, R. H.,** Laxative effect of sorbitol, *Br. Med. J.*, 2, 677, 1958.
17. **Hyams, J. S., Etienne, N. L., Leightner, A. M., and Theuer, R. C.,** Carbohydrate malabsorption following fruit juice ingestion in young children, *Pediatrics*, 82, 64, 1988.

18. **Schulte, M. J. and Lenz, W.,** Fatal sorbitol infusion in patients with fructose-sorbitol intolerance, *Lancet,* 2, 188, 1977.

19. **Winter, S., Frankel, H., Ribot, S., and Kirschner, M. A.,** Sorbitol-induced coma in uremic patients undergoing peritoneal dialysis, *J. Newark Beth Israel Med. Ctr.,* 23, 175, 1973.

20. **Zijlstra, F. J.,** Sorbitol, prostaglandins, and ulcerative colitis enemas, *Lancet,* 2, 815, 1981.

21. **Lillemoe, K. D., Romolo, J. L., Hamilton, S. R., Pennington, L. R., Burdick, J. F., and Williams, G. M.,** Intestinal necrosis due to sodium polystyrene (Kayexalate) in sorbitol enemas: clinical and experimental support for the hypothesis, *Surgery,* 101, 267, 1987.

22. **Farley, T. A.,** Severe hypernatremic dehydration after use of an activated charcoal-sorbitol suspension, *J. Pediatr.,* 109, 719, 1986.

23. **Klein, S. K., Levinsohn, M. W., and Blumer, J. L.,** Accidental chlorpromazine ingestion as a cause of neuroleptic malignant syndrome in children, *J. Pediatr.,* 107, 970, 1985.

24. **McCord, M. M.,** Toxicity of sorbitol-charcoal suspension, *J. Pediatr.,* 111, 307, 1987.

25. **Caldwell, J. W., Nava, A. J., and DeHaas, D. D.,** Hypernatremia associated with cathartics in overdose management, *West. J. Med.,* 147, 593, 1987.

26. **Hill, R. E. and Ramananda Kamath, K.,** "Pink" diarrhoea: osmotic diarrhoea from a sorbitol-containing vitamin C supplement, *Med. J. Aust.,* 1, 387, 1982.

27. **Rumessen, J. J. and Gudmand-Hoyer, E.,** Functional bowel disease: malabsorption and abdominal distress after ingestion of fructose, sorbitol, and fructose-sorbitol mixtures, *Gastroenterology,* 95, 694, 1988.

28. **American Academy of Pediatrics,** Policy Statement. The use of fruit juice in the diets of young children, *AAP News,* 7, 11, 1991.

SOYA LECITHIN

I. REGULATORY CLASSIFICATION

Soya lecithin is classified as an emulsifying and/or solubilizing agent.

$$CH_2OCOR$$
$$CHOCOR$$
$$CH_2O-P-OCH_2CH_2\overset{+}{N}(CH_3)_3$$

FIGURE 64. Lecithin.

II. AVAILABLE FORMULATIONS

A. Foods

Lecithin is a common, naturally occurring food component. It is well-tolerated orally and when administered in intravenous parenteral nutrition formulations. Soybean protein was found in six of seven samples of soya lecithin analyzed, in amounts of 1.03 to 27.2 mg/g.[1]

B. Cosmetics

Lecithin is commonly used as an emulsifying agent and skin conditioner. Types of products include shampoos, makeup foundations, blushes, lipsticks, and moisturizing creams or lotions.[2]

III. TABLE OF COMMON PRODUCTS

A. Inhalation Aerosols

Trade name	Manufacturer
Atrovent inhalation aerosol	Boehringer Ingelheim

IV. HUMAN TOXICITY DATA

Paradoxical bronchospasm was reported in 23 of 1450 (1.6%) asthmatics treated with a metered-dose albuterol inhaler containing soya lecithin. The reactions were immediate in onset and usually lasted less than 3 min.[3] A follow-up study was done to investigate the role of the excipients in this reaction. A metered-dose metaproterenol product containing soya lecithin was used in 900 well-controlled asthmatics and compared to a placebo metered-dose product containing only inert ingredients (oleic acid) and propellants in 175 asthmatics. Bronchoconstriction was observed in 52 subjects, with an incidence of 4.4% in the drug group and 6.9% in the placebo group. Patients with bronchoconstriction from the placebo system did not improve after receiving the active metered-dose inhaler, but did improve after nebulized metaproterenol, implicating one or more of the excipients as causing the adverse response. The investigators did not determine which of the excipients was responsible.[4]

In 1989, the metaproterenol metered-dose inhaler, Alupent/Metaprel, was reformulated to contain soya lecithin as the suspending agent in place of sorbitan trioleate. Within 1 month, escalating reports of adverse reactions resulted in withdrawal of the new formulation. The reactions virtually all occurred with the first puff of the new product and consisted of coughing, gagging sensation, or exacerbation of asthma symptoms.[5]

Fine reported a case of a 40-year-old woman with chronic asthma and extreme soybean hypersensitivity who developed increased wheezing, laryngospasm, and a tingling sensation within several minutes of using an ipratropium bromide inhaler (Atrovent®), which contains soya lecithin in concentrations four times less than the former Alupent formulation.[6]

V. CLINICAL RELEVANCE

A. Bronchodilators

Although soya lecithin has been replaced by other excipients, such as sorbitan trioleate or oleic acid, in bronchodilator metered-dose inhalers, it is probable that paradoxical bronchoconstriction will still continue to be observed in asthmatics using these products, with an incidence of 1.5 to 4%, from one or more of these other excipients. In addition to soya lecithin, other inhaled excipients, such as chlorofluorocarbons, need to be investigated for bronchoconstrictive properties before this problem can be resolved. Soya lecithin remains in one inhalation aerosol product, which may exacerbate symptoms in allergic asthmatics.

REFERENCES

1. **Porras, O., Carlsson, B., Fallstrom, S. P., and Hanson, L. A.,** Detection of soy protein in soy lecithin, margarine and occasionally soy oil, *Int. Arch. Allergy Appl. Immunol.*, 78, 30, 1985.
2. **Nikitakis, J. M.,** *CTFA Cosmetic Ingredient Handbook*, 1st ed., The Cosmetic, Toiletry and Fragrance Association, Washington, D.C., 1988.
3. **Yarbrough, J., Mansfield, L. E., and Ting, S.,** Immediate Bronchoconstrictive Response to Metered Dose Albuterol (MD-A), Abstract presented at the 39th Annual Congress of the American College of Allergy, New Orleans, January 1983.
4. **Yarborough, L., Mansfield, L., and Ting, S.,** Metered dose inhaler induced bronchospasm in asthmatic patients, *Ann. Allergy*, 55, 25, 1985.
5. **Wilms, H. G.,** Market Removal of Prescription Asthma Metered Dose Inhaler ... Alupent ... A Brand of Metaproterenol, Food and Drug Administration FAX to Poison Control Centers, October 27, 1989.
6. **Fine, S. R.,** Possible reactions to soya lecithin in aerosols, *J. Allergy Clin. Immunol.*, 87, 600, 1991.

SOYBEAN OIL

I. REGULATORY CLASSIFICATION

Soybean oil is classified as an oleaginous vehicle.

II. SYNONYMS

Oleum sojae
Soya oil
Soja bean oil
Soya bean oil

III. AVAILABLE FORMULATIONS

A. Constituents

Soybean oil is the refined fixed oil obtained from the seeds of *Glycine soja*. Constituents are primarily triglycerides of oleic, linoleic, linolenic, and saturated acids. Soy protein residues are not consistently present in refined soybean oil. Less purified derivatives, such as some cold-pressed soybean oils, may contain soybean protein.[1] Analysis of eight commercial Swedish soy oil samples revealed soy protein in three samples in amounts ranging from 0.11 to 3.3 mg/g.[2]

B. Foods

Soybean oil is widely used in foods. It is utilized, along with soy protein, in infant formulas for those infants intolerant of cow's milk protein.

C. Cosmetics

This oil is used for its emollient properties in bath oils, shampoos and hair conditioners, cleansing products, creams and lotions, and suntan products.[3]

IV. TABLE OF COMMON PRODUCTS

A. Parenteral Drug Products

Trade name	Manufacturer
Diprivan injection	Stuart
Intralipid 10, 20%	Kabivitrum

V. HUMAN TOXICITY DATA

A. Foods

Patients with immediate hypersensitivity reactions to soy protein are common, particularly in children. Seven adults with soy protein hypersensitivity documented by a history of anaphylaxis and positive RAST assay and skin tests were able to tolerate up to 15 ml of four different soybean oils sold commercially, given by oral double-blind challenge. Skin tests were negative to the oils.[1]

Some soybean oils contain measurable amounts of allergenic soy protein fractions.[2] A positive skin test to soy oil was observed in an asthmatic patient with hypersensitivity to soy protein.[4]

B. Drugs

Intravenous fat emulsions used in parenteral nutrition contain 10% soybean oil, 1.2% egg yolk phospholipids, 2.25% glycerol, and water for injection. Several cases of immediate hypersensitivity reactions have been reported in adults and children receiving this therapy.

An atopic child developed wheezing, erythema, wheals, and urticaria after 11 d of parenteral soybean oil emulsion therapy; this was the only hypersensitivity reaction reported in 159 children treated. No mention of skin testing or RAST assays to soybean oil was made.[5]

A 9-year-old boy developed generalized pruritic urticaria 1 h after administration of a dose of parenteral nutrition with Intralipid 10% after resumption of therapy. He had tolerated the infusion for 19 d before a brief interruption during surgery. The symptoms abated following discontinuation and returned within 10 min after restarting the infusion 10 h later. Skin testing to Intralipid was negative, but the reaction was reproducible after a 1 ml test intravenous dose.[6]

A 41-year-old male with alcoholic pancreatitis and a history of legume allergy was given parenteral nutrition with 10% Intralipid. Dyspnea and diffuse urticaria began 5 h after initiation of the infusion and resolved rapidly after discontinuation and treatment with diphenhydramine. A RAST test was positive to soybean protein, and negative for the other constituents of Intralipid.[7]

VI. CLINICAL RELEVANCE

Soybean oil is generally well tolerated. In patients with hypersensitivity to soybean protein, small residues in some oil products may rarely produce allergic reactions. It would seem prudent for soybean-sensitive patients to undergo skin testing and oral challenge with the particular oil they intend to use in cooking before adopting this food into the diet.

A history of soybean sensitivity should be ruled out prior to initiation of intravenous fat emulsion-containing nutrition.

REFERENCES

1. **Bush, R. K., Taylor, S. S., Nordlee, J. A., and Busse, W. W.,** Soybean oil is not allergenic to soybean-sensitive individuals, *J. Allergy Clin. Immunol.,* 76, 242, 1985.
2. **Porras, O., Carlsson, B., Fallstrom, S. P., and Hanson, L. A.,** Detection of soy protein in soy lecithin, margarine and occasionally soy oil, *Int. Arch. Allergy Appl. Immunol.,* 78, 30, 1985.
3. **Nikitakis, J. M.,** *CTFA Cosmetic Ingredient Handbook,* 1st ed., The Cosmetic, Toiletry and Fragrance Association, Washington, D.C., 1988.
4. **Duke, W. W.,** Soy bean as a possible important source of allergy, *J. Allergy,* 5, 300, 1934.
5. **Hansen, L. M., Hardie, W. R., and Hidalgo, J.,** Fat emulsion for intravenous administration: clinical experience with Intralipid® 10%, *Ann. Surg.,* 184, 80, 1976.
6. **Kamath, K. R., Berry, A., and Cummins, G.,** Acute hypersensitivity reaction to Intralipid, *N. Engl. J. Med.,* 304, 360, 1981.
7. **Hiyama, D. T., Griggs, B., Mittman, R. J., Lacy, J. A., Benson, D. W., and Bower, R. H.,** Hypersensitivity following lipid emulsion infusion in an adult patient, *JPEN,* 13, 318, 1989.

SUCROSE

I. REGULATORY CLASSIFICATION

Sucrose is classified as a coating agent and tablet and/or capsule diluent. Compressible and confectioner's sugar are classified as sweetening agents and tablet and/or capsule diluents.

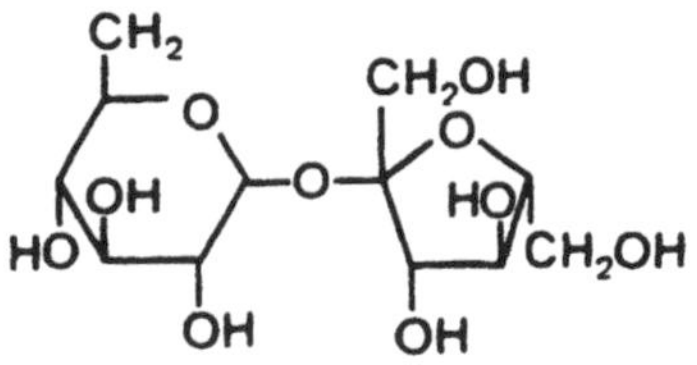

FIGURE 65. Sucrose.

II. SYNONYMS

Azucar
Cane sugar
Refined sugar
Saccharose
Sucre

III. AVAILABLE FORMULATIONS

A. Drugs

Compressible sugar NF is used in tabletting and contains 95 to 98% sucrose; additives may include starch, maltodextrin, invert sugar, and lubricants. Confectioner's sugar contains not less than 95% sucrose in a fine powder with corn starch. Syrup NF contains 85% sucrose. The FDA lists 73 approved pharmaceutical syrup formulations containing 24 to 99% sucrose, 52 oral suspensions containing 6 to 75% sucrose, 30 elixirs containing 12.5 to 60% sucrose, and 47 powders for reconstitution containing up to 80% sucrose. Other pharmaceutical products include buccal tablets, sublingual tablets, chewable tablets, capsules, oral concentrates, oral drops, and oral granules.[1] Pediatric medications commonly contain 30 to 70% sucrose.[2]

IV. TABLE OF COMMON PRODUCTS

A. Oral Liquid Pediatric Medications

Trade name	Manufacturer
Achromycin V oral suspension	Lederle
Actifed syrup	Burroughs Wellcome
Ambenyl cough syrup	Forest
Amoxil oral suspension	Beecham
Anacin-3 Children's liquid	Whitehall
Anspor oral suspension	Smith Kline & French
Asbron G elixir	Sandoz
Atarax syrup	Roerig
Azulfidine oral suspension	Pharmacia
Bactrim suspension	Roche
Calcidrine syrup	Abbott
Carnitor oral solution	Sigma-Tau
Ceclor suspension	Lilly
Chloromycetin palmitate suspension	Parke-Davis
Children's Advil suspension	Whitehall
Choledyl elixir	Parke-Davis
Cleocin pediatric granules	Upjohn
Colace syrup	Mead Johnson
Coly-Mycin S suspension	Parke-Davis
Comtrex liquid	Bristol-Myers
Comtrex cough formula	Bristol-Myers
Deconamine syrup	Berlex
Depakene syrup	Abbott
Dexedrine elixir	Smith Kline & French
Dilantin suspension	Parke-Davis
Diuril oral suspension	Merck Sharp & Dohme
Dramamine liquid	Richardson-Vicks
Duricef oral suspension	Mead Johnson
E.E.S liquids	Abbott
Entex liquid	Norwich Eaton
EryPed drops	Abbott
Feosol elixir	Smith Kline Beecham
Gantanol suspension	Roche
Gantrisin pediatric suspension	Roche
Ilosone suspension	Dista
Keflex suspension, pediatric drops	Dista
Lanoxin elixir pediatric	Burroughs Wellcome
Levsin elixir, drops	Schwarz Pharma
Marax syrup	Roerig
Mestinon syrup	ICN
Mycostatin oral suspension	Squibb

Trade name (cont'd)	Manufacturer (cont'd)
Naldecon DX children's syrup	Bristol
Neo-Calglucon syrup	Sandoz
Nucofed expectorant, syrup	Beecham
Nucofed pediatric expectorant	Beecham
PediaProfen suspension	McNeil
Pediazole	Ross
Pentids syrup	Squibb
Periactin syrup	Merck Sharp & Dohme
Phenobarbital elixir	Lilly
Principen suspension	Squibb
Quibron liquid	Bristol
Retrovir syrup	Burroughs Wellcome
Rynatan pediatric suspension	Wallace
Rynatuss pediatric suspension	Wallace
Suprax suspension	Lederle
Tacaryl syrup	Westwood-Squibb
Tegretol suspension	Geigy
Temaril syrup	Herbert
Theolair liquid	3M Riker
Triaminic oral infant drops	Sandoz
Trimox suspension	Squibb
Tylenol Children's liquid, elixir	McNeil
Ultracef suspension	Bristol
Unipen solution	Wyeth-Ayerst
Veetids solution	Squibb
Velosef suspension	Squibb
Zarontin syrup	Parke-Davis

V. HUMAN TOXICITY DATA

A. Dental Caries

Children receiving daily liquid medications containing sucrose for 6 months or longer had a significantly higher incidence of dental caries (244 caries in 44 children) compared to children of similar age receiving no medication or tablet formulations (65 caries in 47 children). Gingivitis was also more prevalent in the sucrose-treated children. This study was done in a nonfluoridated community, and dental evaluations did not include radiographs.[3] Localized caries has been described, limited to the area of the mouth where liquid sucrose-containing medications were administered.[4]

Dental plaque pH was found to be significantly lower in patients given 5 ml of various sucrose-containing medications and instructed to swish in the mouth for 60 s. The product with the least amount of sucrose, in a concentration of 12.8%, had the least effect on plaque pH.[4]

Chronic exposure to Lanoxin elixir, which contains 30% sucrose, to 40 children living in a fluoridated community was associated with a higher incidence of decayed, missing, or filled primary tooth surfaces compared to a control group of children with no history of chronic medication. In the medicated group, 67.5% had decay scores of 0 to 2 compared to 82.5% in the control group. Of seven children with decay scores greater than 10, six were in the medicated group.[2]

A study of 20 chronically ill children under the age of 3 years with rampant dental caries combined a retrospective chart review with parental interviews to assess the quantity and type of sucrose consumption. The mean daily sucrose load from medications was 17 g (range 1 to 46 g) and mean total sucrose load since birth 8696 g (range 48 to 20,482 g). Commonly, at least two doses of medication were given before or during nap or bedtime without rinsing or toothbrushing.[5]

B. Behavior

Anecdotal stories of exacerbation of adverse behavior in children following consumption of sucrose are common, particularly in hyperkinetic children. One such child underwent a blind challenge with 20 g of sucrose mixed with lemonade, along with control challenges of saccharin, glucose, and lactose. All three challenges with sucrose produced symptoms of frustration, hyperactivity, and uncontrollable behavior within 5 to 10 min, lasting about 4 h. After evaluation of this patient, double-blind challenges with a lemonade mixture containing 75 g of sucrose were done in 50 hyperkinetic children; none reacted adversely to sucrose. Behavioral evaluation was done by subjective parental observation.[6]

Several small double-blind crossover studies done in more controlled circumstances evaluated a heterogenous group of psychiatric inpatients and children with various subjective behavior diagnoses. The results were inconsistent, with positive findings of increased total motor activity in one,[7] decreased fine motor activity in another,[7] and a slight decrease in motor activity in the third study.[8] Another small study of 16 hyperactive boys admitted to a clinical research facility for 3 d found no effect of sucrose on behavioral ratings, learning, or memory.[7]

In the largest study to date, 119 male juvenile delinquents were given 78 g of sucrose in a double-blind design. While overall results showed no evidence of behavior or performance impairment by sucrose and improved performance in some groups (white subjects), analysis of subgroups did show some differences. Youths with a high degree of underlying hyperactivity benefited from the sucrose load, while those with less pronounced problems showed impaired performance.[9] Reactive hypoglycemia following a sucrose load was not demonstrated in this population, although the number of subjects exhibiting serum glucose levels of less than 50 mg/dl during the postprandial portion of the test was significantly greater than that found in a control group of nondelinquent male adolescents (29 vs. 7%).[10]

C. Sucrose Malabsorption

Congenital deficiency of sucrase in the small intestine is an inherited trait, resulting in osmotic diarrhea following sucrose ingestion. The prevalence of this trait is low, with about 250 cases reported in the world literature. Ingestion of sucrose-sweetened medications has been reported to provoke or exacerbate symptoms in these children.[11]

VI. CLINICAL RELEVANCE

A. Dental Caries

With adequate fluoridation, the incidence of significant caries in children receiving chronic medication with sucrose-based products is about 32%.[2] While most children are unaffected, those who must be treated with sucrose-containing medications should be instructed to clear the medication from the mouth with a drink of water or preferably follow administration by toothbrushing. Medication should preferably be given while the child is awake. These measures have not been proven to ameliorate the risk associated with these medications, and pharmaceutical manufacturers should be encouraged to eliminate sucrose and other fermentable carbohydrates from pediatric formulations.

Simply reducing the concentration of sucrose may ameliorate the extent of the plaque pH decrease, but will result in decreased salivary flow rates, prolonging contact with the sucrose solution.[12] Xylitol has been shown to be the least cariogenic sugar substitute.[13]

B. Behavior

There is no convincing evidence that the majority of hyperactive children or adolescents exhibit adverse behavior or performance following ingestion of sucrose. There is some suggestion that sucrose may benefit performance in more severely affected children, perhaps by providing a source of readily available energy.

REFERENCES

1. **Food and Drug Administration,** *Inactive Ingredients Guide.* FDA. Washingon. D.C.. March 1990.
2. **Feigal, R. J., Gleeson, M. C., Beckman, T. M., and Greenwood, M. E.,** Dental caries related to liquid medication intake in young cardiac patients. *J. Dent. Child..* Sept.–Oct.. 360, 1984.
3. **Roberts, I. F. and Roberts, G. J.,** Relation between medicines sweetened with sucrose and dental disease, *Br. Med. J..* 2, 15, 1979.
4. **Feigal, R. J., Jensen, M. E., and Mensing, C. A.,** Dental caries potential of liquid medications, *Pediatrics,* 68, 416, 1981.
5. **Kenny, D. J. and Somaya, P.,** Sugar load of oral liquid medications on chronically ill children, *Can. Dent. Assoc. J..* 55, 43, 1989.
6. **Gross, M. D.,** Effect of sucrose on hyperkinetic children, *Pediatrics.* 74, 876, 1984.
7. **Wolraich, M., Milich, R., Stumbo, P., and Schultz, F.,** Effects of sucrose ingestion on the behavior of hyperactive boys. *J. Pediatr..* 106, 675, 1985.
8. **Behar, D., Rapoport, J., Adams, A., Berg, C., and Cornblath, M.,** Sugar challenge testing with children considered behaviorally "sugar reactive". *J. Nutr. Behav..* 1, 277, 1984.
9. **Bachorowski, J., Newman, J. P., Nichols, S. L., Gans, D. A., Harper, A. E., and Taylor, S. L.,** Sucrose and delinquency: behavioral assessment. *Pediatrics,* 86, 244, 1990.
10. **Gans, D. A., Harper, A. E., Bachorowski, J., Newman, J. P., Shrago, E. S., and Taylor, S. L.,** Sucrose and delinquency: oral sucrose tolerance test and nutritional assessment. *Pediatrics,* 86, 254, 1990.
11. **Gudmand-Hoyer, E.,** Sucrose malabsorption in children: a report of thirty-one Greenlanders. *J. Pediatr. Gastroenterol. Nutr..* 4, 873, 1985.
12. **Lagerlof, F. and Dawes, C.,** Effect of sucrose as a gustatory stimulus on the flow rates of parotid and whole saliva. *Caries Res..* 19, 206, 1985.
13. **Hart, A.,** Sweeteners in medications: a risk to healthy teeth. *Can. Med. Assoc. J..* 131, 806, 1984.

SULFITES

I. REGULATORY CLASSIFICATION

Sulfites are classified as antioxidants and are commonly found in sympathomimetic and aminoglycoside medications, which are very susceptible to oxidation. Concentrations range from 0.3 to 0.75%.

II. SYNONYMS

A. Sodium Sulfite

E221
Exsiccated sodium sulphite
Sodium sulphite

$$Na_2SO_3$$

FIGURE 66. Sodium sulfite.

B. Sodium Bisulfite

E222
Sodium Bisulphite
Sodium hydrogen sulfite

$$NaHSO_3$$

FIGURE 67. Sodium bisulfite.

C. Sodium Metabisulfite

Disodium pyrosulphite
E223

Sodium metabisulphite
Sodium pyrosulphite

Na$_2$S$_2$O$_5$

FIGURE 68. Sodium metabisulfite.

D. Potassium Bisulfite

E228
Potassium bisulphite
Potassium hydrogen sulphite

E. Potassium Metabisulfite

Dipotassium pyrosulphite
E224
Potassium metabisulphite
Potassium pyrosulphite

III. AVAILABLE FORMULATIONS

A. Foods

The list of foods potentially containing sulfiting agents is extensive. Six agents are approved for use in foods: sodium bisulfite, potassium bisulfite, sodium metabisulfite, potassium metabisulfite, sodium sulfite, and sulfur dioxide. Sulfites may be added to foods prior to packaging and distribution or added by the food service industry to keep lettuce and other vegetables crisp and to prevent discoloration. Sulfur dioxide is used to sanitize food containers and fermentation equipment. Sulfites are prohibited from meats or other foods that are sources of thiamine because of instability of the vitamin in their presence.[1]

GRAS status was revoked in 1990 for sulfites used on fresh potato products (any form other than frozen, canned, or dehydrated) intended to be served or sold unpackaged and unlabeled to the consumer.[2] GRAS status was previously revoked for the use of sulfites on raw fruits and vegetables intended to be served to consumers. Other nonstandardized food products have been required to list sulfiting agents present at a concentration of 10 ppm or greater since 1986.[3] A similar proposal has been published to require labeling for standardized foods.[4]

Although the above inorganic sulfites are added to foods, the final form present in the food is frequently a bound form. For example, the major form in wines is acetaldehyde hydroxysulfonate.[5] Lettuce is an exception, with the sulfite present exclusively as the unbound inorganic form.

The form of sulfite present is important in predicting the likelihood of an adverse reaction in a sulfite-intolerant patient. Of eight asthmatics demonstrated to be sulfite-sensitive by double-blind capsule and/or oral solution challenges, only four reacted to a challenge with sulfited lettuce, two with dried apricots, two with white grape juice, one with dehydrated potatoes, and none with sulfited shrimp.[6] In another small study of five sulfite-sensitive asthmatics, all five reacted to a sulfited lettuce challenge.[7] Sensitive patients are diagnosed with a double-blind challenge with an oral solution of inorganic sulfite, usually in an acidic vehicle. Such challenges may not accurately predict subsequent reactions to sulfites in foods containing different forms of sulfite.

Fine et al.[8] demonstrated that the pH of the solution influenced the severity of the reaction in asthmatics. At an acidic pH, sulfite is present primarily as bisulfite ion and results in a potent bronchoconstrictive stimulus. Conversely, at a basic pH, the ionic form is predominantly sulfite, and a very weak bronchoconstrictive response was observed.

A commercial test strip (Sulfitest) is available to detect the presence of sulfites in foods, but has low sensitivity for acidic or ascorbic acid-containing foods. In one study, false-negative results were found in five of nine acidic sulfited foods.[9] Because of the possibility of false-negative and false-positive reactions with this test, it is not recommended by the FDA.[10]

B. Drugs

In 1987, the FDA began requiring labeling of all sulfite-containing drugs, with the exception of injectable epinephrine, with a warning regarding the potential for allergic reactions. In addition to this warning, injectable epinephrine products were required to contain a statement that the presence of sulfite should not deter the administration of the drug for treatment of serious allergic or other emergency situations.[11]

C. Cosmetics

Home permanent wave products may contain either sodium or ammonium bisulfite, alone or in combination.[12]

IV. TABLE OF COMMON PRODUCTS

A. Foods

The following foods may contain sulfites.[13]

Type of product	Estimated residual SO_2 (ppm)
Baked Goods	
Baking mixes with dry fruit or vegetables	Up to 200
Cookies	5
Crackers	1
Pie crust	5
Pizza crust	5
Quiche crust	5
Wheat tortillas	Up to 2,500
Beverages	
Beer	10
Red wine	150
White wine	150
Fruit drinks	100
Instant tea	6
Cola drinks	4

Type of product (cont'd)	Estimated residual SO$_2$ (ppm) (cont'd)
Condiments	
Olives	Up to 210
Onion relish	Up to 30
Pickle relish	Up to 30
Pickles	Up to 30
Salad dressing mixes	Up to 50
Confections	
Brown sugar	5
Powdered sugar	5
Granulated sugar	5
Dairy Analogs	
Filled milk	200
Fish Products	
Clams	Unknown
Crab	Unknown
Dried cod	Up to 200
Lobster	Unknown
Scallops	Unknown
Shrimp	Unknown
Fresh Fruit	
Fruit salads	Up to 500
Grapes	1
Fresh Vegetables	
Mushrooms	50
Salad bars	Up to 3,400
Tomatoes	Up to 10,000
Gelatins/Puddings/Fillings	
Apple filling	Up to 35
Flavored gelatin	Up to 5
Unflavored gelatin	Up to 40
Pectin jelling agents	Up to 100
Grain Products	
Corn starch	40
Food starches	Up to 80
Jam/Jelly	
Commercial	500
Processed Fruits	
Dietetic fruit or juice	1,000
Dried fruit	100
Fruit juice	50
Juice concentrate	10
Marashino cherries	150

Type of product (cont'd)	Estimated residual SO$_2$ (ppm) (cont'd)
Processed Vegetables	
Canned vegetables	1
Dried vegetables	1,000
Canned potatoes	1
Frozen potatoes	10
Instant potatoes	50
Vegetable juice	1,000
Snack Foods	
Apple bits	Up to 275
Filled crackers	Up to 80
Potato/tortilla chips	100
Soups	
Canned soups	1
Dried soup mix	1
Sweet Sauces	
Corn syrup	
Dextrose monohydrate	
Fruit topping	
Glucose syrup	
High fructose corn syrup	
Maple syrup	
Molasses	
Pancake syrup	

B. Inhalation Drug Products

Trade name	Manufacturer
Adrenalin chloride solution for nebulization	Parke-Davis
Arm-A-Med Isoetharine inhalation solution	Armour
Arm-A-Med metaproterenol sulfate inhalation solution	Armour
AsthmaNefrin	Norcliff-Thayer
Bronkosol	Winthrop
Isuprel solution for inhalation	Winthrop

C. Parenteral Drug Products

Trade name	Manufacturer
Adrenalin chloride	Parke-Davis
Aldomet	Merck Sharp & Dohme
Amikin	Bristol
Antilirium	Forest
Aramine	Merck Sharp & Dohme

Trade name (cont'd)	Manufacturer (cont'd)
Bactrim IV	Roche
Bronkephrine	Winthrop
Carbocaine 2% with Neo-Cobefrin	Cook-Waite
Celestone Phosphate	Schering
Compazine 2 ml ampule	Smith Kline & French
Dalalone D.P.	Forest
Decadron L.A.	Merck Sharp & Dohme
Decadron phosphate	Merck Sharp & Dohme
Decadron phosphate with xylocaine	Merck Sharp & Dohme
Dobutrex	Lilly
Enlon	Astra
Duranest with epinephrine	Astra
Epipen/Epipen Jr.	Center
Garamycin	Schering
Garamycin pediatric	Schering
Hydeltrasol	Merck Sharp & Dohme
Hydrocortone phosphate	Merck Sharp & Dohme
Inocor lactate	Winthrop
Intropin	DuPont
Kantrex	Bristol
Levophed	Winthrop
Levsin	Schwarz Pharma
Marcaine 0.5% with Neo-Cobefrin	Cook-Waite
Mepergan	Wyeth-Ayerst
Nebcin 1.6 ADD-Vantage/3.2 multiple-dose	Lilly
Neo-Synephrine	Winthrop
Netromycin	Schering
Norflex	3M Riker
Norzine	Purdue Frederick
Novocain	Winthrop
Numorphan (sodium dithionate)	DuPont
Phenergan ampule	Wyeth-Ayerst
Pontocaine	Winthrop
Pronestyl	Princeton
Sensorcaine with epinephrine	Astra
Septra IV	Burroughs Wellcome
Solganal	Schering
Synkayvite	Roche
Talwin	Winthrop
Tensilon	ICN
Thorazine	Smith Kline & French
Tofranil	Geigy
Torecan	Roxane
Trilafon	Schering

Trade name (cont'd)	Manufacturer (cont'd)
Vasoxyl	Burroughs Wellcome
Xylocaine 1% with epinephrine 1:100,000	Astra
Xylocaine 1.5% with epinephrine 1:200,000	Astra
Xylocaine 2% with epinephrine 1:200,000	Astra
Yutopar	Astra

Note: A preservative-free, sulfite-free epinephrine 1:1,000 solution for injection is now available from American Regent Laboratories.

D. Parenteral Nutrition Products

Trade name	Manufacturer
Dextrose 5%/electrolyte No. 48	Travenol
Dextrose 5%/electrolyte No. 75	Travenol
FreAmine 111 3% amino acid injection/electrolytes	American McGaw
FreAmine 111 8.5% amino acid injection	American McGaw
FreAmine 111 10% amino acid injection	American McGaw
HepatAmine 8% amino acid injection	American McGaw
Nephramine 5.4% amino acid injection	American McGaw
ProcalAmine 3% amino acid/3% glycerin/electrolytes	American McGaw
Travasol 5.5%/electrolytes	Travenol
Travasol 5.5%	Travenol
Travasol 8.5%/electrolytes	Travenol
Travasol 8.5%	Travenol
Travasol 10%	Travenol
Travasol M 3.5%/electrolyte No. 48	Travenol
Travert 5%/electrolyte No. 2	Travenol
Travert 5%/electrolyte No. 4	Travenol
Travert 10%/electrolyte No. 1	Travenol
Travert 10%/electrolyte No. 2	Travenol
Travert 10%/electrolyte No. 3	Travenol
TrophAmine 6% amino acid injection	American McGaw

E. Ophthalmic Products

Trade name	Manufacturer	Therapeutic class
AK-Dilate	Akorn	Mydriatic
Baldex solution	Bausch & Lomb	Corticosteroid
Betagan	Allergan	Beta-blocker
Decadron solution	Merck Sharp & Dohme	Corticosteroid
Epifrin	Allergan	Antiglaucoma
Epitrate	Wyeth-Ayerst	Antiglaucoma
Glaucon	Alcon	Antiglaucoma

Trade name (cont'd)	Manufacturer (cont'd)	Therapeutic class (cont'd)
Murocoll 2	Bausch & Lomb	Cycloplegic
NeoDecadron	Merck Sharp & Dohme	Corticosteroid/antibiotic
Pred-Forte	Allergan	Corticosteroid
Pred-Mild	Allergan	Corticosteroid
Prefrin-A	Allergan	Decongestant/antihistamine
Propine	Allergan	Antiglaucoma
Sulphrin	Bausch & Lomb	Corticosteroid/antibiotic
Sulten-10 solution	Bausch & Lomb	Antibiotic

F. Topical Drug Products

Trade name	Manufacturer
Caldecort cream	Fisons
Carmol HC	Syntex
Eldopaque cream	Elder
Eldopaque forte cream	Elder
Eldoquin cream	Elder
Eldoquin forte cream	Elder
Esoterica Medicated Fade cream	Norcliff Thayer
Lasan cream	Stiefel
Meclan cream	Ortho
Neodecadron cream	Merck Sharp & Dohme
Nizoral cream	Janssen
Nupercainal cream	Ciba Consumer
Solaquin cream	Elder
Solaquin forte cream	Elder
Sulfacet R	Dermik
Sulfamylon cream	Winthrop
Topicycline	Norwich Eaton

G. Otic Drug Products

Trade name	Manufacturer
Cortisporin otic solution	Burroughs Wellcome
Otocort ear drops	
Tympagesic otic drops	Adria

V. HUMAN TOXICITY DATA

A. Inhalation

The incidence of hypersensitivity reactions to sulfiting agents depends greatly on the route of exposure. Inhalation of sulfites in solution will provoke a reproducible bronchoconstriction in virtually all asthmatics if present in adequate concentrations. A sensitive asthmatic may experience exercise-induced bronchospasm after inhalation of as little as 0.1 ppm of sulfur dioxide.[14] Sulfited bronchodilator solutions have been shown to contain 0.1 to 6 ppm.[15-17]

Patients sensitive to oral administration of sulfites may experience similar reactions with inhaled doses that are 0.4% of the provoking oral capsule dose.[18] Inhalation of high concentrations of sodium metabisulfite, greater than 20 mg/ml, results in a primary irritant response with coughing.[19]

The onset of bronchoconstriction is rapid following inhalation, occurring within 1 min, peaking between 2 to 5 min, and declining to baseline within 30 min.[20]

Initial reports of paradoxical bronchoconstriction, following use of nebulized bronchodilators containing sulfites, primarily concerned steroid-dependent asthmatics.[21-24]

B. Oral Solutions

Following inhalation, exposure to sulfited oral solutions is most likely to provoke a reaction in a sensitive individual, presumably due to inhalation of variable amounts of the solution during ingestion.[25] The amount of sulfiting agent in an acidic oral solution that will provoke a reaction ranges from one fourth to one half of the oral capsule amount.[18]

C. Subcutaneous Injection

The possibility that subcutaneous injection of sulfite-containing injectables, particularly local anesthetic solutions, may cause hypersensitivity reactions is controversial. Subcutaneous injection of 0.6 to 0.9 mg of sodium metabisulfite has been reported to produce hypersensitivity reactions in a few cases.[26-29] Reported reactions have included generalized pruritic urticaria;[28] rapid onset of facial edema without dermatologic effects;[30] severe airway obstruction with generalized pruritus and urticaria;[31] and severe asthma, urticaria, and syncope.[32]

Other studies demonstrated tolerance of up to ten times the usual therapeutic dose of sulfited epinephrine injection.[18] Some of the reported reactions have been attributed to anxiety or vasomotor effects of epinephrine,[33] while others appear to be true hypersensitivity reactions documented by positive patch tests or subcutaneous injection challenge to the sulfiting agent, with a negative response to active ingredients.[30,31]

One author has recommended pretreating sulfite-sensitive individuals with prednisone 40 mg and Chlortrimeton® 12 mg the night before and prednisone 40 mg and terfenadine 60 mg 2 h before administration of a local anesthetic containing sulfites.[32] The safety and efficacy of this procedure has not been evaluated by controlled studies.

D. Intravenous Injection

Bronchospasm has been reported after intravenous administration of medications containing metabisulfite.[34] Total parenteral nutrition solutions containing sulfites were implicated in causing neurological problems in another case.[35]

E. Intramuscular Injection

Injection of a vitamin preparation containing metabisulfite was implicated in causing 41 cases of anaphylaxis, 13 cases of bronchospasm, and 22 dermatological reactions during 18 years of monitoring by the Committee on Safety of Medicines in the U.K.[36]

F. Ophthalmic

An adult nonsteroid-dependent asthmatic was reported to develop bronchospasm following administration of an ophthalmic dipivefrin solution containing sodium metabisulfite.[37]

G. Topical Dermal

Occupational contact dermatitis has been attributed to delayed Type IV hypersensitivity to sulfites.[38-40] Delayed eczematous reactions have also been reported in patients using a sodium sulfite-containing antifungal cream[30,41] and in several patients who applied a sulfite-containing home permanent wave solution.[12]

Immediate reactions, such as pruritus, urticaria, syncope, and bronchospasm, have also been reported following topical exposure in sulfite-sensitive asthmatics.[42]

The appropriate patch test concentration has not been determined. Mild irritant reactions, consisting of slight erythema, occur with concentrations of 5%. Hypersensitive patients have demonstrated positive reactions to concentrations as low as 0.2%.[30]

H. Foods

Adverse reactions following ingested sulfiting agents in foods typically occur within 2 to 15 min and consist of severe flushing, itching, urticaria, syncope, hypotension, angioedema, or asthma.[43] Potato products were responsible for 12% of all adverse reactions and for 4 of 17 deaths attributable to sulfiting agents in a monitoring system maintained by the FDA.[2] A fatality was reported following ingestion of a white wine containing sulfites at a concentration of 92 ppm.[44]

I. Mechanism

The primary mechanism appears to be stimulation of pulmonary cholinergic receptors, eliciting a vagal reflex response. Successful prophylaxis has been observed following administration of inhaled and oral anticholinergic agents, such as atropine and doxepin.[45] Other mechanisms may be infrequently responsible in some individuals.

Administration of cyanocobalamin prevented reactions in four of six patients, presumably by catalyzing oxidation of sulfite to sulfate. A subclinical sulfite oxidase deficiency may be interfering with the enzymatic route of detoxification of sulfite in these patients.[45-47] Cromolyn sodium has also been shown to be protective.[45-48]

An IgE-mediated mechanism has been demonstrated in sporadic case reports.[24,33,49-51] Positive skin prick tests or intradermal tests were found in 5 of 53 sulfite-sensitive patients tested with skin prick tests in one series. Passive transfer was positive in the two subjects in which this was attempted.[52] Attempts to desensitize sulfite-sensitive individuals using alum-precipitated sulfite injections was successful in 83% of patients in an anecdotal series.[53]

Mast cell degranulation was proven not to be the mechanism in five patients sensitive to oral challenge with negative skin prick tests.[54]

J. Intrathecal Exposure

Inadvertent subarachnoid injection of 2-chloroprocaine preserved with sodium bisulfite was associated with paralyses, consisting of sensorimotor loss in the legs and impaired rectal sphincter control in several human patients.[55-58]

These effects were reproduced in subsequent animal experiments, which demonstrated sodium bisulfite to cause spinal reflex activity blockade at concentrations of 0.6%, which manifested as a prolonged motor block.[57,60-62] These studies prompted removal of bisulfite from solutions intended for epidural injection.

VI. CLINICAL RELEVANCE

A. Asthma

The primary population at risk for sulfite-induced adverse reactions is the chronic asthmatic, particularly patients with moderate or severe bronchial hyperreactivity who are dependent on corticosteroids. Only nine anecdotal reports have been collected by the FDA concerning nonasthmatic individuals.[2] The incidence of sulfite-induced bronchoconstriction was higher in steroid-dependent adult asthmatics (8.4%) than in nonsteroid-dependent asthmatics (0.8%) in a study of 203 patients.[63]

Sulfite-induced bronchoconstriction has been provoked by practically all routes of exposure, including inhalation, ingestion of oral solutions, subcutaneous injection, intravenous injection, dermal, and ophthalmic application. No reactions have been reported from ingestion of commercial solid pharmaceutical dosage forms, presumably due to the small amount contained in these products. There are no pharmaceutical products containing more than 3.2 mg per tablet.[64]

Symptoms in sensitive individuals can include chest pain, bronchoconstriction, pruritus, tingling, urticaria, angioedema, and hypotension.[65]

Elucidation of the incidence of sulfite intolerance in chronically asthmatic children has been confounded by potential selection bias in populations containing more severe cases. In these studies, the incidence of adverse responses was 35.3 and 65.5%, [66,67] a rate seemingly higher than that reported in adult asthmatics, which is estimated to be in the range of 3.5 to 8%.[48,63,68,69] In one of the pediatric studies, 96.5% of the children were receiving inhaled corticosteroids and 34.4% oral steroids.[67]

In a study performed in a residential house for mild nonsteroid-dependent perennial asthmatic children, the incidence was 7.1% following oral capsule challenge and 3.5% following oral solution challenge. In this study, reactions were only seen in children with moderate to severe bronchial reactivity, determined by methacholine challenge.[70]

In a study of 1544 asthmatic patients, aged 8 to 71 years, the incidence of sulfite sensitivity, determined by nonacidic oral solution challenge, was 3.4%. In contrast to the previous study in children, there was no relationship between the underlying airway reactivity, assessed by exercise testing and bronchodilator responsiveness, and the severity of metabisulfite sensitivity.[48]

B. Oral Exposure

The amount of oral encapsulated sulfite required to provoke a reaction in a sensitive asthmatic ranges from 5 to 200 mg.[65]

REFERENCES

1. **Lecos, C.,** Food preservatives: a fresh report, *FDA Consumer,* April 1984.
2. **Food and Drug Administration,** Sulfiting agents: revocation of GRAS status for use on fresh potatoes served or sold unpackaged and unlabeled to consumers and request for data on use of sulfites on frozen potatoes; rule and proposed rule, *Fed. Reg.,* 55, 9826, 1990.
3. **Food and Drug Administration,** Food labeling: declaration of sulfiting agents, *Fed. Reg.,* 51, 25012, 1986.
4. **Food and Drug Administration,** Sulfiting agents in standardized foods; labeling requirements, *Fed. Reg.,* 53, 51062, 1988.

5. **Burroughs, L. F. and Sparks, A. H.,** Sulphite-binding power of wines of ciders. III. Determination of carbonyl compounds in a wine and calculation of its sulphite-binding power, *J. Sci. Food Agric.,* 24, 207, 1973.

6. **Taylor, S. L., Bush, R. K., Selner, J. C., Nordlee, J. A., Wiener, M. B., Holden, K., Koepke, J. W., and Busse, W. W.,** Sensitivity to sulfited foods among sulfite-sensitive subjects with asthma, *J. Allergy Clin. Immunol.,* 81, 1159, 1988.

7. **Howland, W. C., III and Simon, R. A.,** Sulfite-treated lettuce challenges in sulfite-sensitive subjects with asthma, *J. Allergy Clin. Immunol.,* 83, 1079, 1989.

8. **Fine, J. M., Gordon, T., and Sheppard, D.,** The roles of pH and ionic species in sulfur dioxide-and sulfite-induced bronchoconstriction, *Am. Rev. Respir. Dis.,* 136, 1122, 1987.

9. **Wanderer, A. A. and Solomons, C.,** Detection characteristics of a commercially available sulfite detection test (Sulfitest): problems with decreased sensitivity and false negative reactions, *Ann. Allergy,* 58, 41, 1987.

10. **Simon, R. A.,** Sulfite sensitivity, *Ann. Allergy,* 59, 100, 1987.

11. **Food and Drug Administration,** Sulfiting agents; labeling in drugs for human use; warning statement, *Fed. Reg.,* 51, 43900, 1986.

12. **Fisher, A. A.,** Dermatitis due to sulfites in home permanent preparations. Part II, *Cutis,* 44, 108, 1989.

13. **Anon.,** Sulfite use and residual levels foods, comments filed in response to Docket No. 81N-0314, Sulfiting agents: proposed affirmation of GRAS status with specific limitations; removal from GRAS status as direct human food ingredient, *Fed. Reg.,* 47, 29956, 1982.

14. **Sheppard, D., Saisho, A., Nadel, J. A., and Boushey, H. A.,** Exercise increases sulfur dioxide-induced bronchoconstriction in asthmatic subjects, *Am. Rev. Respir. Dis.,* 123, 486, 1981.

15. **Koepke, J. W., Selner, J. C., and Dunhill, A. L.,** Presence of sulfur dioxide in commonly used bronchodilator solutions, *J. Allergy Clin. Immunol.,* 72, 504, 1983.

16. **Schwartz, H. J. and Chester, E. H.,** Bronchospastic responses to aerosolised metabisulfite in asthmatic subjects: potential mechanisms and clinical implications, *J. Allergy Clin. Immunol.,* 74, 15, 1984.

17. **Witek, T. J. and Schachter, E. N.,** Detection of sulfur dioxide in bronchodilator aerosols, *Chest,* 86, 592, 1984.

18. **Goldfarb, G. and Simon, R.,** Provocation of sulfite sensitive asthma, *J. Allergy Clin. Immunol.,* 73, 135, 1984.

19. **Nichol, G. M., Nix, A., Chung, K. F., and Barnes, P. J.,** Characterisation of bronchoconstrictor responses to sodium metabisulphite aerosol in atopic subjects with and without asthma, *Thorax,* 44, 1009, 1989.

20. **Wright, W., Zhang, Y. G., Salome, C. M., and Woolcock, A. J.,** Effect of inhaled preservatives on asthmatic subjects. I. Sodium metabisulfite, *Am. Rev. Respir. Dis.,* 141, 1400, 1990.

21. **Jamieson, D. M., Guill, M. F., Wray, B. B., and May, J. R.,** Metabisulfite sensitivity: case report and literature review, *Ann. Allergy,* 54, 115, 1985.

22. **Koepke, J. W., Christopher, K. L., Chai, H., and Selner, J. C.,** Dose-dependent bronchospasm from sulfites in isoetharine, *JAMA,* 251, 2982, 1984.

23. **Sher, T. H. and Schwartz, H. G.,** Bisulfite sensitivity manifesting as an allergic reaction to aerosol therapy, *Ann. Allergy,* 54, 224, 1985.

24. **Twarog, F. J. and Leung, D. Y. M.,** Anaphylaxis to a component of isoetharine (sodium bisulfite), *JAMA,* 248, 2030, 1982.

25. **Delohery, J., Simmul, R., Castle, W. D., and Allen, D. H.,** The relationship of inhaled sulfur dioxide reactivity to ingested metabisulfite sensitivity in patients with asthma, *Am. Rev. Respir. Dis.,* 130, 1027, 1984.

26. **Huang, A. S. and Fraser, W. M.,** Are sulfite additives really safe?, *N. Engl. J. Med.,* 311, 542, 1984.

27. **Riggs, B. S., Harchelroad, F. P., and Poole, C.,** Allergic reaction to sulfiting agents, *Ann. Emerg. Med.,* 15, 77, 1986.

28. **Schwartz, H. J. and Sher, T. H.,** Bisulfite sensitivity manifesting as allergy to local dental anesthesia, *J. Allergy Clin. Immunol.,* 75, 525, 1985.

29. **Stevenson, D. D. and Simon, R. A.,** Sensitivity to ingested metabisulfites in asthmatic subjects, *J. Allergy Clin. Immunol.,* 68, 26, 1981.

30. **Dooms-Goossens, A., Gide de Alam, A., Degreef, H., and Kochuyt, A.,** Local anesthetic intolerance due to metabisulfite, *Contact Dermatitis,* 20, 124, 1989.

31. **Schwartz, H. J., Gilbert, I. A., Lenner, K. A., Sher, T. H., and McFadden, E. R.,** Metabisulfite sensitivity and local dental anesthesia, *Ann. Allergy,* 62, 83, 1989.

32. **Fisher, A. A.,** Reactions to injectable local anesthetics. Part IV. Reactions to sulfites in local anesthestics, *Cutis,* 44, 283, 1989.

33. **Simon, R. A.,** IgE mediated sulfite sensitive asthma: a case report, *J. Allergy Clin. Immunol.,* 77, 157, 1986.

34. **Baker, G. J., Collett, P., and Allen, D. H.,** Bronchospasm induced by metabisulphite-containing foods and drugs, *Med. J. Aust.,* 2, 614, 1981.

35. **Abumrad, N. N., Schneider, A. J., Steel, D., and Rogers, L. S.,** Amino acid intolerance during prolonged total parenteral nutrition reversed by molybdate therapy, *Am. J. Clin. Nutr.,* 34, 2551, 1981.

36. **Committee on Safety of Medicines,** Injectable vitamin preparations: serious allergic reactions, *WHO Drug Information,* 3, 69, 1989.

37. **Schwartz, H. J. and Sher, T. H.,** Bisulfite intolerance manifest as bronchospasm following topical dipivefrin hydrochloride therapy for glaucoma, *Arch. Ophthalmol.,* 103, 14, 1985.

38. **Apetato, M. and Marques, M. S. J.,** Contact dermatitis caused by sodium metabisulphite, *Contact Dermatitis,* 14, 194, 1986.

39. **Epstein, E.,** Sodium sulfite, *Contact Dermatitis Newsletter,* 7, 155, 1970.

40. **Nater, J. P.,** Allergic contact dermatitis caused by potassium metabisulphite, *Dermatologica,* 136, 477, 1968.

41. **Vissers-Croughs, K. J. M., Van der Kley, A. M. J., Vulto, A. G., and Hulsmans, R. F. H. J.,** Allergic contact dermatitis from sodium sulfite, *Contact Dermatitis,* 18, 252, 1988.

42. **Fisher, A. A.,** Urticaria, asthma, and anaphylaxis due to sodium sulfite in an antifungal cream complicated by treatment with aminophylline in an ethylenediamine-sensitive person. Part I, *Cutis,* 44, 19, 1989.

43. **Fisher, A. A.,** Reactions to sulfites in foods: delayed eczematous and immediate urticarial, anaphylactoid, and asthmatic reactions. Part III, *Cutis,* 44, 187, 1989.

44. **Tsevat, J., Gross, G. N., and Dowling, G. P.,** Fatal asthma after ingestion of sulfite-containing wine, *Ann. Intern. Med.,* 107, 263, 1987.

45. **Simon, R., Goldfarb, G., and Jacobson, D.,** Blocking studies in sulfite sensitive asthmatics, *J. Allergy Clin. Immunol.,* 73, 136, 1984.

46. **Jacobsen, D. W., Simon, R. A., and Singh, M.,** Sulfite oxidase deficiency and cobalamin protection in sulfite-sensitive asthmatics (SSA), *J. Allergy Clin. Immunol.,* 73, 135, 1984.

47. **Schwartz, H. J.,** Observations on the use of oral sodium cromoglycate in a sulfite-sensitive asthmatic patient, *Ann. Allergy,* 57, 36, 1986.

48. **McClellan, M. D., Wanger, J. S., and Cherniack, R. M.,** Attenuation of the metabisulfite-induced bronchoconstrictive response by pretreatment with cromolyn, *Chest,* 97, 826, 1990.

49. **Prenner, B. M. and Stevens, J. J.,** Anaphylaxis after ingestion of sodium bisulfite, *Ann. Allergy,* 37, 180, 1976.

50. **Simon, R. A. and Wasserman, M. D.,** IgE mediated sulfite sensitive asthma, a case report, *J. Allergy Clin. Immunol.,* 77, 157, 1986.

51. **Meggs, W. J., Atkins, F. M., Wright, R. H., et al.,** Sulfite challenges in patients with systemic mastocytosis or unexplained anaphylaxis, *J. Allergy Clin. Immunol.,* 75, 144, 1985.

52. **Yang, W. H., Purchase, E. C. R., and Rivington, R. N.,** Positive skin tests and Prausnitz-Kustner reactions in metabisulfite-sensitive subjects, *J. Allergy Clin. Immunol.,* 78, 443, 1986.

53. **Hosen, H.,** Specific immunologic therapy with the sulfite chemicals, *J. Asthma,* 24, 219, 1987.

54. **Sprenger, J. D., Altman, L. C., Marshall, S. G., Pierson, W. E., and Koenig, J. Q.,** Studies of neutrophil chemotactic factor of anaphylaxis in metabisulfite sensitivity, *Ann. Allergy,* 62, 117, 1989.

55. **Covino, B. G., Marx, G. F., Finster, M., and Zsigmond, E. K.,** Prolonged sensory/motor deficits following inadvertent spinal anesthesia, *Anesth. Analg.,* 59, 399, 1980.

56. **Moore, D. C., Spierdijk, J., Van Kleef, J. D., Coleman, R. L., and Love, G. F.,** Chloroprocaine neurotoxicity: four additional cases, *Anesth. Analg.,* 61, 155, 1982.

57. **Ravindran, R. S., Bond, V. K., Tasch, M. O., Gupta, C. D., and Leurssen, T. G.,** Prolonged neural blockade following regional analgesia with 2-chloroprocaine injection, *Anesth. Analg.,* 59, 446, 1980.

58. **Reisner, L. S., Hochman, B. N., and Plumer, M. H.,** Persistent neurological deficit and adhesive arachnoiditis following intrathecal 2-chloroprocaine injection, *Anesth. Analg.,* 59, 452, 1980.

59. **Ravindran, R. S., Turner, M. S., and Muller, J.,** Neurological effects of subarachnoid administration of 2-chloroprocaine-CE, bupivacaine, and low pH normal saline in dogs, *Anesth. Analg.,* 61, 279, 1982.

60. **Wang, B. C., Hillman, D. E., Spielholz, N. I., and Turndorf, H.,** Chronic neurological deficits and Nesacaine-CE: an effect of the anesthetic, 2-chloroprocaine, or the antioxidant, sodium bisulfite, *Anesth. Analg.,* 63, 445, 1984.

61. **Ready, L. B., Plumer, M. H., Haschke, R. H., Austin, E., and Sumi, M.,** Neurotoxicity of intrathecal local anesthetics in rabbits, *Anesthesiology,* 63, 364, 1985.

62. **Hersh, E. V., Condouris, G. A., and Havelin, D.,** Actions of intrathecal chloroprocaine and sodium bisulfite on rat spinal reflex function utilizing a noninvasive technique, *Anesthesiology*, 72, 1077, 1990.

63. **Bush, R. K., Taylor, S. L., Holden, K., Nordlee, J. A., and Busse, W. W.,** Prevalence of sensitivity to sulfiting agents in asthmatic patients, *Am. J. Med.*, 81, 818, 1986.

64. **Golightly, L. K., Smolinske, S. C., Bennett, M. L., Sutherland E. W., and Rumack, B. H.,** Pharmaceutical excipients; adverse effects associated with "inactive" ingredients in drug products (Part II), *Med. Toxicol.*, 3, 209, 1988.

65. **Bush, R. K., Taylor, S. L., and Busse, W.,** A critical evaluation of clinical trials in reactions to sulfites, *J. Allergy Clin. Immunol.*, 78, 191, 1986.

66. **Friedman, M. E. and Easton, J. G.,** Prevalence of positive metabisulfite challenges in children with asthma, *Pediatr. Asthma Allergy Immunol.*, 1, 53, 1987.

67. **Towns, S. J. and Mellis, C. M.,** The role of acetylsalicylic acid and sodium metabisulfite in chronic childhood asthma, *Pediatrics*, 73, 631, 1984.

TALC

I. REGULATORY CLASSIFICATION

Talc is classified as a glidant and/or anticaking agent and a tablet and/or capsule lubricant.

$$Mg_6(Si_2O_5)_4(OH)_4$$

FIGURE 69. Talc.

II. SYNONYMS

Powdered talc
Purified French chalk
Purified talc
Soapstone
Steatite
Talcum

III. AVAILABLE FORMULATIONS

A. Constituents

Talc consists of purified native, hydrated magnesium silicate. Impurities vary with the natural origin of the talc and may include aluminum silicate, aluminum oxide, calcium carbonate, and iron oxide. Purification of the native hydropolysilicate is performed by pulverization and flotation processes, followed by powdering and treatment with dilute hydrochloric acid.

B. Drugs

Talc is present in over 680 FDA-approved oral solid dosage forms, including capsules, tablets, buccal tablets, and sublingual tablets. Amounts range from 0.003 to 220.4 mg. Six topical powders and three topical creams, lotions, or ointments are also listed.[1]

C. Cosmetics

Talc is used in cosmetics as an absorbent, anticaking agent, bulking agent, and opacifier in powdered eye shadows, makeup foundations, powders, blushers, and dusting powders.[2]

IV. TABLE OF COMMON PRODUCTS

A. Oral Drug Products with Abuse Potential

Trade name	Manufacturer
Demerol tablets	Winthrop
Desoxyn tablets	Abbott
Dexedrine tablets	Smith Kline & French
Dolophine tablets	Lilly
Fiorinal with codeine capsules	Sandoz
Levo-Dromoran tablet	Roche
MS Contin	Purdue Frederick
MSIR tablet	Purdue Frederick
PBZ tablets	Geigy
Prelu-2 tablet	Boehringer Ingelheim
Preludin tablet	Boehringer Ingelheim
Tenuate Dospan tablet	Marion Merrell Dow

V. HUMAN TOXICITY DATA

A. Pneumonoconiosis

Intravenous abuse of suspensions made from talc-containing drugs results in pneumoconiosis resembling that seen in talc miners with inhalation exposure to unpurified talc. In a case report of an addicted physician who injected Dolophine® for 5 years, pulmonary granulomas, diffuse panacinar emphysema, and cor pulmonale were noted at autopsy. Talc crystals were found in every organ examined, including the spleen, heart, liver, kidney, brain, adrenal glands, and thyroid.[3]

An 18-year-old girl who died following intravenous abuse of Ritalin® had morphologic evidence of severe pulmonary hypertension, resulting in cor pulmonale. Talc granulomas blocking the pulmonary arteries were numerous. In a review of 16 autopsies in patients with talc granulomas from injected dissolved tablets, all had talc granulomas in the pulmonary interstitium (the predominant lesion in 11) and pulmonary artery lumina (the predominant lesion in 4). Patients with a shorter history of abuse tended to have arterial lesions which led to pulmonary hypertension. Those with long-standing abuse tended to have interstitial fibrosis.[4]

Chest radiographs done in 17 living patients with a history of intravenous injection of 2,500 to 50,000 methadone tablets over a 1- to 9-year period were normal in 10 patients and showed a diffuse pin-point, micronodular pattern in 7. Two patients also had evidence of massive progressive fibrosis. Dyspnea was present in 58%.[5] A follow-up study of three of these patients and three additional cases was published 10 years after the first report. Despite discontinuation of drug abuse, gradual coalescence of the nodules developed, resulting in large opacities in the parahilar regions and upper lobes. Rapid evolution of functional impairment with severe airflow limitation, air trapping, hyperinflation, and decreased diffusing capacity with characteristics of emphysema occurred in the three patients from the original study and in the three additional patients.[6]

Talc granulomas are recognized by plate-like birefringent crystals visible under polarizing light. Talc particles may be found in the retinal vessels.[6]

Pathological examination of seven Ritalin® abusers who died of profound obstructive lung disease showed severe panlobular emphysema that was more severe in the lower lung zones. Alpha 1-antitrypsin deficiency was ruled out in five patients, and smoking history ranged from 5 to 100 pack-years. In contrast to the previous report of methadone abusers, interstitial fibrosis was not a predominant finding in these patients.[7]

VI. CLINICAL RELEVANCE

Progressive panacinar emphysema has been a consistent finding in autopsy studies of intravenous drug abusers who inject talc-containing drug products. Variable degrees of talc granulomas, inflammatory infiltrates, and pulmonary vascular occlusion may be present. The mechanism has not been elucidated, but it appears not to involve alpha 1-antitrypsin deficiency. Most patients have been cigarette smokers, but have developed the disease at a younger age than classic smoke-related emphysema. In contrast to the centrilobular distribution in smokers, the drug abusers invariably had a diffuse panlobular pattern.

REFERENCES

1. **Food and Drug Administration,** *Inactive Ingredients Guide,* FDA, Washington, D.C., March 1990.
2. **Nikitakis, J. M.,** *CTFA Cosmetic Ingredient Handbook,* 1st ed., The Cosmetic, Toiletry and Fragrance Association, Washington, D.C., 1988.
3. **Groth, D. H., Mackay, G. R., Crable, J. V., and Cochran, T. H.,** Intravenous injection of talc in a narcotics addict, *Arch. Pathol.,* 94, 171, 1972.
4. **Waller, B. F., Brownlee, W. J., and Roberts, W. C.,** Structure-function correlations in cardiovascular and pulmonary diseases (CPC): self-induced pulmonary granulomatosis, *Chest,* 78, 90, 1980.
5. **Pare, J. A., Fraser, R. G., Hogg, J. C., Howlett, J. G., and Murphy, S. B.,** Pulmonary "mainline" granulomatosis: talcosis of intravenous methadone abuse, *Medicine,* 58, 229, 1979.
6. **Pare, J. P., Cote, G., and Fraser, R. S.,** Long-term follow-up of drug abusers with intravenous talcosis, *Am. Rev. Respir. Dis.,* 139, 233, 1989.
7. **Schmidt, R. A., Glenny, R. W., Godwin, J. D., Hampson, N. B., Cantino, M. E., and Reichenbach, D. D.,** Panlobular emphysema in young intravenous Ritalin® abusers, *Am. Rev. Respir. Dis.,* 143, 649, 1991.

THIMEROSAL

I. REGULATORY CLASSIFICATION

Thimerosal, also known as thiomersal, sodium ethylmercurithiosalicylate, and Merthiolate®, is classified as an antimicrobial preservative and is used in concentrations of 0.002 to 0.01%.[1]

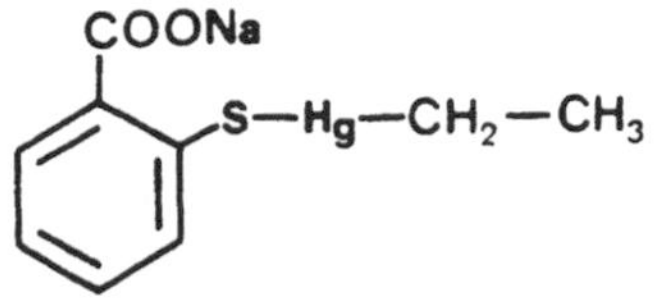

FIGURE 70. Thimerosal.

II. AVAILABLE FORMULATIONS

A. Drugs
Thimerosal is infrequently used in ophthalmologic products due to concerns over adverse reactions. A 1989 FDA list cited the availability of thimerosal in 12 ophthalmic, 3 topical, and 10 parenteral approved drug products.[1]

B. Cosmetics
Thimerosal is used as a preservative in eye makeup, such as eye shadow, mascara, eyeliners, and makeup removers.[2]

III. TABLE OF COMMON PRODUCTS AND ALTERNATIVES

A. Ophthalmic Products

Trade name	Manufacturer	Therapeutic use
Adsorbonac	Alcon	Hypertonic agent
Adsorbotear	Alcon	Artificial tear
AK-Spore	Akorn	Antibiotic
Bleph-10 solution	Allergan	Antibiotic
Collyrium for Fresh Eyes	Wyeth-Ayerst	Irrigant

411

Trade name (cont'd)	Manufacturer (cont'd)	Therapeutic use (cont'd)
Cortisporin suspension	Burroughs Wellcome	Corticosteroid/antibiotic
Fluoracaine	Akorn	Disclosing agent
Gonioscopic prism solution	Alcon	Bonding agent
Hydrocare Cleaning/Disinfect	Allergan	Contact lens solution
Hydrocare Preserved Saline	Allergan	Contact lens solution
LC-65 Daily Cleaner	Allergan	Contact lens solution
Lensrins Preserved Saline	Allergan	Contact lens solution
Lens-Wet Lubricatin/Rewetting	Allergan	Contact lens solution
Liquifilm forte	Allergan	Artificial tear
Neosporin ophthalmic solution	Burroughs Wellcome	Antibiotic
Neotricin solution	Bausch & Lomb	Antibiotic
Ocufen solution	Allergan	NSAID
Poly-Pred	Allergan	Corticosteroid/antibiotic
Rose Bengal solution	Akorn	Diagnostic agent
Stoxil solution	Smith Kline & French	Antiviral
Vasocidin solution	IOLAB	Corticosteroid/antibiotic
Viroptic	Burroughs Wellcome	Antiviral

B. Parenteral Drug Products

Trade name	Manufacturer
Antivenin crotalidae polyvalent	Wyeth-Ayerst
Antivenin micrurus fulvius	Wyeth-Ayerst
Diphtheria & Tetanus Toxoids Adsorbed	Lederle, Wyeth-Ayerst
Diuril sodium IV	Merck Sharp & Dohme
Flu-Imune	Lederle
Gammar	Armour
Gammulin RhO(D)	Armour
Hep-B-Gammagee	Merck Sharp & Dohme
Heptavax-B	Merck Sharp & Dohme
HibTITER	Lederle
HypRho-D Mini-Dose	Miles-Cutter
HypRho-D	Miles-Cutter
Lactrodectus mactans Antivenin	Merck Sharp & Dohme
MICRhoGAM	Johnson & Johnson
Mini-Gamulin RhO	Armour
PNU-Imune 23	Lederle
ProHiBIT	Connaught
Recombivax HB	Merck Sharp & Dohme
RhoGAM	Johnson & Johnson
Sodium Edecrin	Merck Sharp & Dohme
Tetanus Toxoids Adsorbed	Lederle, Wyeth-Ayerst
Tetanus Toxoids Fluid	Lederle, Wyeth-Ayerst
Tri-Immunol	Lederle

IV. ANIMAL TOXICITY DATA

Most of the animal literature regarding thimerosal deals with the potential ocular toxicity. *In vitro* studies using concentrations of 0.01% on corneal endothelium have demonstrated cytotoxic effects;[3,4] however, *in vivo* topical instillation of 2% drops to rabbits did not result in detectable endothelial damage.[5] Continuous application of 0.004% thimerosal hourly for 2 d to healthy rabbit corneas produced mild, and probably reversible, changes to the endothelial cells as demonstrated by electron microscopy, thus patients with continuous exposure to thimerosal-preserved eyedrops may be at some risk for corneal ultrastructural changes.[6]

V. HUMAN TOXICITY DATA

A. Ophthalmic Products

The primary concern over the use of thimerosal in ophthalmic products is the frequent development of delayed contact hypersensitivity reactions. As many as 10 to 70% of soft contact lens wearers may be hypersensitive to thimerosal-preserved solutions as demonstrated by positive patch tests.[7,8] Patch tests to thimerosal 0.1% in petrolatum were positive in 37% of 174 patients with allergic contact conjunctivitis in one series. All patients had clinical signs attributed to the use of eyedrops or soft contact lens solutions containing thimerosal.[9]

Conjunctivitis is the most common adverse reaction. Redness of the eye, eye irritation, photophobia, blurred vision, decreased lens tolerance, and reversible corneal infiltrates have been reported.[10-12] The reactions are usually, but not always, confined to the ocular surface, sparing the surrounding skin and eyelids.[12-14] Combination conjunctivitis and eyelid dermatitis has been described.[15] Isolated eyelid dermatitis, without conjunctival involvement, was described in one case of occupational exposure in an ophthalmological assistant who demonstrated insertion of soft contact lenses without actually contact with the ocular surface.[16]

Positive skin patch and intradermal tests have confirmed a T-cell-mediated etiology for these reactions.[7,17,18]

An unusual drug interaction involving ophthalmic administration of thimerosal and oral administration of tetracycline was described in nine patients and confirmed by an animal study in rabbits. All patients had been using a thimerosal-containing contact lens solution for at least 6 months and developed sudden onset of conjunctival hyperemia, irritation, and blepharitis shortly after beginning oral tetracycline therapy. The adverse effects resolved within a few days after discontinuation of either thimerosal or tetracycline. The mechanism of this interaction was not determined.[19]

B. Topical Products

Thimerosal may cause two distinct types of cutaneous reactions. The most common type is a typical eczematous delayed contact hypersensitivity, analogous to the ophthalmologic reactions. Hand eczemas are particularly implicated, often manifesting as the pompholyx variety.[20] The frequency of positive patch tests to thimerosal 0.1% in petrolatum in eczema patients has ranged from 8 to 13% in North American subjects[21-23] and up to 19% in Austrian subjects.[24] A patch test series done in 593 healthy Italian subjects, with no history of dermatitis or ocular defects, demonstrated thimerosal to be the most frequent allergen with an incidence of 4.7%.[25]

The incidence of positive patch tests in another Italian series involving 2163 subjects varied widely depending on the type of prior exposure to thimerosal. The incidence was 5% in healthy male military recruits with prior exposure to eyedrops in six and recent vaccinations in all, 2% in healthy subjects without dermatological disease, 19% of patients treated with immunotherapy using standard allergen extracts containing thimerosal, 9% of patients with contact dermatitis, and 37% with allergic contact conjunctivitis.[9]

The second type of cutaneous reaction is a tuberculin-type delayed hypersensitivity, presumably related to prior vaccination or skin testing with thimerosal-containing agents. Histologic examination of tuberculin-like reactions have shown typical lymphocytic infiltrates with some eosinophils. Epicutaneous testing confirmed hypersensitivity to thimerosal in 7% of 105 recently vaccinated military recruits[26] and in 15% of Swedish patients tested with a thimerosal-containing tuberculin product.[27] The typical reaction consists of redness and swelling at the injection site. These reactions may be mistaken for positive tuberculin or coccidioidin results in some cases.[27,28]

The allergenic determinant responsible for hypersensitivity reactions may be either of the two moieties present in the thimerosal molecule, the ethylmercury radical or the thiosalicylate radical. Older studies showed strong sensitization potential from the thiosalicylate moiety,[29,30] while more recent studies have usually implicated the organic mercury component.[31,32] One recent Austrian study found cross-reactions to thiosalicylate in 21% of patients with a positive patch test to thimerosal.[33] Thimerosal readily decomposes in unstabilized saline solutions into ethylmercury chloride and thiosalicylate. The former compound is a strong skin sensitizing agent and was demonstrated to be the hapten binding to ocular proteins in a rabbit model.[34] Cross-reactions may occur to other mercury salts in patients sensitive to the mercury radical.[23,35]

A serious delayed contact reaction was reported in a 58-year-old man following the use of a sore throat spray containing 0.033% thimerosal. Within 24 h he developed life-threatening laryngeal obstruction with aphonia and crowing respirations, requiring emergency tracheostomy. Patch testing confirmed a severe delayed response to thimerosal.[36]

An exothermic reaction between thimerosal and aluminum foil was responsible for a thermal burn in a patient treated preoperatively with a thimerosal tincture prior to undergoing hysterectomy. A 5 cm blister, surrounded by erythema, was found at the site of contact of an aluminum foil diathermy electrode and the disinfected area. Reenactment of the thimerosal-aluminum combination in volunteers confirmed development of heat generation and a burning sensation. It was thought that mercury acted as a catalyst in the oxidation of the aluminum.[37]

C. Injectable Products

In addition to the tuberculin-type reactions described above, repeated long-term intramuscular injection of thimerosal-containing immune serum globulin has resulted in accumulation of mercury and subsequent mercury toxicity in patients with hypogammaglobulinemia. Elevated urine mercury concentrations were found in 19 of 26 patients with hypogammaglobulinemia receiving weekly intramuscular injections of 25 to 50 mg/kg gamma globulin for 6 months to 17 years. The total dose of mercury administered ranged from 4 to 734 mg. No clinical evidence of toxicity was observed.[38]

In a case report of a 20-year-old man who had received immune globulin injections for 15 years, mercury toxicity "acrodynia" was demonstrated as documented by slowed nerve conduction velocity; sensory neuropathy in the fingertips; a pruritic, pink, scaling rash on the palms and soles; photophobia; irritability; and tremor. The total estimated elemental mercury load was 40 to 50 mg. One month following injection of immune globulin, his blood and urine mercury concentrations were elevated at 8 mcg/l and 80 mcg/d, respectively.[39]

Systemic hypersensitivity has been reported following vaccination with thimerosal-containing hepatitis, influenza, and tetanus immune globulin products. The reactions were delayed, with an onset of 1 to 3 d following injection, and consisted of acute urticaria in two cases and a generalized exanthematic eruption in another case. Patch tests were positive in all three cases to thimerosal 0.1% in petrolatum.[40]

D. Oral Disease

Although the role of allergic contact dermatitis in the pathogenesis of oral diseases, such as lichen planus, is controversial, studies have demonstrated a high incidence of positive patch tests to mercury compounds in such patients, 28% in one study of 53 patients with biopsy proven oral lichenoid reactions.[41] Significant improvement has been shown when these patients undergo amalgam replacement or avoid use of thimerosal-preserved contact lens solutions.[42]

VI. CLINICAL RELEVANCE

A. Ocular Hypersensitivity

A significant number of patients using soft contact lenses complaining of conjunctivitis will demonstrate hypersensitivity to thimerosal, ranging from 25 to 37%.[9,18] It is clear that these patients should switch to other types of preservatives. It is not clear, however, to what extent these sensitized patients should avoid other routes of thimerosal administration.

One manufacturer of hepatitis B vaccine receives one or more telephone calls daily questioning whether ocular hypersensitivity is a contraindication to receiving the vaccine. The FDA has received two spontaneous reports of apparent reactions to thimerosal in hepatitis vaccine in these types of patients; one patient developed transient faintness and nausea, and the other patient developed a delayed skin rash. A small study of nine patients with a history of ocular hypersensitivity to thimerosal showed no adverse reactions to hepatitis B vaccination.[43]

B. Skin Hypersensitivity

The frequency of positive patch tests to thimerosal 0.1% in petrolatum in eczema patients has ranged from 8 to 13% in North American subjects. There is some indication that this patch test concentration may be too high and induces a mixed allergic-irritant reaction, which makes interpretation of clinical relevance difficult. Of 40 subjects reacting to a 0.1% concentration, sensitivity was clinically relevant in 4 patients who were soft contact lens wearers with eyelid or hand dermatitis. Patch testing with 0.05% in petrolatum was more specific. Patients reacting to this concentration had stronger responses, were more likely to have a positive intradermal test, and cross-reacted to other mercurials more frequently.[32] In another series of 113 patients, clinically relevant problems were associated with only 3 patients (2.6%) with positive patch tests.[33]

C. Vaccination Reactions

Both delayed-type hypersensitivity and generalized urticarial reactions have been described following the administration of thimerosal-preserved vaccines to sensitized individuals. The risk is apparently low. A small study of nine patients with a history of ocular hypersensitivity to thimerosal showed no adverse reactions to hepatitis B vaccination.[43] In a prospective series of 113 patients with positive patch tests to thimerosal, only 2 reported massive local reactions (redness, swelling, and pain) 1 to 2 d following subcutaneous vaccinations.[33] At this time, there is insufficient evidence to enforce a strict contraindication or to proclaim the procedure to be without risk. Each case should be evaluated for the risk-benefit ratio.

D. Hypogammaglobulinemia

Patients with congenital hypogammaglobulinemia, treated with long-term therapy with immune serum globulin, are at risk of accumulating an increased body burden of mercury. Despite documentation of high blood and urine mercury levels in such patients, only a small percentage will go on to develop the idiosyncratic mercury toxicity syndrome known as acrodynia.

REFERENCES

1. **Weiner, M. and Bernstein, I. L.,** *Adverse Reactions to Drug Formulation Agents,* Marcel Dekker, New York, 1989.
2. **Nikitakis, J. M.,** *CTFA Cosmetic Ingredient Handbook,* 1st ed., The Cosmetic, Toiletry and Fragrance Association, Washington, D.C., 1988.
3. **Collin, H. B., Grabsch, B. E., Carroll, N., and Hammond, V. E.,** Morphological changes to keratocytes and endothelial cells of the isolated guinea pig cornea due to thimerosal, *Int. Contact Lens Clin.,* 9, 275, 1982.
4. **Van Horn, D. L., Edelhauser, H. F., Prodanovich, G., Eiferman, R., and Pederson, H. J.,** Effect of the ophthalmic preservative thimerosal on rabbit and human corneal endothelium, *Invest. Ophthalmol. Vis. Sci.,* 16, 273, 1977.
5. **Gassett, A. R., Ishii, Y., Kaufman, H. E., and Miller, T.,** Cytotoxicity of ophthalmic preservatives, *Am. J. Ophthalmol.,* 78, 98, 1974.
6. **Collin, H. B. and Carroll, N.,** In vivo effects of thimerosal on the rabbit corneal endothelium: an ultrastructural study, *Am. J. Optometry Physiol. Optics,* 64, 123, 1987.
7. **Rietschel, R. L. and Wilson, L. A.,** Ocular inflammation in patients using soft contact lenses, *Arch. Dermatol.,* 118, 147, 1982.
8. **Miller, J. R.,** Sensitivity to contact lens solutions, *West. J. Med.,* 140, 791, 1984.
9. **Tosti, A., Guerra, L., and Bardazzi, F.,** Hyposensitizing therapy with standard antigenic extracts: an important source of thimerosal sensitization, *Contact Dermatitis,* 20, 173, 1989.
10. **Mondino, B. J. and Groden, L. R.,** Conjunctival hyperemia and corneal infiltrates with chemically disinfected soft contact lenses, *Arch. Ophthalmol.,* 98, 1767, 1980.
11. **Sendele, D. D., Kenyon, K. R., Mobilia, E. F., Rosenthal, P., Steinert, R., et al.,** Superior limbic keratoconjunctivitis in contact lens wearers, *Ophthalmology,* 90, 616, 1983.
12. **Fisher, A. A.,** Allergic reactions to contact lens solutions, *Cutis,* 21, 209, 1985.
13. **van Ketel, W. G. and Melzer-van Riemsduk, F. A.,** Conjunctivitis due to soft lens solutions, *Contact Dermatitis,* 6, 321, 1980.
14. **Pedersen, N. B.,** Allergic conjunctivitis from Merthiolate in soft contact lenses, *Contact Dermatitis,* 4, 165, 1978.
15. **Fregert, S. and Hjorth, H.,** *Textbook of Dermatology,* Blackwell, Oxford, 1972.
16. **De Groot, A. C., van Wijnen, W. G., and van Wijnen-Vos, M.,** Occupational contact dermatitis of the eyelids, without ocular involvement, from thimerosal in contact lens fluid, *Contact Dermatitis,* 23, 195, 1990.
17. **Mondino, B. J., Salamon, S. M., and Zaidman, G. W.,** Allergic and toxic reactions in soft contact lens wearers, *Surv. Opthalmol.,* 26, 337, 1982.
18. **Wilson, L. A., McNatt, J., and Reitschel, R.,** Delayed hypersensitivity to thimerosal in soft contact lens wearers, *Ophthalmology,* 88, 804, 1981.
19. **Crook, T. G. and Freeman, J. J.,** Reactions induced by the concurrent use of thimerosal and tetracycline, *Am. J. Optometry Physiol. Optics,* 60, 759, 1983.
20. **Moller, H.,** Merthiolate allergy: a nationwide iatrogenic sensitization, *Acta Derm. Venereol.,* 57, 509, 1977.
21. **Emmons, W. W. and Marks, J. G.,** Immediate and delayed reactions to cosmetic ingredients, *Contact Dermatitis,* 13, 258, 1985.

22. **Rudner, E. J., Clendenning, W. E., Epstein, E., et al.,** Epidemiology of contact dermatitis in North America: 1972, *Arch. Dermatol.,* 108, 537, 1973.

23. **Epstein, S.,** Sensitivity to merthiolate: a cause of false delayed intradermal reactions, *J. Allergy,* 34, 225, 1963.

24. **Wekkeli, M., Hippmann, G., Rosenkranz, A. R., Jarisch, R., and Gotz, M.,** Mercury as a contact allergen, *Contact Dermatitis,* 22, 295, 1990.

25. **Seidenari, S., Manzini, B. M., Danese, P., and Motolese, A.,** Patch and prick test study of 593 healthy subjects, *Contact Dermatitis,* 23, 162, 1990.

26. **Forstrom, L., Hannuksela, M., Kousa, M., and Lehmuskallio, E.,** Merthiolate hypersensitivity and vaccination, *Contact Dermatitis,* 6, 241, 1980.

27. **Hansson, H. and Moller, H.,** Intracutaneous test reactions to tuberculin containing merthiolate as a preservative, *Scand. J. Infect. Dis.,* 3, 169, 1971.

28. **Sbarbaro, J. A.,** Skin test antigens: an evaluation whose time has come, *Am. Rev. Respir. Dis.,* 118, 1, 1978.

29. **Ellis, F. A.,** The sensitizing factor in merthiolate, *J. Allergy,* 18, 212, 1947.

30. **Gaul, L. E.,** Sensitizing component in thiosalicylic acid, *J. Invest. Dermatol.,* 31, 91, 1958.

31. **Takino, C.,** Thiomersal contact dermatitis, *Jpn. J. Clin. Dermatol.,* 25, 1175, 1971.

32. **Lisi, P., Perno, P., Ottaviani, M., and Morelli, P.,** Minimum eliciting patch test concentration of thimerosal, *Contact Dermatitis,* 24, 22, 1991.

33. **Aberer, W.,** Vaccination despite thimerosal sensitivity, *Contact Dermatitis,* 24, 6, 1991.

34. **Cai, F., Backman, H. A., and Baines, M. G.,** Thimerosal: an ophthalmic preservative which acts as a hapten to elicit specific antibodies and cell mediated immunity, *Curr. Eye Res.,* 7, 341, 1988.

35. **Sertoli, A., DiFonzo, E., Spallanzani, P., and Panconesi, E.,** Allergic contact dermatitis from thimerosal in a soft contact lens wearer, *Contact Dermatitis,* 6, 292, 1980.

36. **Maibach, H.,** Acute larngeal obstruction presumed secondary to thiomersal (Merthiolate) delayed hypersensitivity, *Contact Dermatitis,* 1, 221, 1975.

37. **Jones, T. H.,** Danger of skin burns from thiomersal, *Br. Med. J.,* 2, 504, 1972.

38. **Haeney, M. R., Carter, G. F., Yeoman, W. B., and Thompson, R. A.,** Long-term parenteral exposure to mercury in patients with hypogammaglobulinaemia, *Br. Med. J.,* 2, 12, 1979.

39. **Matheson, D. S., Clarkson, T. W., and Gelfand, E. W.,** Mercury toxicity (acrodynia) induced by long term injection of gammaglobulin, *J. Pediatr.,* 97, 153, 1980.

40. **Tosti, A., Melino, M., and Bardazzi, F.,** Systemic reactions due to thiomersal, *Contact Dermatitis,* 15, 187, 1986.

41. **Todd, P., Garioch, J., Lamey, P. J., Lewis, M., Forsyth, A., and Rademaker, M.,** Patch testing in lichenoid reactions of the mouth and oral lichen planus, *Br. J. Dermatol.,* 123 (Suppl. 37), 26, 1990.

42. **Garioch, J., Todd, P., Lamey, P. J., Lewis, M., Forsyth, A., and Rademaker, M.,** The significance of a positive patch test to mercury in oral disease, *Br. J. Dermatol.,* 123 (Suppl. 37), 25, 1990.

43. **Kirkland, L. R.,** Ocular sensitivity to thimerosal: a problem with Hepatitis B vaccine?, *South. Med. J.,* 83, 497, 1990.

TINCTURE OF ORANGE

I. REGULATORY CLASSIFICATION

Tincture of orange peel is one of several compounds used as flavoring agents in oral pharmaceuticals. Official USP/NF preparations that contain orange oil include aromatic elixir USP, orange oil USP, sweet orange peel tincture NF, compound orange spirit NF, and orange syrup NF. These flavorings may be present in commercially available products and are also used in extemporaneous compounding.

II. AVAILABLE FORMULATIONS

A. Drugs

Orange oil, sweet orange peel, or orange peel extract are listed in 14 oral liquids and 1 oral effervescent granule product in the FDA list of NDA-approved drugs. Concentrations listed ranged from 0.0075 to 0.9%.[1]

III. TABLE OF COMMON PRODUCTS

Because current guidelines, published by the Pharmaceutical Manufacturing Association and followed by most pharmaceutical companies, state that flavoring agents do not have to be specifically identified due to concerns over their proprietary nature, it is extremely difficult to prepare a list of agents containing a particular flavor. The following list includes products listed by the manufacturer as "orange-flavored" and may include tincture of orange peel as well as other orange flavorings. Individual manufacturers should be contacted in case of a suspected adverse reaction to the flavoring.

Trade name	Manufacturer
Aldomet oral suspension	Merck Sharp & Dohme
Augmentin oral suspension 250 mg/5 ml	Beecham
Caltrate Jr. chewable tablet	Lederle
Dexedrine elixir	Smith Kline & French
Dilantin suspension	Parke-Davis
Duricef oral suspension	Mead Johnson

419

Trade name (cont'd)	Manufacturer (cont'd)
Ilosone oral suspension 125 mg/5 ml	Dista
Ilosone pediatric drops	Dista
Klor-Con/EF	Upsher-Smith
K-lyte orange effervescent tablet	Bristol
K-lyte DS orange effervescent tablet	Bristol
Lasix oral solution	Hoechst-Roussel
Luride lozi-tab 1 mg	Colgate-Hoyt
Mandelamine granules	Parke-Davis
Metamucil powder orange flavor	Procter & Gamble
Mintezol chewable tablet	Merck Sharp & Dohme
Naprosyn suspension	Syntex
Phos-Flur oral rinse	Colgate-Hoyt
Reglan syrup	Robins
Thorazine syrup	Smith Kline & French
Vermox chewable tablet	Janssen

IV. HUMAN TOXICITY DATA

A. Gastrointestinal Intolerance

Tincture of orange peel present as a flavoring in erythromycin oral suspensions was associated with the development of severe abdominal pain, nausea, and vomiting in a patient seen by a physician in a pediatric consulting practice. Substitution of the formulation with a different strength of the same brand (which contained a different flavoring agent) at the same dosage did not result in adverse reaction. Inadvertent rechallenge with the original preparation reproduced the gastrointestinal effects. This same formulation was responsible for gastrointestinal complaints in another child treated for pertussis, in a 6-week-old infant treated with an extemporaneous spironolactone-hydrochlorothiazide suspension formulated with tincture of orange, and in 16 patients in a retrospective review of the previous 15 years by the same physician.[2]

B. Hypersensitivity Reactions

Orange oil contains more than 90% limonene, an essential oil that is a primary irritant and sensitizer. Allergic cheilitis has been reported in patients with a history of removing orange peel with their teeth.[3] Cheilitis has not been described from the use of orange oil in pharmaceutical products.

V. CLINICAL RELEVANCE

Based on anecdotal reports by a Canadian physician and validated by rechallenge in some cases, tincture of orange flavoring may be responsible for some cases of adverse gastrointestinal effects from erythromycin suspensions. This association should be confirmed by controlled studies; however, it may be worthwhile to consider substitution of an alternate flavored product in patients with gastrointestinal intolerance of an orange-flavored product.

REFERENCES

1. **Food and Drug Administration,** *Inactive Ingredient Guide,* FDA, Washington, D.C., March 1990.
2. **Napke, E. and Stevens, D. G. H.,** Excipients and additives: hidden hazards in drug products and in product substitution, *Vet. Hum. Toxicol.,* 32, 253, 1990.
3. **Fisher, A. A.,** *Contact Dermatitis,* 3rd ed., Lea & Febiger, Philadelphia, 1986.

TRAGACANTH

I. REGULATORY CLASSIFICATION

Tragacanth is classified as suspending and/or viscosity-increasing agent.

II. SYNONYMS

E413
Goma alcatira
Gum dragon
Gum tragacanth

III. AVAILABLE FORMULATIONS

A. Constituents

Tragacanth is a dried gummy exudate from cut trunk and branches of *Astragalus gummifer* and other species in this genus. A water-soluble fraction, consisting mostly of uronic acid and arabinose, and a mucilage-forming fraction, consisting of mostly bassorin, have been identified. Gum tragacanth swells into a jelly when mixed with water and forms a thick, transparent solution after addition of sufficient amounts of water. Tragacanth may be adulterated with karaya, India gum, or acacia.[1]

B. Foods

Tragacanth is widely used in foods with no established acceptable dietary intake limit.[2]

C. Drugs

Tragacanth is listed in 21 FDA-approved drug products, including nasal solutions, buccal or sublingual tablets, oral suspensions, syrups, and tablets. The concentration in oral liquids ranges from 4.8 to 6%; tablets contain 0.42 to 100 mg.[3]

D. Cosmetics

Tragacanth is used as a binder, emulsion stabilizer, film former, and viscosity-increasing agent in face powders, mud packs, and eye shadows.[4]

423

IV. TABLE OF COMMON PRODUCTS

A. Foods

Tragacanth is found in many foods including chewing gum, French dressings, frozen fruits, dehydrated juices, commercial gravies, mayonnaise, syrups, toppings, catsup, tomato puree, and whipped cream.[5]

B. Drugs

Trade name	Manufacturer
Aci-jel vaginal jelly	Ortho
Agoral liquid	Parke-Davis
Choledyl tablets	Parke-Davis
Diuril oral suspension	Merck Sharp & Dohme
Halotestin tablets	Upjohn
Indocin oral suspension	Merck Sharp & Dohme
Mintezol suspension	Merck Sharp & Dohme
PBZ tablet 50 mg	Geigy
Ritalin tablets	Ciba
Serpasil-Apresoline tablets	Ciba
Serpasil tablets	Ciba
Sumycin syrup	Squibb
Ultracef suspension	Bristol

V. HUMAN TOXICITY DATA

A. Immediate Hypersensitivity Reactions
1. Occupational Exposure

Occupational asthma was described in a 26-year-old worker in a gum factory. Initial symptoms of rhinitis and sneezing began about 1 year after starting employment at the factory, which progressed to severe asthma 3 months later after a particularly large inhalation exposure. A skin prick test with tragacanth was markedly positive, and passive transfer was documented for both tragacanth and gum arabic.[1]

2. Foods

Life-threatening angioedema, accompanied by pruritus, dyspnea, and abdominal pain, developed after ingestion of a "Big Mac" hamburger. Intradermal testing confirmed an immediate reaction to gum tragacanth.[6]

3. Drugs

Three kidney transplant patients receiving long-term therapy with prednisone tablets containing acacia and tragacanth were reported to develop hypersensitivity reactions consisting of rash, pruritus, fever, and arthralgia. One patient had a positive scratch test to acacia, one had a positive test to tragacanth, and the other patient was not tested. Subsequent testing of 11 other transplant recipients showed a reaction to tragacanth in 1 patient, but not associated with any clinical manifestations.[7]

Asthma and generalized urticaria were attributed to tragacanth in a tripelennamine tablet in one reported case. An IgE-mediated mechanism was suggested by demonstration of intracutaneous testing and passive transfer.[8]

B. Allergic Contact Dermatitis

A 4-year-old boy developed allergic contact dermatitis following the use of an electrode jelly used during an electrocardiogram procedure 2 months earlier. Patch testing was positive to the intact jelly and to gum tragacanth 1% aqueous solution.[9]

VI. CLINICAL RELEVANCE

Immediate IgE-mediated hypersensitivity reactions, consisting of angioedema, asthma, rhinitis, or urticaria, have been rarely described. Only one case of delayed-type hypersensitivity has been reported.

REFERENCES

1. **Gelfand, H. H.,** The allergenic properties of the vegetable gums, *J. Allergy,* 14, 203, 1942.
2. **World Health Organization,** Twenty-Ninth Report of the Joint FAO/WHO Expert Committee on Food Additives, Tech Rep Ser No. 723, WHO, Geneva, 1986.
3. **Food and Drug Administration,** *Inactive Ingredients Guide,* FDA, Washington, D.C., March 1990.
4. **Nikitakis, J. M.,** *CTFA Cosmetic Ingredient Handbook,* 1st ed., The Cosmetic, Toiletry and Fragrance Association, Washington, D.C., 1988.
5. **Nilsson, D. C.,** Sources of allergenic gums, *Ann. Allergy,* 18, 518, 1960.
6. **Danoff, D., Lincoln, L., Thomson, D. M. P., and Gold, P.,** "Big Mac attack", *N. Engl. J. Med.,* 298, 1095, 1978.
7. **Rubinger, D., Friedlander, M., and Superstine, E.,** Hypersensitivity to tablet additives in transplant recipients on prednisone, *Lancet,* 2, 689, 1978.
8. **Brown, E. B. and Crepea, S. B.,** Allergy (asthma) to ingested gum tragacanth, *J. Allergy,* 18, 214, 1947.
9. **Coskey, R. J.,** Contact dermatitis caused by ECG electrode jelly, *Arch. Dermatol.,* 113, 839, 1977.

TRANSDERMAL SYSTEMS

I. REGULATORY CLASSIFICATION

Transdermal drug delivery systems are currently available for five drugs (clonidine, estradiol, fentanyl, nitroglycerin, and scopolamine). These systems allow controlled continuous administration of drugs through the skin, avoiding side effects associated with oral administration, such as gastrointestinal discomfort.

Drugs being studied in transdermal formulations include antihistamines, coumarin, indomethacin, isoproterenol, isosorbide dinitrate, nicotine, pyridostigmine, testosterone, and timolol.

II. AVAILABLE FORMULATIONS

Transdermal systems are multilayered formulations containing a variety of inactive ingredients. The basic design consists of four layers.

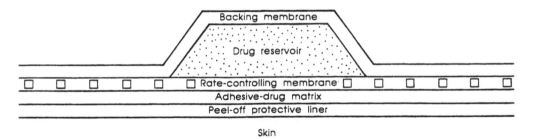

FIGURE 71. Transdermal delivery system.

A. Adhesive Layer
This layer is in direct contact with the skin and contains active drug dispersed in an adhesive matrix. Inactive matrix ingredients include acrylate polymers, polyisobutylene, mineral oil, lactose, resinous cross-linking agents, and colloidal silicone.

B. Rate-Controlling Membrane
This layer contains a semiporous membrane that allows gradual controlled diffusion of

the active drug into the adhesive matrix. Inactive membrane ingredients include ethylene-vinyl acetate copolymers and polypropylene. Some formulations do not contain this layer as a separate layer, but incorporate rate-controlling materials into the adhesive layer.

C. Drug Reservoir

This layer contains the bulk of the active drug. Inactive reservoir ingredients include alcohol USP, hydroxypropyl cellulose, mineral oil, polyisobutylene, lactose, and colloidal silicon dioxide.

D. Backing Membrane

This is the outermost layer, which generally contains an aluminized polyester film, with no active drug.

III. TABLE OF COMMON PRODUCTS

A. Transdermal Drug Delivery Systems

Trade name	Active ingredient	Manufacturer	Adhesive type
Catapres-TTS	Clonidine	Boehringer Ingelheim	Polyisobutylene
Deponit NTG	Nitroglycerin	Schwarz Pharma	Polyisobutylene
Duragesic	Fentanyl	Janssen	Silicon-based
Estraderm	Estradiol	Ciba	Polyisobutylene
Minitran	Nitroglycerin	3M	Acrylate-based polymer
Nitrodisc	Nitroglycerin	Searle	Acrylate-based polymer
Nitro-Dur	Nitroglycerin	Key	Acrylate-based polymer
Transderm-Nitro	Nitroglycerin	Summit	Silicon-based
Transderm-Scop	Scopolamine	Ciba	Polyisobutylene

IV. HUMAN TOXICITY DATA

A. Occlusive Effects

Because transdermal systems provide an occlusive barrier left on the skin for 1 to 7 d, transient adverse effects of skin occlusion are common. Occlusive effects are maximized in warm or humid climates. These include sweat duct occlusion, miliaria rubra, and bacterial and yeast overgrowth. Alternating the site of application is recommended to minimize these reactions.

Modifications of the systems design have been investigated in attempts to reduce these undesirable effects of the delivery system. Attachment of a hydrogel system to the adhesive layer reduced skin irritation from sweat duct occlusion, but resulted in intense bacterial growth. Bacterial growth was inhibited by pretreatment of the skin with a topical chlorhexidine gluconate solution.[1]

B. Thermal Burns

A second-degree burn conforming to the size and shape of a transdermal nitroglycerin patch was described in a 51-year-old man who sat next to a microwave oven which was later determined to have a leak. The aluminum in the patch was presumed to have been heated by the radiation leaking from the oven.[2]

C. Primary Irritant Reactions

Nonimmunologic irritant erythema may transiently occur after removal of the adhesive from the skin. Erythema is common following use of nitroglycerin transdermal systems due to the vasodilatory effects of the active drug.

Primary irritant effects, consisting of mild erythema at the application site, was reported in 4 of 11 healthy volunteers following repeated application to the same site.[3]

More severe primary irritant effects have been reported after a longer duration of use. A sharply demarcated macular erythematous burn-like rash at the site of application of both an active nitroglycerin transdermal system and a placebo patch was reported in a 63-year-old man previously treated with nitroglycerin ointment. The reaction faded, but was still visible, 1 month after discontinuation of the patch. Because of the absence of involvement of the surrounding tissue, this may have represented a primary irritant dermatitis.[4]

Irritant reactions (excluding simple erythema) were reported in 15% of patients during the first 70 d of use of a transdermal nitroglycerin patch in a prospective study of 33 patients.[5]

Two other cases of a burn-like primary irritant reaction were reported in patients who consistently applied the patches to the same sites or anatomic area.[6] These irritant reactions have been suggested to be related to glycerin contamination of the nitroglycerin component, which is chemically modified by the growth of bacteria at the occluded site to produce acrylic aldehyde, a strong irritant.[7]

D. Allergic Contact Dermatitis

1. Clonidine

A prospective open study of 29 patients treated with clonidine patches found delayed-type hypersensitivity reactions in 11 (38%). Patch testing with individual components of the transdermal system showed positive reactions to the active ingredient in six of seven patients tested; the remaining patient was allergic to polyisobutylene.[8]

Consistent with the finding in this series that the majority of the allergic reactions were to the active ingredient are several case reports documenting contact sensitivity to clonidine during the first 20 weeks of transdermal product use. In one series, 22 of 35 (63%) cases developed positive reactions to patch testing with clonidine 9% in petrolatum.[9] All but 3 of 260 patients subsequently challenged with oral clonidine have been able to tolerate the drug. One patient developed a localized flare-up at the site of the original dermatitis, and two patients developed a generalized maculopapular rash and pruritus.[9,10]

The characteristic lesion produced by the Type IV allergic reaction to this transdermal system includes severe irritation, induration, erythema, and occasionally vesiculation. Lesions increase in severity for 48 h after removal of the patch. Because the clonidine patch is designed to be left on the skin for a longer period of time (7 d), allergic reactions may be more frequent than with patches designed to be removed daily or every third day, such as nitroglycerin or scopolamine.[9]

2. Estradiol

During a therapeutic trial of 124 postmenopausal women treated with estradiol transdermal patches for 3 weeks, 6 had severe skin reactions to the placebo patch system, consisting of rash, redness, or itching.[11]

A case report described a 35-year-old woman who developed pruritus, edema, and hyperpigmentation 2 weeks after using an estradiol transdermal system twice weekly. Extensive patch testing showed positive results only for hydroxypropyl cellulose, an ingredient of the drug reservoir. The reaction was vehicle-dependent, with positive results in an ethanol and mineral oil vehicle and negative results in an aqueous vehicle.[12]

An investigation of nine patients with delayed hypersensitivity reactions to the estradiol transdermal patch found four patients sensitive to the adhesive. Four other patients reacted to a component in the drug reservoir; this was shown to be hydroxypropyl cellulose in two patients, and an unknown allergen in the other two patients. One patient reacted to both the active drug and to the adhesive. These reactions worsened during the 48 h following removal and began 2 to 8 weeks after first using the patches in eight of nine cases.[13]

3. Nitroglycerin

A case of severe contact dermatitis occurred 4 months after switching from nitroglycerin ointment, used for 1 year, to the transdermal patch and was described as a localized, painful erythematous eruption with a few ruptured vesicles. The lesions disappeared when the patch was discontinued and recurred within 6 to 12 h after application. Three weeks after restarting the ointment, the rash recurred and was demonstrated to be related to the active drug after patch testing.[14]

Continued application of a nitroglycerin transdermal patch to the same site produced severe erythema, edema, and crusting in two patients who were also demonstrated to have delayed hypersensitivity reactions to the nitroglycerin component.[7] Application to variable sites has also been reported to result in allergic contact dermatitis, subsequently shown to occur only with the active patch.[15]

Allergic contact dermatitis with postinflammatory hyperpigmentation was described in a 59-year-old man following use of a nitroglycerin patch for 3 months. At the time of evaluation, he presented with 45 to 50 brown-blue patches corresponding to previous sites of application. Diagnostic patch testing revealed a 2+ reaction after 24 h to crushed nitroglycerin tablets in petrolatum and a 1+ reaction after 96 h to the placebo transdermal device. This case was the only dermatologic reaction to nitroglycerin patches observed in 3273 patients treated in the same department.[16]

A similar case of allergic dermatitis with hyperpigmentation was reported 13 months after using a transdermal system correctly applied to variable sites. Patch testing was positive to the intact complete system, but negative to nitroglycerin in petrolatum, a placebo transdermal system, and one component of the adhesive layer. The reaction was attributed to one of the untested excipients or to a complex mixture of more than one component.[17]

Another reaction attributed to one or more of the excipients in the adhesive bandage has been documented.[18]

4. Scopolamine

The scopolamine patches are intended for short-term use, therefore allergic hypersensitivity reactions have been less frequently reported. Allergic contact dermatitis has been reported after 3 weeks of use in one case.[18] A 53-year-old woman who used the patch continuously for 3 months for the treatment of persistent vertigo developed contact dermatitis, which was attributed to the active ingredient after extensive patch testing with the individual components.[19]

A similar case, in a 10-year-old mentally retarded child with contact dermatitis after 2 weeks of continuous use for control of ptyalism, was also attributed to scopolamine hypersensitivity.[20]

Long-term use for 1.5 to 15 months in 164 naval crew workers was associated with development of allergic contact dermatitis in 10%. All patients had negative reactions to placebo patches.[21]

E. Immediate Hypersensitivity Reactions

One case of angioedema and seven cases of contact urticaria have been reported following use of the clonidine transdermal system.[9]

V. CLINICAL RELEVANCE

A. Immunologic Reactions

Hypersensitivity reactions occur usually to the active ingredient or to one of the inactive ingredients in the adhesive or reservoir layer. Sensitization to these constituents is facilitated by the irritant and occlusive properties of the formulations. The incidence and allergen involved depends on the type of active drug, the duration the patch is left on the skin uninterrupted, and the inactive components. With some patches, such as clonidine and nitroglycerin, hypersensitivity appears to be primarily related to the active ingredient; clonidine is also the patch with the greatest skin contact duration. For other patches, such as estradiol, hypersensitivity is usually related to one of the inactive ingredients. Excipients involved in these delayed Type IV hypersensitivity reactions have included polyisobutylene and hydroxypropyl cellulose.

Patients with consistent cutaneous side effects to transdermal patches should consult their physician immediately. Evidence of decreased absorption of the active drug when applied to an inflamed site has been documented by the occurrence of rebound phenomenon or a decreased therapeutic response.[15,22]

B. Nonimmunologic Reactions

Transient erythema is a fairly frequent side effect observed in all types of transdermal therapy. The symptoms are generally mild and do not result in discontinuation of therapy. Occlusive effects, such as miliaria rubra and microbial overgrowth, are maximized in warm or humid climates and minimized by alternation of the site of application. Patients using the nitroglycerin patches may be more prone to severe irritant reactions due to the tendency to place the patches over the heart with restricted site rotation. Irritant reactions are characterized by erythema and induration, which is maximal at the margins. A sharply demarcated lesion that does not extend beyond the dimensions of the patch is typical.

REFERENCES

1. **Hurkmans, J. F. G. M., Bodde, H. E., Van Driel, L. M. J., Van Doorne, H., and Junginger, H. E.,** Skin irritation caused by transdermal drug delivery systems during long-term (5 days) application, *Br. J. Dermatol.,* 112, 461, 1985.
2. **Murray, K. B.,** Hazard of microwave ovens to transdermal delivery system, *N. Engl. J. Med.,* 310, 721, 1984.
3. **Utian, W. H.,** Transdermal estradiol overall safety profile, *Am. J. Obstet. Gynecol.,* 156, 1335, 1987.
4. **Letendre, P. W., Barr, C., and Wilkens, K.,** Adverse dermatologic reaction to transdermal nitroglycerin, *DICP,* 18, 69, 1984.
5. **Vaillant, L., Biette, S., Machet, L., Constans, T., and Monpere, C.,** Skin acceptance of transcutaneous nitroglycerin patches: a prospective study of 33 patients, *Contact Dermatitis,* 23, 142, 1990.
6. **Fischer, R. G. and Tyler, M.,** Severe contact dermatitis due to nitroglycerin patches. *South. Med. J.,* 78, 1523, 1985.
7. **Topaz, O. and Abraham, D.,** Severe allergic contact dermatitis secondary to nitroglycerin in a transdermal therapeutic system, *Ann. Allergy,* 59, 365, 1987.
8. **Groth, H., Vetter, H., Knuesel, J., and Vetter, W.,** Allergic skin reactions to transdermal clonidine, *Lancet,* 2, 850, 1983.

9. Maibach, H. I., Oral substitution in patients sensitized by transdermal clonidine treatment, *Contact Dermatitis,* 16, 1, 1987.

10. Grattan, C. E. H. and Kennedy, C. T. C., Allergic contact dermatitis to transdermal clonidine, *Contact Dermatitis,* 12, 225, 1985.

11. Place, V. A., Powers, M., Darley, P. E., Schenkel, L., and Good, W. R., A double-blind comparative study of Estraderm and Premarin in the amelioration of postmenopausal symptoms, *Am. J. Obstet. Gynecol.,* 152, 1092, 1985.

12. Schwartz, B. K. and Clendenning, W. E., Allergic contact dermatitis from hydroxypropyl cellulose in a transdermal estradiol patch, *Contact Dermatitis,* 18, 106, 1988.

13. McBurney, E. I., Noel, S. B., and Collins, J. H., Contact dermatitis to transdermal estradiol system, *J. Am. Acad. Dermatol.,* 20, 508, 1989.

14. Rosenfeld, A. S. and White, W. B., Allergic contact dermatitis secondary to transdermal nitroglycerin, *Am. Heart J.,* 108, 1061, 1984.

15. Carmichael, A. J. and Foulds, I. S., Allergic contact dermatitis from transdermal nitroglycerin, *Contact Dermatitis,* 21, 113, 1989.

16. Harari, Z., Sommer, I., and Knobel, B., Multifocal contact dermatitis to Nitroderm TTS 5 with extensive postinflammatory hypermelanosis, *Dermatologica,* 174, 249, 1987.

17. Di Landro, A., Valsecchi, R., and Cainelli, T., Contact dermatitis from Nitroderm, *Contact Dermatitis,* 21, 115, 1989.

18. Fisher, A. A., Dermatitis due to transdermal therapeutic systems, *Cutis,* 34, 526, 1984.

19. Trozak, D. J., Delayed hypersensitivity to scopolamine delivered by a transdermal device, *J. Am. Acad. Dermatol.,* 13, 247, 1985.

20. van der Willigen, A. H., Oranje, A. P., Stolz, E., and van Joost, Th., Delayed hypersensitivity to scoplamine in transdermal therapeutic systems, *J. Am. Acad. Dermatol.,* 18, 146, 1988.

21. Gordon, C. R., Shupak, A., Doweck, I., and Spitzer, O., Allergic contact dermatitis caused by transdermal hyoscine, *Br. Med. J.,* 298, 1220, 1989.

22. White, T. M. and Guidry, J. R., Rebound hypertension associated with transdermal clonidine and contact dermatitis, *West. J. Med.,* 145, 104, 1986.

UROCANIC ACID

I. REGULATORY CLASSIFICATION

Urocanic acid is used as a skin conditioning agent in cosmetics.

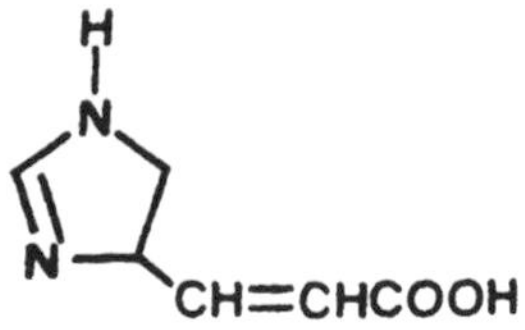

FIGURE 72. Urocanic acid.

II. SYNONYMS

Deaminated histidine
4-imidazoleacrylic acid
Imidazole-4-acrylic acid
Imidazoleacrylic acid
Urocaninic acid

III. AVAILABLE FORMULATIONS

Urocanic acid is a naturally occurring substance found in high concentrations in human stratum corneum. In 1989, there were 15 cosmetic products listed with the FDA containing urocanic acid. These products included ten sunscreens, three body lotions, one makeup base, and one makeup foundation. Most of these products have been reformulated.

IV. TABLE OF COMMON PRODUCTS

The following list of cosmetics includes those formerly containing urocanic acid. Some of these products may remain available to consumers until existing store stocks are depleted.

Trade name	Manufacturer
Facial moisturizing lotion	Shiseido
Facial nourishing cream rich	Shiseido
Germaine Monteil Pre-tan conditioner	Revlon
Oil-free tanning formula	Estee Lauder
Overnight Pre-Tan Accelerator	Estee Lauder
Pre-makeup cream base	Shiseido
Self-tanning formula	Clinique
Sun Face Block for Sensitive Skin	Estee Lauder
Waterworld sunscreen	Estee Lauder

V. ANIMAL TOXICITY DATA

A. Immunosuppression

The naturally occurring *trans*-isomer of urocanic acid, present in about 0.7% of dry weight of the epidermis, has been shown to act as a receptor for ultraviolet light, resulting in conversion to the *cis*-isomer. It was originally thought to act as a natural sunscreen or photoprotective agent. An immunosuppressive effect was found in irradiated skin pretreated with urocanic acid, possibly due to generation of suppressor T-cells.[1]

B. Carcinogenicity

Topical application of a commercial water-resistant sunscreen base containing *trans*-urocanic acid 0.2% to hairless mice exposed chronically to daily minimal erythemal doses of ultraviolet light resulted in an increase in the number of overt UV-initiated tumors from 1.7 to 3.3 tumors per mouse. The urocanic-acid treated mice also had a greater percentage of malignant types of tumors. The proposed mechanism is evasion of ultraviolet-initiated tumor cells from the immunosurveillance mediated by T-cells in the epidermis. Urocanic acid shares a 4-substituted imidazole ring structure with azathioprine, an immunosuppressant that increases susceptibility to malignant skin tumors in renal transplant patients.[2]

VI. CLINICAL RELEVANCE

There is no evidence of skin cancer-promoting activity of urocanic acid in humans. This excipient has been voluntarily withdrawn from most or all of the cosmetics involved.

REFERENCES

1. **Norval, M., Simpson, T. J., and Ross, J. A.,** Urocanic acid and immunosuppression, *Photochem. Photobiol.,* 50, 267, 1989.
2. **Reeve, V. E., Greenoak, G. E., Canfield, P. J., Boehm-Wilcox, C., and Gallagher, C. H.,** Topical urocanic acid enhances UV-induced tumour yield and malignancy in the hairless mouse, *Photochem..Photobiol.,* 49, 459, 1989.

INDEX

Milton Keynes UK
Ingram Content Group UK Ltd.
UKHW052022071024
449327UK00027B/2389